Aircraft Systems Classifications

Aerospace Series

Aircraft System Classifications: A Handbook of Characteristics and Design Guidelines
Allan Seabridge and Mohammad Radaei

UAS Integration into Civil Airspace: Policy, Regulations and Strategy
Douglas M. Marshall

Introduction to UAV Systems, Fifth Edition
Paul G. Fahlstrom, Thomas J. Gleason, Mohammad H. Sadraey

Introduction to Flight Testing
James W. Gregory, Tianshu Liu

Foundations of Space Dynamics
Ashish Tewari

Essentials of Supersonic Commercial Aircraft Conceptual Design
Egbert Torenbeek

Design of Unmanned Aerial Systems
Mohammad H. Sadraey

Future Propulsion Systems and Energy Sources in Sustainable Aviation
Saeed Farokhi

Flight Dynamics and Control of Aero and Space Vehicles
Rama K. Yedavalli

Design and Development of Aircraft Systems, 3rd Edition
Allan Seabridge, Ian Moir

Helicopter Flight Dynamics: Including a Treatment of Tiltrotor Aircraft, 3rd Edition
Gareth D. Padfield CEng, PhD, FRAeS

Space Flight Dynamics, 2nd Edition
Craig A. Kluever

Performance of the Jet Transport Airplane: Analysis Methods, Flight Operations, and Regulations
Trevor M. Young

Small Unmanned Fixed-wing Aircraft Design: A Practical Approach
Andrew J. Keane, András Sóbester, James P. Scanlan

Advanced UAV Aerodynamics, Flight Stability and Control: Novel Concepts, Theory and Applications
Pascual Marqués, Andrea Da Ronch

Differential Game Theory with Application to Missiles and Autonomous Systems Guidance
Farhan A. Faruqi

Introduction to Nonlinear Aeroelasticity
Grigorios Dimitriadis

Introduction to Aerospace Engineering with a Flight Test Perspective
Stephen Corda

Aircraft Control Allocation
Wayne Durham, Kenneth A. Bordignon, Roger Beck

Remotely Piloted Aircraft Systems: A Human Systems Integration Perspective
Nancy J. Cooke, Leah J. Rowe, Winston Bennett Jr., DeForest Q. Joralmon

Theory and Practice of Aircraft Performance
Ajoy Kumar Kundu, Mark A. Price, David Riordan

Adaptive Aeroservoelastic Control
Ashish Tewari

The Global Airline Industry, 2nd Edition
Peter Belobaba, Amedeo Odoni, Cynthia Barnhart, Christos Kassapoglou

Introduction to Aircraft Aeroelasticity and Loads, 2nd Edition
Jan R. Wright, Jonathan Edward Cooper

Theoretical and Computational Aerodynamics
Tapan K. Sengupta

Aircraft Aerodynamic Design: Geometry and Optimization
András Sóbester, Alexander I J Forrester

Stability and Control of Aircraft Systems: Introduction to Classical Feedback Control
Roy Langton

Aerospace Propulsion
T.W. Lee

Civil Avionics Systems, 2nd Edition
Ian Moir, Allan Seabridge, Malcolm Jukes

Aircraft Flight Dynamics and Control
Wayne Durham

Modelling and Managing Airport Performance
Konstantinos Zografos, Giovanni Andreatta, Amedeo Odoni

Advanced Aircraft Design: Conceptual Design, Analysis and Optimization of Subsonic Civil Airplanes
Egbert Torenbeek

Design and Analysis of Composite Structures: With Applications to Aerospace Structures, 2nd Edition
Christos Kassapoglou

Aircraft Systems Integration of Air-Launched Weapons
Keith A. Rigby

Understanding Aerodynamics: Arguing from the Real Physics
Doug McLean

Aircraft Design: A Systems Engineering Approach
Mohammad H. Sadraey

Theory of Lift: Introductory Computational Aerodynamics in MATLAB/Octave
G.D. McBain

Sense and Avoid in UAS: Research and Applications
Plamen Angelov

Morphing Aerospace Vehicles and Structures
John Valasek

Spacecraft Systems Engineering, 4th Edition
Peter Fortescue, Graham Swinerd, John Stark

Unmanned Aircraft Systems: UAVS Design, Development and Deployment
Reg Austin

Visit www.wiley.com to view more titles in the Aerospace Series.

Aircraft Systems Classifications

A Handbook of Characteristics and Design Guidelines

Allan Seabridge and Mohammad Radaei

WILEY

This edition first published 2022
© 2022 John Wiley & Sons, Inc.

All rights reserved. No part of this publication may be reproduced, stored in a retrieval system, or transmitted, in any form or by any means, electronic, mechanical, photocopying, recording or otherwise, except as permitted by law. Advice on how to obtain permission to reuse material from this title is available at http://www.wiley.com/go/permissions.

The right of Allan Seabridge and Mohammad Radaei to be identified as the authors of this work has been asserted in accordance with law.

Registered Office
John Wiley & Sons, Inc., 111 River Street, Hoboken, NJ 07030, USA

Editorial Office
111 River Street, Hoboken, NJ 07030, USA

For details of our global editorial offices, customer services, and more information about Wiley products visit us at www.wiley.com.

Wiley also publishes its books in a variety of electronic formats and by print-on-demand. Some content that appears in standard print versions of this book may not be available in other formats.

Limit of Liability/Disclaimer of Warranty
The contents of this work are intended to further general scientific research, understanding, and discussion only and are not intended and should not be relied upon as recommending or promoting scientific method, diagnosis, or treatment by physicians for any particular patient. In view of ongoing research, equipment modifications, changes in governmental regulations, and the constant flow of information relating to the use of medicines, equipment, and devices, the reader is urged to review and evaluate the information provided in the package insert or instructions for each medicine, equipment, or device for, among other things, any changes in the instructions or indication of usage and for added warnings and precautions. While the publisher and authors have used their best efforts in preparing this work, they make no representations or warranties with respect to the accuracy or completeness of the contents of this work and specifically disclaim all warranties, including without limitation any implied warranties of merchantability or fitness for a particular purpose. No warranty may be created or extended by sales representatives, written sales materials or promotional statements for this work. The fact that an organization, website, or product is referred to in this work as a citation and/or potential source of further information does not mean that the publisher and authors endorse the information or services the organization, website, or product may provide or recommendations it may make. This work is sold with the understanding that the publisher is not engaged in rendering professional services. The advice and strategies contained herein may not be suitable for your situation. You should consult with a specialist where appropriate. Further, readers should be aware that websites listed in this work may have changed or disappeared between when this work was written and when it is read. Neither the publisher nor authors shall be liable for any loss of profit or any other commercial damages, including but not limited to special, incidental, consequential, or other damages.

Library of Congress Cataloging-in-Publication Data applied for:
ISBN: 9781119771845

Cover Design: Wiley
Cover Image: © Wu Hao/Getty Images

Set in 9.5/12.5pt STIXTwoText by Straive, Chennai, India

10 9 8 7 6 5 4 3 2 1

Contents

About the Authors *ix*
Acknowledgements *xi*
Sources of Background Information *xiii*
Glossary *xv*

1   Introduction *1*
    Further Reading *4*

2   The Airframe and Systems Overview *5*
2.1 Introduction *5*
2.2 The Airframe *6*
2.3 The Aircraft Systems *10*
2.4 Classification of Aircraft Roles *14*
2.5 Classification of Systems *25*
2.6 Stakeholders *26*
2.7 Example Architectures *27*
2.8 Data Bus *29*
2.9 Summary and Conclusions *34*
    References *34*
    Exercises *35*

3   Vehicle Systems *37*
3.1 Propulsion System *38*
3.2 Fuel System *44*
3.3 Electrical Power Generation and Distribution *49*
3.4 Hydraulic Power Generation and Distribution *53*
3.5 Bleed Air System *56*
3.6 Secondary Power Systems *59*
3.7 Emergency Power Systems *61*

3.8	Flight Control System	65
3.9	Landing Gear	68
3.10	Brakes and Anti-skid	71
3.11	Steering System	73
3.12	Environmental Control System	76
3.13	Fire Protection System	79
3.14	Ice Detection	82
3.15	Ice Protection	84
3.16	External Lighting	86
3.17	Probe Heating	89
3.18	Vehicle Management System (VMS)	91
3.19	Crew Escape	93
3.20	Canopy Jettison	97
3.21	Oxygen	99
3.22	Biological and Chemical Protection	102
3.23	Arrestor Hook	104
3.24	Brake Parachute	107
3.25	Anti-spin Parachute	110
3.26	Galley	112
3.27	Passenger Evacuation	115
3.28	In-Flight Entertainment	117
3.29	Toilet and Water Waste	119
3.30	Cabin and Emergency Lighting	122
	References	123
	Exercise	126

4	**Avionic Systems**	**127**
4.1	Displays and Controls	127
4.2	Communications	131
4.3	Navigation	134
4.4	Example Navigation System Architecture	135
4.5	Flight Management System (FMS)	138
4.6	Weather Radar	140
4.7	Air Traffic Control (ATC) Transponder	143
4.8	Traffic Collision and Avoidance System (TCAS)	146
4.9	Terrain Avoidance and Warning System (TAWS)	149
4.10	Distance Measuring Equipment (DME)/TACAN	152
4.11	VHF Omni-Ranging (VOR)	154
4.12	Automatic Flight Control System	156
4.13	Radar Altimeter (Rad Alt)	160
4.14	Automated Landing Aids	163

Contents | vii

4.15	Air Data System (ADS) 168
4.16	Accident Data Recording System (ADRS) 172
4.17	Electronic Flight Bag (EFB) 174
4.18	Prognostics and Health Management System (PHM) 178
4.19	Internal Lighting 181
4.20	Integrated Modular Architecture (IMA) 183
4.21	Antennas 185
	References 189

5	**Mission Systems** 191
5.1	Radar System 192
5.2	Electro-optical System 197
5.3	Electronic Support Measures (ESM) 200
5.4	Magnetic Anomaly Detection (MAD) 202
5.5	Acoustic System 205
5.6	Mission Computing System 207
5.7	Defensive Aids 209
5.8	Station Keeping System 212
5.9	Electronic Warfare System 214
5.10	Camera System 217
5.11	Head Up Display (HUD) 220
5.12	Helmet Mounted Systems 222
5.13	Data Link 224
5.14	Weapon System 227
5.15	Mission System Displays and Controls 230
5.16	Mission System Antennas 234
	References 237
	Further Reading 239
	Exercises 239

6	**Supporting Ground Systems** 241
6.1	Flight Test Data Analysis 243
6.2	Maintenance Management System 246
6.3	Accident Data Recording 248
6.4	Mission Data Management (Mission Support System) 250
6.5	UAV Control 252
	References 254
	Exercises 255

7	**Modelling of Systems Architectures** 257
7.1	Introduction 257
7.2	Literature Survey of Methods 259

7.3	Avionics Integration Architecture Methodology *277*
7.4	Avionics Integration Modelling of Optimisation *292*
7.5	Simulations and Results for a Sample Architecture *297*
7.6	Conclusion *300*
	References *300*

8 Summary and Future Developments *305*
8.1	Introduction *305*
8.2	Systems of Systems *305*
8.3	Architectures *314*
8.4	Other Considerations *315*
8.5	Conclusion *323*
8.6	What's Next? *323*
	Exercise *327*

Index *329*

About the Authors

Allan Seabridge was until 2006 the Chief Flight Systems Engineer at BAE SYSTEMS at Warton in Lancashire in the United Kingdom. In over 50 years in the aerospace industry, his work has included the opportunity to work on a wide range of BAE Systems projects including Canberra, Jaguar, Tornado, EAP, Typhoon, Nimrod, and an opportunity for act as reviewer for Hawk, Typhoon, and Joint Strike Fighter, as well being involved in project management, R&D, and business development. In addition, Allan has been involved in the development of a range of flight and avionics systems on a wide range of fast jets, training aircraft, and ground and maritime surveillance projects. From experience in BAE Systems with a Systems Engineering education, he is keen to encourage a further understanding of integrated engineering systems. An interest in engineering education continues since retirement with the design and delivery of systems and engineering courses at a number of UK universities at undergraduate and postgraduate level including: the Universities of Bristol, Cranfield, Lancaster, Loughborough, Manchester, and the University of the West of England. Allan has been involved at Cranfield University for many years and has served as an external examiner for the M.Sc course in Aerospace Vehicle Design, and as external examiner for MSc and PhD students.

Allan has co-authored a number of books in the Aerospace Series with Ian Moir, all published by John Wiley. He is currently a member of the BAE Systems Heritage Department at Warton and is fully involved in their activities, working closely with a colleague to produce a project history book published by the Heritage Group: EAP: The Experimental Aircraft Programme by Allan Seabridge and Leon Skorzcewski, which was published in 2016.

Mohammad Radaei has got a PhD in aerospace engineering specialized in avionics systems integration from Cranfield University, United Kingdom. He obtained his BSc in aeronautical engineering from Air University, and MSc in aerospace engineering, flight dynamics, and control from National University of Iran, Tehran. He also holds a commercial pilot license. Mohammad has been involved in two EU-funded projects including FUCAM and GAUSS during his PhD at Cranfield. His research interests are aircraft systems design, avionics systems integration and systems architecting, aircraft and avionics systems flight testing, applied mathematics, flight dynamics and control of manned and unmanned aircraft as well as Human-machine interaction. He is currently lecturing in avionics systems at a number of universities.

Acknowledgements

This work is the culmination of many years of work in the field of military and civil aircraft systems engineering. My work experience has been enriched by the opportunity to work with a number of universities at undergraduate and postgraduate level to develop and add to degree courses, where the delegates unwittingly became critics and guinea pigs for my subject matter. Discussions during the courses with the academics and the students have broadened my knowledge considerably. In particular I would like to mention the Universities of Manchester, Loughborough, Cranfield, Bristol, University of the West of England and Lancaster for their MSc, and short courses attended by students and engineers from industry.

My experience at Cranfield has played a big part in encouraging me to acquire information about aircraft systems that will be of use to engineers studying at undergraduate and post graduate level as well as those entering the workplace. Special thanks must go to Dr Craig Lawson, Dr Huamin Jia, and Professor Shijun Guo for inviting me to participate in their MSc modules in Air Vehicle Design and short courses in Aircraft Systems Design at Cranfield University. Their international students have been most attentive and have made significant contributions to my knowledge.

My thanks as always to Ian Moir, he and I worked on many books and courses. I have raided our past collaborations for information in order to produce a book that brings together information for all aircraft systems that is not based on implementation, but generic information about the interactions between systems that typifies modern complex aircraft.

We have received considerable help from the staff at Wiley especially Laura Poplawski and Sarah Lemore, as well as their proof readers, copy editors, and publishing and production staff.

Dent, Cumbria *Allan Seabridge*
UK, November 2021

I have been dreaming to write a book since I started my professional education in aerospace engineering and I should confess that writing a technical book is harder than I thought. Honestly, this would not be possible without Allan's endless support. I would like to thank Allan Seabridge who provided this opportunity for me and from whom I learned a lot in avionics data networking, hardware integration, and testing course at Cranfield University. He also supported me during my PhD as well as writing this book. Moreover, I would like to thank my PhD supervisors, Dr Huamin Jia and Dr Craig Lawson, for all their great advice and recommendations. I have benefited from their supervisions in many aspects including the method and attitude of scientific research as well as hard-working. Last but not least, I would like to thank my family and friends for their endless love and support.

November 2021   *Mohammad Radaei*

Sources of Background Information

In addition to the references included at the end of each chapter, the following sources of information are provided to allow readers to obtain a broader grasp of the topics addressed in this book.

Atmosphere and Climate: A collection of papers on the atmosphere and the effects of aviation on the environment. Part 9 of Encyclopedia of Aircraft Engineering, Green Aviation, Ed Ramesh Agarwal, Fayette Collier, Andreas Schäfer and Allan Seabridge. John Wiley & Sons.

Chapra, S.C. (2017). *Applied Numerical Methods with MATLAB for Engineers and Scientists*, 4th e. Mcgraw Hill.
Farouki, S. (2020). *Future Propulsion Systems and Energy Sources in Sustainable Aviation*. Wiley.
Kluever, C.A. (2018). *Space Flight Dynamics*. Wiley.
The Mathworks Inc. (2005). MATLAB, Simulink. www.mathworks.com.
Padfield, G.D. (2018). *Helicopter Flight Dynamics*, 3rd e. Wiley.
Sadrey, M.H. (2020). *Design of Unmanned Aerial Systems*. Wiley.
Seabridge, A. and Ian, M. (2020). *Design and Development of Aircraft Systems*, 3e. Wiley.
Torenbeek, E. (2020). *Essentials of Supersonic commercial aircraft conceptual design*. Wiley.
Yedavali, R.K. (2020). *Flight Dynamics and Control of Aero and Space Vehicles*. Wiley.

Glossary

This glossary is intended to be of assistance to readers of other documents provided in the references and sources of material in this Handbook. The Glossary contains, therefore, many more entries than the abbreviations, units, and terms used in this book. It will not be complete, terms change and new terms emerge. The Internet is a good place to find many terms, abbreviations, and acronyms in general use.

3D	three dimensional
4D	four dimensional
AAA	anti-aircraft artillery (triple A)
A&AE	Aircraft & Armament Evaluation (Squadron, Boscombe Down) see A&AEE
A&AEE	Aircraft & Armament Experimental Establishment
A4A	Airlines for America
AADL	architecture analysis and design language
ABL	airborne laser
ABS	automatic braking system
AC	airworthiness circular – document offering advice on specific aircraft operations
AC	alternating current
ACA	Agile Combat Aircraft
ACARS	aircraft communications and reporting system
ACARS	ARINC communications and reporting system
ACE	actuator control electronics
ACFD	advanced civil flight deck
ACK	receiver acknowledge
ACM	air cycle machine

ACM	air driven motor pump
ACO	ant colony optimisation
ACP	audio control panel
ACS	active control system
ACT	active control technology
A-D	analogue to digital
Ada	a high order software language
ADC	air data computer
ADC	analogue to digital conversion/converter
ADCN	Aircraft Data Communication Network
ADD	airstream direction detector
ADF	automatic direction finding
ADI	attitude direction indicator
ADIRS	air data and inertial reference system
ADIRU	air data and inertial reference unit (B777)
ADM	air data module
ADMC	actuator drive and monitoring computer
ADN	Avionics Data Network
ADP	air driven pump
ADR	accident data recorder
ADS-A	automatic dependent surveillance - address
ADS-B	automatic dependent surveillance - broadcast
ADU	actuator drive unit
ADV	Air Defence Variant (of Panavia Tornado)
AE	acoustic emission
AESA	active electronically scanned array
AEU	antenna electronic unit
AEW	airborne early warning
AEW&C	airborne early warning and control
AFCS	automatic flight control system
AFDC	autopilot flight director computer
AFDS	autopilot flight director system
AFDX	avionics full-duplex switched Ethernet
AGARD	advisory group for aerospace and development
AGC	automatic gain control
AH	ampere hour
AH	artificial horizon

AHARS	attitude and heading reference system
AI	airborne interception
AI	artificial intelligence
AICS	air intake control system
AIFF	advanced IFF
AIMS	aircraft information management system (B777)
AIT	Aeritalia
Al	aluminium
ALARM	air launched anti-radar missile
ALARP	as low as reasonably practical
ALF	ambient lighting facility
AlGaAs	aluminium gallium arsenide
ALT	barometric altitude
ALU	arithmetic logic unit
AM	amplitude modulation
AMAD	airframe mounted accessory gearbox
AMCC	Applied Micro Circuits Corporation
AMECS	advanced military engine control system
AMLCD	active matrix liquid crystal displays
AMP	air driven motor pump, avionics modification programme
AMRAAM	advanced medium range air to air missile
AMSU	aircraft motion sensor unit
ANO	air navigation order
ANP	actual navigation performance
AoA	angle of attack
AOC	Airline Operational Centre
AOR-E	Azores Oceanic Region - East
AOR-W	Azores Oceanic Region - West
AP	autopilot
APEX	application executive
APGS	auxiliary power generation system
API	application programming interface
APSCU	air supply and pressure control unit
APU	auxiliary power unit
ARI	Air Radio Installation
ARINC 400 series	ARINC specifications providing a design foundation for avionic equipment

ARINC 404	early ARINC standard relating to the packaging of avionic equipment
ARINC 429	widely used civil aviation data bus standard
ARINC 500 Series	ARINC specifications relating to the design of analogue avionic equipment
ARINC 578	ARINC standard relating to the design of VHF omni-range (VOR)
ARINC 579	ARINC standard relating to the design of instrument landing system (ILS)
ARINC 600	later ARINC standard relating to the packaging of avionic equipment
ARINC 600 Series	ARINC specifications relating to enabling technologies for avionic equipment
ARINC 629	ARINC standard relating to a 2 Mbit/s digital data bus
ARINC 664	ARINC standard relating to aircraft full multiplex (AFDX) digital data bus
ARINC 700 Series	ARINC specifications relating to the design of digital avionic equipment
ARINC 708	ARINC Standard relating to the design of weather radar
ARINC 755	ARINC standard relating to the design of multi-mode receivers (MMR)
ARINC	Air Radio Inc.
ARM	anti-radar missile, anti-radiation missile
ARP	aerospace recommended practice (SAE)
ASCB	avionics standard communications bus (Honeywell)
ASCII	American Standard Code for Information Interchange
ASE	aircraft survivability equipment
ASI	aircraft station interface, airspeed indicator
ASIC	application specific integrated circuit
ASPCU	air supply and pressure control unit
ASR	air sea rescue
ASR	anonymous subscriber messaging
ASRAAM	advanced short range air to air missile
AST	air staff target
AST	asynchronous transfer mode
ASTOR	airborne stand-off radar
ASUW	anti-surface unit warfare
ASW	anti-submarine warfare

ATA	advanced tactical aircraft
ATA	Air Transport Association
ATC	air traffic control
A to D	analogue to digital
ATE	automatic test equipment
ATF	advanced tactical fighter
ATF	altitude test facility
ATI	air transport indicator
ATM	air targeting mode
ATM	air transport management, air traffic management
ATN	Aeronautical Telecommunications Network
ATR	Air Transport Radio (LRU form factor or box size)
ATS	air traffic services
ATSU	air traffic service unit – Airbus unit to support FANS
AWACS	airborne warning and command system
AWG	American wire gauge
Az	azimuth
BAC	British Aircraft Corporation
BAe	British Aerospace (now BAE Systems)
BAG	bandwidth allocation group
BAT	battery
BC	bus controller
BCAR	British Civil Airworthiness Requirement
BCD	binary coded decimal
BFL	balanced field length
BFoV	binocular field of view
BGAN	broadcast global area network
BIT	built in test
BIU	bus interface unit
BLC	battery line contactors
BMS	business management system
BP	binary programming
BPCU	brake power control unit
bps	bits per second
BRNAV	basic area navigation in RNP
BSCU	brake system control unit
BTB	bus tie breakers

BTC	bus tie contactor
BTMU	brake temperature monitoring unit
BVR	beyond visual range
BWB	blended wing body
C band	C band (3.90–6.20 GHz)
C++	a programming language
C3, C^3	command, control, and communications
CA	course acquisition – GPS operational mode
CAA	Civil Airworthiness Authority
CAD	computer aided design
CAE	computer aided engineering
CAIV	cost as an independent variable
CAMU	communications and audio management unit
CANbus	controller area network bus
CAP	combat air patrol
CAS	close air support
CAS	calibrated air speed
CAST	Certification Authorities Software Team
Cat I	automatic approach category I, Cat I category I auto-land
Cat II	automatic approach category II, Cat II category II auto-land
Cat III	automatic approach category III
Cat IIIA	category IIIA auto-land
Cat IIIB	category IIIB auto-land
CB	circuit breaker
CBIT	continuous built in test
CBLS	carrier bombs light store
CCA	common cause analysis
CCD	charge coupled device
CCIP	continuously computed impact point
CCR	common computing resource
CCRP	continuously computed release point
CCS	communications control system
CCV	control configured vehicle
CD	collision detection
Cd/m^2	candela per square metre
CDR	critical design review
CDU	control and display unit

Glossary xxi

CEP	circular error probability
CF	constant frequency
CF	course to a fix
CFC	carbon fibre composite
CFC	chloro-fluoro-carbon compounds
CFD	computational fluid dynamics
CFIT	controlled flight into terrain
CFR	Code of Federal Regulations
CG, cg, C of G	centre of gravity
CHBDL	common high band data link
CIFU	cockpit interface unit
CLA	creeping line ahead – a maritime patrol search pattern
CLB	configurable logic block
CMA	Centralised Maintenance Application, common mode analysis
CMD	counter measures dispenser
C-MOS	complementary metal oxide semiconductor
CNI	communications, navigation identification
CNS	communications, navigation, surveillance
CO_2	carbon dioxide
Cold Soak	prolonged exposure to cold temperatures
COM	command channel
COMED	combined map and electronic display
COMINT	communications intelligence
COMMS	communications mode
COMPASS	Chinese equivalent of GPS
CORE/CoRE	controlled requirements expression
COTS	commercial off the shelf systems
CPIOM	central processor input output module
CPM	common processing module, core processing module
CPU	central processing unit
CRC	cyclic redundancy check
CRDC	common remote data concentrator (A350)
CRI	configuration reference item
CRM	crew resource management
CRT	cathode ray tube
CS	certification specification

CSAS	command stability augmentation system
CSD	constant speed drive
CSDB	commercial standard data bus
CSG	computer symbol generator
CSMA	carrier sense multiple access
CSMA/CD	carrier sense multiple access/collision detection
CTC	cabin temperature controller
Cu	copper
CVR	cockpit voice recorder
CVS	combined vision system
CW	continuous wave
CW/FM	continuous wave/frequency modulated
D to A, D-A	digital to analogue
DA	decision altitude
DAC	digital to analogue conversion/converter
DAL	design assurance level
DASS	defensive aids sub-system
dB	decibel
DBS	Doppler beam sharpening
DC	direct current
DCA	data concentration application
DCDU	data link control and display unit (Airbus)
DCMP	DC motor driven pump
DDVR	displays data video recorder
DECU	digital engine control unit
Def Stan	Defence Standard
DefAids	defensive aids sub-system
DF	direct to a fix
DF	direction finding
DFDM	direct force modes
DG	directional gyro
DGPS	differential GPS
DH	decision height
DIMA	distributed integrated modular avionics
DIP	dual in-line package
DIRCM	direct infrared counter measures
DLP	digital light projector

DMA	direct memory access
DMC	display management computer
DMD	digital micro-mirror
DME	distance measuring equipment
DMP	display management computer
DO	design office
DoA, DOA	direction of arrival
DoD	Department of Defence (US)
DOORS	a requirements management tool
Downey Cycle	procurement model once used in the UK MoD
DPX	a style of rear rack connector
D-RAM	dynamic random access memory
DRL	data requirements list
DSM	design structure matrix
DTED	digital terrain elevation data
DTI	Department of Trade and Industry
DTSA	dynamic time slot allocation
DU	display unit
DVI	direct voice input
DVO	direct vision optics
DVOR	Doppler VOR
E	east
EADI	electronic ADI
EAP	Experimental Aircraft Programme
EAS	equivalent airspeed
EASA	European Aviation Safety Administration
EC	European Community
ECA	European Combat Aircraft
ECAM	electronic crew alerting and monitoring (Airbus)
ECC	error correcting code
ECCM	electronic counter-counter measures
ECF	European Combat Fighter
ECL	electronic check list
ECM	electronic counter measures
ECS	environmental control system
ECU	electronic control unit
EDP	engine driven pump

EDR	engineering design requirement
EE	electrical equipment
EEC	electronic engine controller
EEPROM	electrically erasable and programmable read only memory
EEZ	economic exclusion zone
EFA	European Fighter Aircraft
EFB	electronic flight bag
EFIS	electronic flight instrument system
EGI	embedded GPS inertial
EGNOS	European Geostationary Navigation Overlay System
EGPWS	enhanced ground proximity warning system
EHA	electro hydrostatic actuator
EHF	extremely high frequency
EHP	electro-hydraulic pump
EHSI	electronic HSI
EHSV	electro-hydraulic servo valve
EICAS	engine indication and crew alerting system (Boeing)
EIS	entry into service, electronic instrumentation system
ELAC	elevator/aileron computer (A320)
ELCU	electrical load control unit
ELINT	electronic intelligence
ELMS	electrical load management system
EM	electro-magnetic
EMA	electro-mechanical actuator
EMC	electro-magnetic compatibility
EMCON	emission control
EMH	electro-magnetic health
EMI	electro-magnetic interference
EMP	electrical motor pumps, EMP electromagnetic pulse
EMR	electro-magnetic radiation
EO	electro-optical
EOB	electronic order of battle
EOF	end of frame
EOS	electro-optical system
EPB	external power breaker
EPC	electrical power contactor
EPLD	electrically programmable logic device

EPROM	electrically programmable read only memory
EPU	emergency power unit
ESA	electronically steered array, ESA European Space Agency
ESM	electronic support measures
ESS	environmental stress screening
Ess	essential
ETA	estimated time of arrival
ETOPS	extended twin operations
ETOX	erase-through-oxide
EU	European Union
EUROCAE	European Organisation for Civil Aviation Equipment
EVS	enhanced vision system (EASA nomenclature)
EW	electronic warfare
FA	fix to altitude
FAA	Federal Aviation Authority
FAC	flight augmentation computer (Airbus)
FADD	fatigue and defect damage
FADEC	full authority digital engine control
FAF	final approach fix
FANS	future air navigation system
FANS1	future air navigation system implemented by Boeing
FANSA	future air navigation system implemented by Airbus
FAR	Federal Airworthiness Requirements
FAV	first article verification
FBW	fly-by-wire
FCC	flight control computer
FCDC	flight control data concentrator
FCDU	flight control data concentrator unit
FCP	flight control panel
FCPC	flight control primary computer
FCR	fire control radar
FCS	flight control system
FCSC	flight control secondary computer (A330/340)
FCU	flight control unit
FD	flight director
FDDS	flight deck display system
FDX	fast switched Ethernet

FEBA	forward edge of the battle area
FET	field effect transistor
FFD	Ferranti Functional Documentation
FFS	formation flight system
FFT	fast Fourier transform
FGMC	flight management guidance computers – Airbus terminology for FMS
FHA	functional hazard analysis
FIFO	first in, first out
FL	flight level
fL	foot-lambert
FLIR	forward looking infrared
FLOTOX	floating gate tunnel oxide
FM	frequency modulation
FMEA	failure modes and effects analysis
FMECA	failure modes effects and criticality analysis
FMES	failure mode effects summary
FMGC	flight management guidance computer
FMGEC	flight management and guidance envelope computer (A330/340)
FMGU	flight management guidance unit
FMQGC	fuel management and quantity gauging computer
FMS	flight management system
FMS	foreign military sales
FMSP	flight mode selector panel
FOB	fuel on board
FOG	fibre optic gyro
FoR	field of regard
FORTRAN	formula translation, a software language
FoV	field of view
FPA	focal plane array
FPGA	field programmable logic array
FRACAS	failure reporting and corrective action system
FRD	functional requirements document
FRR	final readiness review
fs	sampling frequency
FSCC	flap/slat control computer

FSEU	flap slats electronic unit
FSF	flight safety foundation
FSK	frequency shift key
FTA	fault tree analysis
FTE	flight technical error
FTI	flight test instrumentation
FTP	foil twisted pair
Full duplex	a data bus that passes data in a bi-directional manner
FWC	flight warning computer
G&C	guidance and control
GA	genetic algorithm
GA	general aviation
GaAs	gallium arsenide
Galileo	European equivalent of GPS
GAMA	General Aviation Manufacturers Association
GATM	global air traffic management
GCB	generator control breaker
GCU	generator control unit
GEM	group equipment manufacturer
GEO	geostationary earth orbit
GEOS	geo-stationary satellite
GFE	government furnished equipment
GHz	gigahertz (10^9 Hz)
GINA (bus)	Gestion information numerique avionique
GLONASS	Russian equivalent of GPS
GMR	ground mapping radar
GMTI	ground moving target indicator
GNSS	Global Navigation Satellite System
GP	general purpose
gpm	gallons per minute
GPM	global processing module
GPS	global positioning system
GPWS	ground proximity warning system (see also TAWS)
GTS	ground targeting mode
GUI	graphical user interface
H	Earth's magnetic field

H/W	hardware
H$_2$O	water
Ha	height of aircraft
HALE	high altitude long endurance (UAV)
Half Duplex	a data bus that passes data in a uni-directional manner
HALT	hardware accelerated life test
HAS	hardware accomplishment summary
HDD	head down display(s)
HDMI	high definition multimedia interface
HEPA filter	high efficiency particulate air filter
HF	high frequency
HFDL	high frequency data link
HFDS	head-up flight display system (Thales)
Hg	mercury
HGS	head-up guidance system (Rockwell Collins)
HIRF	high intensity radio field
HISL	high intensity strobe light
HMD	helmet mounted displays
HMI	human–machine interface
HMS	helmet mounted sight
HOL	high order language
HOOD	Hierarchical Object Oriented Design
Hot soak	prolonged exposure to high temperatures
HOTAS	hands on throttle and stick
HP	horse power
HSA	Hawker Siddeley Aviation
HSD	horizontal situation display
HSI	horizontal situation indicator
Ht	height
HUD	head-up display
HVGS	head-up visual guidance system
HVP	hardware verification plan
HX	holding to a fix
H_X	X component of H
H_Y	Y component of H
Hz	Hertz
H_Z	Z component of H

I/O	input/output
IAC	integrated avionics cabinets
IAP	integrated actuator package
IAS	indicated airspeed
IATA	International Air Transport Association
IAWG	Industrial Avionics Working Group
IBIT	interruptive built in test
IC	integrated circuit
ICAO	International Civil Aviation Organisation
ICD	interface control document
ID	identifier
IDG	integrated drive generator
IED	Industrial Engineering Department
IEEE 1398	high speed data bus
IEPG	Independent European Programme Group
IF	initial fix
IFALPA	International Federation of Air Line Pilots Association
IFE	in-flight entertainment
IFF	identification friend or foe (see ADS-B)
IFF/SSR	identification friend or foe/secondary surveillance radar (ADS-B)
IFPCS	integrated flight and propulsion control system
IFR	instrument flight rules
IFSD	in flight shut down
IFSME	in flight structural mode excitation
IFU	interface unit
IFZ	independent fault zone
IGOS	inclined geo-stationary orbits
ILP	integer linear programming
ILS	instrument landing system
IMA	integrated modular architecture
IMINT	image intelligence
In Hg	inches of mercury
IN	inertial navigation
INCOSE	International Council on Systems Engineering
INMARSAT	International Maritime Satellite Organisation
INS	inertial navigation system

INU	inertial navigation unit
INV	inverter
IOC	interim operational clearance
IOR	Indian Ocean Region
IP	integer programming, internet protocol
IPC	initial provision cost
IPFD	Integrated primary flight display (Honeywell SVS)
IPR	intellectual property rights
IPT	Integrated Product Team
IR	infrared, infrared
IRS	inertial reference system
ISAR	inverse synthetic aperture radar
ISDOD	information system design and optimisation system
ISIS	integrated standby instrument system
ISO	International Organisation for Standardisation
IT	information technology
ITAR	International traffic in Arms Regulations
ITCZ	Inter-tropical Convergence Zone
JAA	Joint Airworthiness Authority
JAR	Joint Airworthiness Requirement
JARTS	Joint Aircraft Recovery and Transportation Squadron
JASC	Joint Aircraft System/Component (FAA)
JAST	Joint Advanced Strike Technology
JDAM	Joint Direct Attack Munition
JOVIAL	a high order software language
JSF	Joint Strike Fighter
JTIDS	Joint Tactical Information Distribution System
JTRS	Joint Tactical Radio System
K	Kelvin temperature scale unit
K^1	K^1 band (10.90–17.25 GHz)
Ka	KA band (36.00–46.00 GHz)
kbit	10^3 bit (kilo bit)
kbps	kilo bits per second
km	kilometres
Ku	Ku band (33.00–36.00 GHz)
kVA	kilo volt amps
kW	kilo Watt

L	L band (0.39–1.55 GHz)
LAAS	local area augmentation system
LAN	local area network
LBAS	locally based augmentation system
LCC	leadless chip carrier, life cycle cost
LCD	liquid crystal display
LCoS	liquid crystal on silicon
LE	leading edge
LED	light emitting diode
LF	low frequency
LGB	laser guided bomb
Link 11	naval tactical data link
Link 16	tactical data link (basis of JTIDS)
Link 22	see NILE
LLTI	long lead time items
LLTV	low light TV
LNAV	lateral navigation
LoC	lines of code
LORAN	hyperbolic navigation beacon system
LoS	line of sight
LOX	liquid oxygen
LP	linear programming, low pressure (engine pressure)
LPI	low probability of intercept
LPV	localiser performance with vertical guidance
LRG	laser rate gyro
LRI	line replaceable item
LRM	line replaceable module
LROPS	long range operations
LRU	line replaceable unit
Ls	Ls band (0.90–0.95 GHz)
LSB	least significant bit, lower side band
LSI	large scale integration
LTPB	linear token passing bus
LVDT	linear variable differential transformer
LVN	load classification number (runway)
LWF	light weight fighter (F-16)
LWIR	long wave infrared

MA	Markov analysis
MAC	media access control
Mach	the speed of an aircraft in relation to the speed of sound
MAD	magnetic anomaly detector
MAL	Marconi Avionics Ltd. (now BAE Systems)
MALE	medium altitude long endurance (UAV)
MASCOT	modular approach to software code, operation and test
MASPS	minimum aviation system performance standard
MAU	modular avionic units
MAW	missile approach warning
MBB	Messerschmitt-Bolkow-Blohm (now EADS part of Airbus)
MBD	model based development
Mbit	10^6 bit (mega bit)
Mbps	mega bits per second
MBSE	model based systems engineering
MCDM	multi criteria decision making
MCDU	multi-function control and display unit
MCU	modular concept unit
MDA	minimum descent altitude
MDC	miniature detonating cord
MDD	manufacturing design and development
MDH	minimum descent height
MDP	maintenance data panel
MEA	more electric aircraft
MEL	minimum equipment list
MEOS	medium Earth orbit satellite
MF	medium frequency
MFD	multi-function display
M-GEO	multi-objective generalised extremal optimisation
MHBK	Military Handbook – A US military publication
MHDD	multi-function head down display
MHRS	magnetic heading and reference system
MHz	megahertz, 10^6 Hz
MIL-STD	military standard
MIL-STD-1553	widely used military data bus standard
MIP	mixed integer programming
MIPS	million instructions per second

MISRA	Motor Industry Reliability Association
mK	milli Kelvin
MLS	Microwave Landing System
MMEL	master minimum equipment list
M_{mo}	maximum operating Mach number
MMR	multi-mode receiver
MMW	milli-metric wave
MoD (PE)	Ministry of Defence Procurement Executive
MoD	Ministry of Defence (UK)
Mode A	ATC mode signifying aircraft call sign (range and bearing)
Mode C	ATC mode signifying aircraft call sign (range, bearing, and altitude)
Mode S	ATC mode signifying additional data (range, bearing, altitude, unique ID)
Mon	monitor channel
MOPS	minimum operational performance standards
MOS	metal oxide semiconductor
MOSFET	metal oxide semiconductor field effect transistor
MoU	memorandum of understanding
MPA	maritime patrol aircraft
MPCD	multi-purpose control and display
MPCDU	multi-purpose control and display unit
MPP	master programme plan
mr	milli radian
MRTT	multi role tanker transport
MSI	medium scale integration
MSL	mean sea level
MTBF	mean time between failures
MTBR	mean time between removals
MTI	moving target indicator
MTOW	maximum take-off weight
MTR	marked target receiver
Mux	multiplexed
MVA	mega volt amps
MWIR	medium wave infrared
MWS	missile warning system

N	north
NA	numerical aperture
NACA	National Advisory Committee for Aeronautics
NASA	National Aeronautics and Space Administration
NATO	North Atlantic Treaty Organisation
NATS	National Air Transport System
Nav aids	navigation aids
NAV	navigation mode
NAV-WASS	navigation and weapon aiming sub-system
NBC	nuclear, biological and chemical
NBP	no break power
NC	numerically controlled
ND	navigation display
NDA	non-disclosure agreement
NDB	non directional beacon
NDT	non-destructive testing
NEMP	nuclear electromagnetic pulse
NETD	noise equivalent temperature difference
NextGen	next generation air transport system (USA)
NGL	Normalair Garrett Ltd.
NH	speed of rotation of the HP Turbine expressed in %
NiCd	nickel cadmium (battery)
NILE	NATO improved Link 11 (Link 22, Stanag 5522)
NOTAM	notice to airmen
NOx	nitrogen oxides
NRC	non-recurring costs
NRZ	non-return to zero
NSGA	non-dominated sorting genetic algorithm
NTSB	National Transportation Safety Board
NTSC	National TV Standards Committee
NVG	night vision goggles
NVIS	night vision imaging system
NVRAM	non-volatile random access memory
O_3	ozone
OAT	outside air temperature
OBIGGS	on-board inert gas generation system
OBOGS	on-board oxygen generation system

OEST	outline European staff target
OIC	operational interruption cost
O-LED	organic light emitting diode
OMG	object management group
OMT	object modelling technique
OOA	object oriented analysis
OOD	object oriented design
OOOI	OUT-OFF-ON-IN: the original simple ACARS message format
Op Amp	operational amplifier
P&D Test	Production and Development Test
PAPI	precision approach path indicator
PBIT	power-up built in test
PBN	performance based navigation
PBs	product breakdown structure
PC	personal computer
PCB	Plenum chamber burning
PCI	peripheral component interconnect
PCM	pulse code modulation
PCU	power control unit
PD	product development
P-DME, p-DME	precision DME
PDR	preliminary design review
PDT	pilot demand transmitter
PDU	pilot's display unit
PED	personal electronic device
PFC	primary flight control computer
PFD	primary flight display
PFR	primary flight reference
PHAC	plan for hardware aspects of certification
PHM	prognostics and health management
PIO	pilot induced oscillation
PJND	perceived just noticeable difference
PLB	personal locator beacon
PLD	programmable logic device
PMA	permanent magnet alternator
PMG	permanent magnet generator

POR	Pacific Ocean Region
PoR	point of regulation
PowerPC	Power optimisation with enhanced RISC microprocessor architecture
PPI	plan position indicator
PPS	present positioning service
PRA	particular risks analysis
PRF	pulse repetition frequency
PRNAV	precision area navigation
PROM	programmable read only memory
PRR	production readiness review
PRSOV	pressure reducing shut off valve
Ps	static pressure
PSA	problem statement analyser
PSEU	proximity switch electronic unit
Psi	pounds per square inch
PSL	problem statement language
PSO	particle swarm optimisation
PSR	primary surveillance radar
PSSA	preliminary system safety assessment
PST	post stall technology
PSU	power supply unit
Pt	total pressure
PTU	power transfer unit
q	dynamic pressure
QA	quality assurance
QFE	elevation
QMS	quality management system
QNH	barometric altitude
Quadrax	data bus wiring technique favoured by Airbus
R&D	research and development
RA	radar altimeter, radio altimeter, resolution advisory
Rad Alt	radar altimeter
RADINT	radar intelligence
RAE	Royal Aircraft Establishment
RAeS	Royal Aeronautical Society
RAF	Royal Air Force

RAIM	receiver autonomous integrity monitoring
RAM	radar absorbent material
RAM	random access memory
RASP	recognised air surface picture
RAT	ram air turbine
RCS	radar cross-section
RDC	remote data concentrator
RDP	radar data processor
RF	constant radius to a fix, radio frequency
RFI	request for information
RFP	request for proposal
RFU	radio frequency unit
RHAG	runway hydraulic arrestor gear
RIO	remote input/output
RISC	reduced instruction set computer/computing
RIU	remote interface unit
RLG	ring laser gyro
RM&T	reliability, maintainability and testability
RMI	radio magnetic indicator
RMP	radio management panel
RN	Royal Navy
RNAV (GNSS)	see RNP APCH
RNAV (GPS)	see RNP APCH
RNAV	area navigation
RNP APCH	RNP approach
RNP AR APCH	RNP with authorisation required approach
RNP	required navigation performance
ROE	rules of engagement
ROM	read only memory
RPDU	remote power distribution unit
RPK	rolling piano key
RPV	remotely piloted vehicle
RR	Rolls-Royce
RSAF	requirement specifications for avionics functions
RSRE	Royal Signals and Radar Establishment (UK)
RSS	root sum squares
RT	remote terminal

RTA	required time of arrival
RTCA	Radio Technical Committee Association
RTL	register transfer level
RTOS	real time operating system
RTR	remote transmission request
RTZ	return to zero
RVDT	rotary variable differential transformer
RVR	runway visual range
RVSM	reduced vertical separation minimum
RWR	radar warning receiver
Rx	receiver, receive
S	S band (1.55–5.20 GHz)
S/UTP	shielded unscreened twisted pair
S/W	software
SA	simulated annealing
SAAAR	special aircraft and aircrew authorisation required (US equivalent of RNP APCH)
SAARU	secondary attitude and air data reference unit (B777)
SAE	Society of Automomotive Engineers
SAFRA	semi-automated requirements analysis
SAHRS	stand-by attitude and heading reference system
SAHRU	secondary attitude and heading reference unit
SAM	surface to air missile
SAR	search and rescue, synthetic aperture radar
SARS	severe acute respiratory syndrome
SAS	standard altimeter setting
SAT	static air temperature
SATCOM	satellite communications
SATNAV	satellite navigation
SAW	simple additive weighting, surface acoustic wave
SB	side band
SBAC	Society of British Aerospace Companies (UK)
SBAS	space based augmentation system
SC	Special Committee 213 (RTCA/MASPS)
SDD	system design document
SDR	system design review
SDU	satellite drive unit

SEAD	suppression of enemy air defence
SEC	spoiler elevator computer
SELCAL	selective calling
SEP	specific excess power
SESAR JU	SESAR Joint Understanding
SESAR	Single European Sky ATM Research
SFC, sfc	specific fuel consumption
SFCC	slat/flap control computer (A330/340)
SG	synchronisation gap
SGU	signal generator unit
SH	sample and hold
SHF	super high frequency
SI	Smiths Industries (now GE Aviation)
SIAP	standard instrument approach procedure
SID	standard instrument departure
SIFT	synthetic in-flight training
SIGINT	signals intelligence
SIM	serial interface module
SiO_2	silicon dioxide
SKE	station keeping equipment
SLC	side lobe clutter
SLR	sideways looking radar
SMD	surface mount device
SMS	stores management system
SMT	surface mount technology
SNMP	simple network management protocol
SOF	start of frame
SOIC	small outline integrated circuit
SOV	shut off valve
SOW	statement of work
SPC	statistical process control
SPFDB	super plastic formed diffusion bonding
SPS	secondary power system
S-RAM	static random access memory
SRR	system requirements review
SSA	system safety assessment
SSB	single side band

SSC	ship set cost
SSD	solid state device
SSI	small scale integration
SSPC	solid state power controller
SSR	secondary surveillance radar
SST	supersonic transport
SSTP	shielded screen twisted pair
Stanag	standardisation agreement (NATO)
STAR	standard terminal arrival route
STC	supplementary type certificate
Stn	station
STOVL	short take-off vertical landing
STP	screened twisted pair
STR	sustained turn rate
SV	servo valve
SVS	synthetic vision system
SWIR	short wave infrared
SWR	software requirement
SysML	Systems Modelling Language
System of systems	a systems embracing a collection of other systems
T3CAS	traffic/terrain/transponder collision avoidance system
TA	traffic advisory
TACAN	tactical air navigation
TACCO	tactical commander
TAS	true air speed
TAT	total air temperature
TAWS	terrain avoidance warning system
TBD	to be determined
TBT	turbine blade temperature
TCAS	traffic collision avoidance system
TCP	tri-cresyl phosphate
TCP/IP	transport control procedure/internet protocol
TCV	terminally configured vehicle
TDMA	time division multiplex access
TDZ	touchdown zone
Terprom	terrain profile mapping
TF	track to a fix

TFR	terrain following radar
TFTP	trivial file transfer protocol
TG	terminal gap
THS	tailplane horizontal stabilator
TI	transmission interval
TIALD	thermal imaging and laser designation
TIR	total internal reflection
Tp	total pressure
TPMU	tyre pressure monitoring unit
TR, T/R	transmitter/receiver, transmit/receive
TRL	technology readiness level
TRU	transformer rectifier unit
TSO	technical standards order
TTL	transistor–transistor logic
TTP	time triggered protocol
TV	television
TWT	travelling wave tube
Tx	transmit, transmitter
UART	universal asynchronous receiver transmitter
UAV	unmanned air vehicle
UCAV	unmanned combat air vehicle
UCS	utilities control system
UDP	user datagram protocol
UHF	ultra high frequency
UK	United Kingdom
ULA	uncommitted logic array
UML	Unified Modelling Language
UOR	unplanned operational requirements
US, USA	United States of America
USB	upper side band
USMS	utility systems management system
UTP	un-shielded twisted pair
UV	ultra violet
V	velocity
V/STOL	vertical/short take-off and landing
V/UHF	combined VHF and UHF radio

V/UHF	very/ultra high frequency
VAC	volts AC
VALID	variability and life data
VAPS	a tool for modelling displays and controls formats
VDC	volts DC
VDR	VHF digital radio
VF	variable frequency
VFR	visual flight rules
VGA	video graphics adapter
VGS	vertical guidance system (Honeywell/BAE Systems)
VHDL	very high speed integrated hardware description language
VHF	very high frequency
VHFDL	very high frequency data link
VIFF	vectoring in forward flight
VL	virtual link
VLF	very low frequency
VLSI	very large scale integration
VMC	visual meteorological conditions
V_{mo}	maximum operating speed
VMS	vehicle management system
VNAV	vertical navigation
VOC	volatile oil compound
VOR	VHF omni-ranging
VORTAC	VOR TACAN
VS	vertical speed
VSCF	variable speed constant frequency
VSD	vertical situation display
VSI	vertical speed indicator
VTOL	vertical take-off and landing
W	Watt
WAAS	wide area augmentation system
WBS	work breakdown structure
WFG	waveform generator
WGS 84	World Geodetic System of 1984
WOW	weight on wheels
WSM	weighted sum method

WWII	World War II
X	X axis
X	X band (5.20–10.90 GHz)
Xb	Xb band (6.25–6.90 GHz)
Y	Y axis
Z	Z axis
ZOH	zeroth order hold
ZSA	zonal safety analysis
ρ	air density (rho)

1

Introduction

Design and Development of Aircraft Systems 3rd edition by Allan Seabridge and Ian Moir covers this material in a single chapter as a series of brief tables. This handbook is intended to provide more detailed information and some historical information about the development of each system and also to indicate how the systems will develop in the future. The reason for this is to give information to people working on heritage or older aircraft still in use, to span current types in operation and to provide assistance to those looking at future projects. The intention is to provide a full set of information for all aircraft systems in one single volume which complements other books in this field and is suitable for suitable for practitioners, students, graduates, and apprentices in aerospace.

People in the aircraft industry will expect to work on aircraft on all types and ages in their career. As a result they will be exposed to different technologies, different design methods, and different certification procedures. Typical aircraft will include

- Aircraft on display in museums
- Flying aircraft in heritage fleets
- Aircraft in private ownership
- Aircraft in routine operation by state owned airlines
- Aircraft in operation with commercial business organisations
- Aircraft in military air force operations
- Aircraft close to end of life
- Newly developed aircraft about to enter into service

The life span of aircraft in these classifications ranges from over a hundred years to a few years and spans many technologies. Some aircraft have a relatively low utilisation rate, whereas others are in operation on a daily basis. Their daily flight may be uncomplicated and stress free, or they may be consuming structural life. This means that the wear and tear of system components will vary and that many aircraft will be subject to routine maintenance and repairs. Knowledge of these scenarios is important to anyone engaged in support of the aircraft.

This book is to provide information for people researching or working with heritage aircraft, current in-service aircraft and future projects and to provide a broad but brief description of each of the aircraft systems likely to be encountered. There are references to enable the reader to find more detailed information, suggestions for further reading to expand their horizons, and naturally there is always Google to rely on – other search engines are available.

Each chapter of this book contains the following information for each aircraft system:

- Key characteristics of all aircraft systems
- Clear descriptions with diagrams where appropriate
- Full glossary and bibliography

Readers preferably will require knowledge of the aircraft industry and its products at introduction level with sufficient links to encourage reading to seek more detailed information. The book is intended to help people who do not have that knowledge but need to know more in certain areas.

The content is intended to be of interest to people intending to join or already working in

- Organisations directly involved in the design, development, and manufacture of manned and unmanned, fixed-wing and rotary-wing aircraft – both military and commercial.
- Systems and equipment supply companies involved in providing services, sub-systems, equipment, and components to the manufacturers of aviation products.
- Organisations involved in the repair, maintenance, and overhaul of aircraft for their own use or on behalf of commercial or military operators.
- Commercial Airlines and armed forces operating their own or leased aircraft on a daily basis.
- Organisations involved in the training of personnel to work on aircraft.

The book is also aimed at educational establishments involved in the teaching of systems engineering, aerospace engineering, or specialist branches of the topic such as avionics or equipment engineering at high school, university undergraduate, or postgraduate level. It is also suitable for short courses intended for the professional development of industry professionals and practitioners. It will be useful for those who do not work directly in engineering, but have a role to play in the industry. Examples include the following:

- Support departments in the aerospace industry such as commercial, contracts, procurement, legal, training, communications, and public relations.

- Supply chain company staff
- Colleges, training academies, universities.
- Journalists, aerospace magazine/journal writers
- Practitioners in the aerospace industry – prime contractors and suppliers
- Graduates and apprentices in the aerospace industry – prime contractors and suppliers
- Undergraduates and postgraduates at universities offering aerospace related courses

These are the sort of people who will be found in the broad range of stakeholders in complex aerospace projects, illustrated in Figure 2.5. This gives an example of the aviation system and some of the people and groups affected by the systems or directly affecting the system. This diagram has been developed to illustrate the stakeholders in the development of a particular aircraft solution as an example to show that each specific project and indeed each system will have its own specific set of stakeholders.

The intention of this book is to provide a basic understanding of the principles of practical systems engineering, not to justify or to recommend specific processes or tools. Examples will be used to illustrate the principles; however, it is important to note that there is not one single "right" approach to an engineering process –nor need there be. As long as there is consistency of approach in the partners in a project, and as long as the process works, then that is the correct approach for that project. This understanding will be particularly useful to engineers designing systems or equipment and will provide essential background information for engineers or technicians using or maintaining the systems.

What this book aspires to do is to create an open-minded approach so that systems engineers feel comfortable that the process they have chosen will produce a safe and successful result. It will also serve to introduce people to the language, jargon, and terms used in industry. Hence, there are no solutions in this book, no definitive architectures. One reason for this is that such architectures go out of date quickly, and they are usually proprietary information so they will be sharply focused on the aims of the company designing the aircraft from which the solution is taken. Another reason is that it is not always good design practice to be overly influenced by previous designs. This book provides an opportunity for readers to understand their system and its requirements and ensure that the system is not only fit for purpose in its own right but also fits into other systems in the total aircraft architecture.

Exercises have been included at the end of most chapters to encourage readers to develop their reading of the chapter. There are no answers given; in many cases, there are no 'right answers', but doing the work, alone or in groups, will help to develop the skills of understanding a system and developing it to a firm solution.

Many references and suggestions for further reading have been provided to assist in this process, and the Internet serves as a source of further information.

Chapter 2 addresses the general nature of an aircraft system and leads to a definition of such systems in the context of a physical application. Some characteristics of systems and their environments will be introduced to encourage the reader to adopt a behavioural skill of broad systems thinking when addressing the analysis and design of systems. This description will include the associated ground systems such as those required for the support and logistics organisations to analyse fault and prognostic information, as well as the systems required to operate and analyse the information collected by Unmanned Air Systems for real-time operations.

Chapter 3 presents the key characteristics of all aircraft systems providing a brief summary of what each system is and to providing references to source material for further detailed descriptions.

Chapter 4 presents the key characteristics of all avionic systems providing a brief summary of what each system is and to providing references to source material for further detailed descriptions.

Chapter 5 presents the key characteristics of all mission systems providing a brief summary of what each system is and to providing references to source material for further detailed descriptions.

Chapter 6 briefly describes some of the ground-based systems which link directly to the operation of the aircraft. Interfaces with these systems must be considered in the design of the airborne system. Some of the functions of ground-based systems will be found in the system diagrams in Chapters 4, 5, and 6.

Chapter 7 provides a literature review of research in progress to provide a mechanism for modeling complex system architectures in order to find an optimum solution. A recently developed model is described in detail using proprietary modeling tools.

Chapter 8 presents a summary of the book, some discussion and poses some questions for the future.

Further Reading

Marais, K., Wolfe, P.J., and Waitz, L.A. (2016). *Air Transport and the Environment*
    (ed. P. Belobaba, A. Odone and C. Barnhardt). The Global Airline Industry. Wiley.
Mitkoff, A., Hansome, J., and Reynolds, T.G. (2016). *Airline Flight Operations*
    (ed. P. Belobaba, A. Odone and C. Barnhardt). The Global Airline Industry. Wiley.
Odoni, A. (2016). *The International Institutional and Regulatory Environment*
    (ed. P. Belobaba, A. Odone and C. Barnhardt). The Global Airline Industry. Wiley.

2

The Airframe and Systems Overview

2.1 Introduction

This chapter introduces the airframe and its systems to start to explore the interactions between them. The chapter also introduces the mechanism by which the key characteristics of each will be described in the following chapters. An important point to be understood is how the systems exist as individuals for the purpose of design responsibility, but they are also inextricably linked as a result of interfaces and interactions resulting from their own functional behaviours, and also by the actions of the humans in the system – the aircrew, the cabin crew, and ground operations. There are factors that act in common on all systems directly and indirectly. They may occur randomly throughout the life of the system and although they may appear at first glance to be unpredictable, they are, nevertheless, well understood and must be included in design considerations. These factors include environmental conditions such as temperature, vibration, and electromagnetic effects. It is important to keep this in view at every stage of the design because these interactions are not always readily apparent.

The aircraft is a combination of the airframe and its systems that give the aircraft a purpose and the means to achieve that purpose with the aid of human operators. It is therefore a combination that requires many interconnections and interactions, some intentional, indeed designed, and others that may not have been intended but happen nevertheless, sometimes known as unintended consequences.

Figure 2.1 shows this combination – a view that is applicable to fixed wing and rotary wing types whatever their purpose in life. This includes aircraft-specific roles such as:

- Commercial long haul passenger
- Commercial regional passenger
- Commercial freight
- Commercial business

Aircraft Systems Classifications: A Handbook of Characteristics and Design Guidelines, First Edition.
Allan Seabridge and Mohammad Radaei.
© 2022 John Wiley & Sons, Inc. Published 2022 by John Wiley & Sons, Inc.

2 The Airframe and Systems Overview

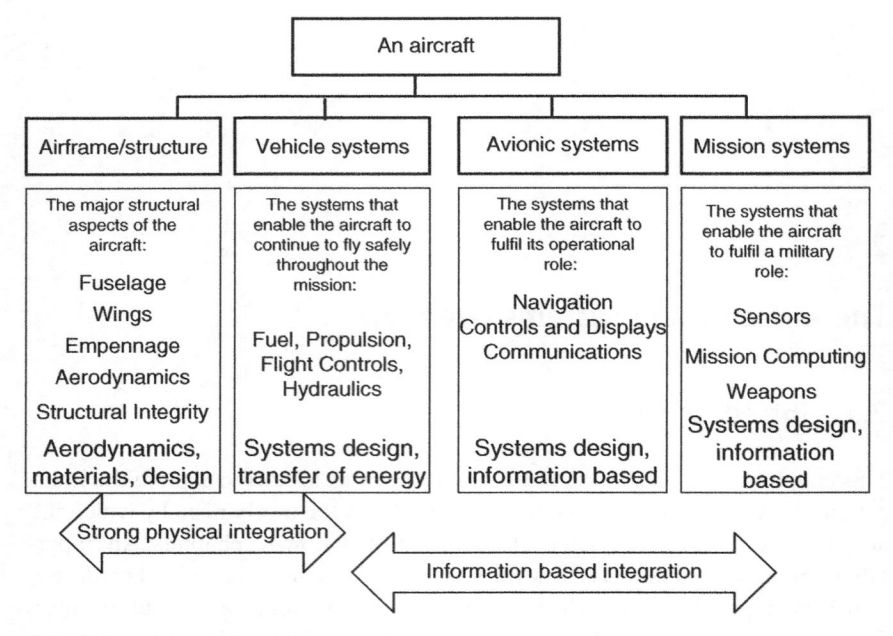

Figure 2.1 The aircraft as a set of systems.

- Military fast jet
- Military surveillance
- Military freight
- Commercial helicopters
- Military helicopters
- Unmanned air vehicles

The detailed implementation of this combination will differ from type to type, and there will be many commonalities between commercial and military and also some essential differences (Figure 2.2).

2.2 The Airframe

Not surprisingly, the airframe is a network of interconnecting and integrated threads. The departmental structure of many aerospace organisations tends to treat these threads as single independent disciplines. However, there is much to be gained by treating them as an integrated entity.

The structure is intended to withstand all the loads associated with flight, with landing and decelerating, with turbulence as well as vibration, climatic conditions,

2.2 The Airframe

Environment
- Aero loads
- Landing loads
- Lightning
- UV
- Cosmic radiation
- Solar radiation

Systems
- Installation
- Mass
- Reaction
- Wiring
- Earth/bond

AIRFRAME STRUCTURE

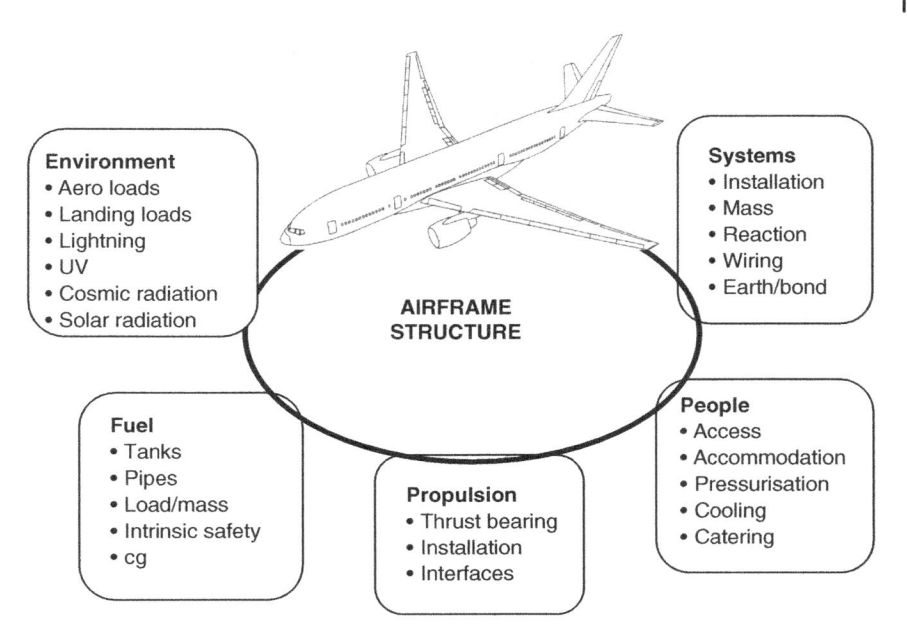

Fuel
- Tanks
- Pipes
- Load/mass
- Intrinsic safety
- cg

Propulsion
- Thrust bearing
- Installation
- Interfaces

People
- Access
- Accommodation
- Pressurisation
- Cooling
- Catering

Figure 2.2 The airframe as an integrated system.

and age. It may be constructed of different materials including aluminium alloys, composites, titanium, and plastics, all with different characteristics of strength, corrosion resistance, fire resistance, and electrical conductivity.

The structural components are combined to form a pressure vessel that is the life-support system for the occupants providing a pressurised environment to allow the human body to absorb oxygen in suitable temperature and humidity conditions.

It is an integral portion of all electrically powered systems since the structure provides the zero potential or 'earth' for all equipment and cable screens. It is also the 'bond' connection for all metallic components such as equipment housings, fuel, and hydraulic pipes – vital to preserve intrinsic safety.

The structure forms a Faraday cage to protect the internal equipment from radio interference effects and also to restrict radio interference escaping from the aircraft to the outside world, an important aspect of some military operations.

Finally, it is the storage location for fuel in wing and fuselage tanks. The continuous movement of fuel and its temperature gradients exert loads on some airframe components. To minimise some load changes, the way in which fuel is moved from tank to tank is scheduled to limit the range of movement of the centre of gravity of fuel, and indeed to maintain the aircraft centre of gravity within precise limits for reasons of fuel economy and for safety.

2 The Airframe and Systems Overview

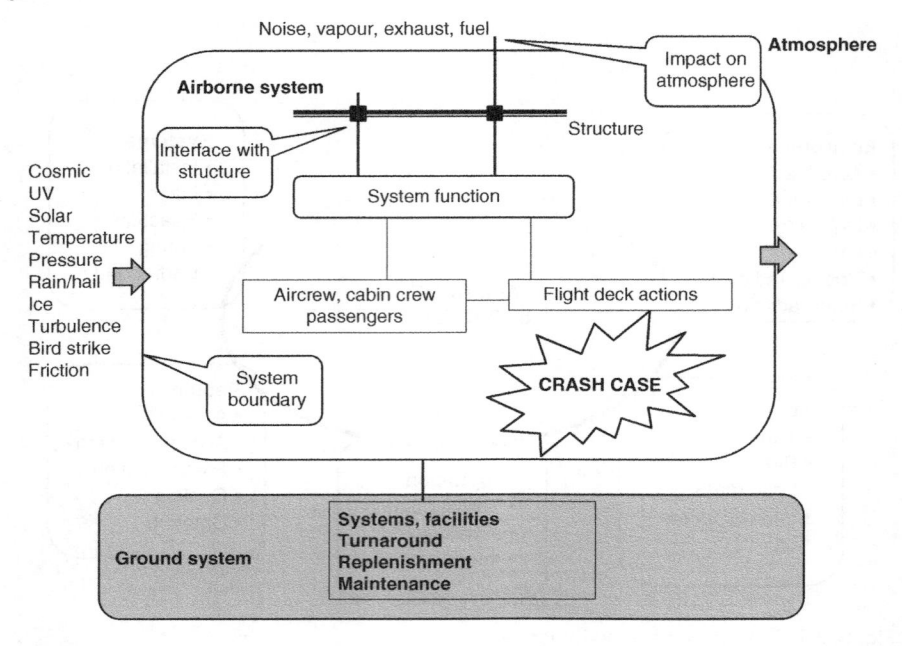

Figure 2.3 Airframe system diagram.

A system diagram for the airframe is shown in Figure 2.3 to show the main interactions within the airframe system and also interaction with the environment or atmosphere, many of which are predictable and known physical entities but are beyond the control of the designer and operator.

2.2.1 Impact on the structure

Loads are exerted by systems such as flight control via the moving control surfaces, the landing gear, and with loads from the oleos and from runway and braking loads. The flight control system on a fast jet aircraft will impose severe transient loads in high g manoeuvres and carrier landings will impose loads through the main wheels and the arrestor hook. A high load is imposed on the nose wheel by the catapult system.

2.2.2 Impact on Atmosphere

The aircraft has a significant impact on the atmosphere. Noise is most likely felt not only by humans in the vicinity of airports but also by the over-flown population, wild life and farm animals on approach paths. Modern engines are considerably better than heritage types, but those affected may declare that further improvement is necessary. Vapour trails (contrails) are said to be the cause of cirrus cloud formation. Engine exhaust contains a number of compounds known to cause atmospheric pollution. Fuel vapour will be present in situations in which fuel vents are opened or in un-burnt fuel from the auxiliary power unit and engines at airports.

2.2.3 Atmosphere Impact on Structure

The airframe is subjected to various external radiation phenomena that have an effect on the aircraft systems. Solar radiation heats the structure and is transmitted to the interior. This is particularly pronounced when the aircraft is parked for long periods in direct sunlight – known as hot soak. This will affect the cabin interior temperatures and rapid cooling is needed for passengers as they embark or for maintenance crews working in the cabin or equipment bays. The fast jet cockpit with a small volume and a large transparency is particularly susceptible and temperatures in desert conditions have been known to be high enough to damage switches and cockpit equipment. Hot soak conditions cause fuel to expand and a fully laden aircraft can damage tarmac by fuel venting to the ground.

Ultra violet radiation has a deteriorating effect on plastic materials and antenna covers are susceptible at long duration high-altitude operations.

Ionising radiation from space in the form of cosmic radiation will penetrate the fuselage and all equipment. It is especially damaging at high altitude where there is no atmospheric attenuation. There is no material that will prevent cosmic ray penetration of aircraft. The impact on crew can be regulated by working to a mandated dose rate which determines how many hours a crew member can fly about a certain altitude. Management of crew rosters can be designed to accommodate this.

Non-ionising radiation from radars must be reduced by careful design of the radar lobes and by installation of radar absorbent material on bulkheads. In some cases, it will be necessary to reduce radar transmission on the ground for the safety of ground crew – often by using the oleo weight on wheels signal.

In dense flash memory, individual cells can be affected since a micro-chip device may be struck by a cosmic neutron every few seconds. This can lead to a temporary change of state that recovers after power down. For critical systems, it may be sensible to implement error detection and recovery routines into software.

The exterior of the aircraft is affected by a number of conditions including driving rain/hail, ice, and bird strike. All of these will cause damage to the exterior finish and may affect sensors such as air data probes and radomes. In conditions where rough runways are frequently used, it may be necessary to install stone deflectors on the wheels to reduce damage to equipment installed in wheel bays.

Turbulence may introduce severe shock to passengers and their seats as well as to equipment mounted in the airframe. Security of attachment of equipment and of modules inside equipment must be addressed in specifications.

Bird strike is an unexpected event but must be included in the aircraft design to reduce damage to structure, canopy, windscreen, probes, and engines.

2.2.4 The Crash Case

The crash case is a specific design case that affects the aircraft, equipment, and human occupants. The case must be included in all specifications. Specifically, it accepts that in some cases a crash or severe deceleration may occur and that certain levels of integrity must be provided to allow safe evacuation of inhabitants and limitation of fire. This will include security of attachment of seats, cabin equipment, and electronic equipment. This is intended to reduce serious injury to passengers and to cabin crew and to maintain exits and pathways clear to allow safe evacuation. To reduce the risk of fire, there may be a mechanism for shutting off fuel to the engines and to isolate all electrical supplies except those needed for emergency lighting.

2.3 The Aircraft Systems

The aircraft systems are described individually in Chapters 3, 4, and 5 in a common format starting with a system diagram and then a list of key characteristics and their explanations for each system individually. Where there are aspects of integration with other systems, they are explained. The true issue of integration can only be determined by examining the entire set of systems installed in the airframe.

2.3.1 System Diagram

Each set of system characteristics is preceded by a system diagram as illustrated in Figure 2.4. This is intended to give a representation of the key functions and interfaces of the system to give an overall impression of the interconnections and integration to be expected. The diagram also gives information on interfaces with ground systems, the aircraft structure, and the atmosphere. In effect, it is a tool that can be used to develop the system conceptual design. In this book, a single diagram model has been used for consistency; however, anyone involved in system design should make their own representation. The diagram acts as a scratch pad to note all aspects of the design which will be added to and amended in all developing discussion – by pencil or a tablet to record all discussions and decisions until the final issue of the system design as part of the design record.

Accidents in modern aviation are commonly the result of failures in one system affecting another system. Therefore, it is important for the systems designer to consider how a malfunction in their system will affect the operation of another

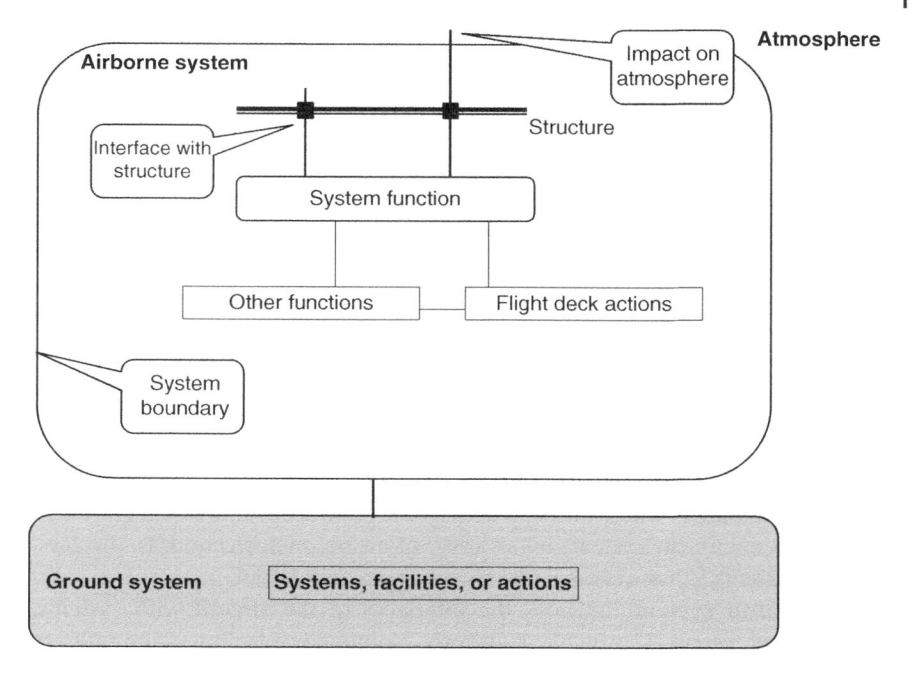

Figure 2.4 Illustration of a system diagram.

system, which requires close coordination across the traditional systems boundaries. The system diagram provides information about inter-system connectivity which can be used to analyse the system behaviour in the event of a failure.

The system boundary is intended to encompass the main aspects of a system and its functions as well as interfaces with other systems that are needed to complete the overall function. The structure is included within the system boundary to show that it is part of the system. Any interfaces with the structure are shown in the diagram to indicate that the structural designers must be consulted.

Interface with the structure is intended to show all interfaces such as an attachment, a breach of the pressure cabin or the provision of doors, hatches, and covers. This is to show that consideration has been given to all such breaches and to indicate where the agreement of structural designers must be obtained and indicated by signature approval on drawings. Note that many interfaces with the structure are contained within the system boundary, even if they protrude into the atmosphere.

Impact on the atmosphere indicates where any aspect of a system has an impact on the environment. This may be in the form of leaks, deliberate discharges such as venting or fuel dumping, emissions from exhausts such as engine or APU, and any other form of consideration. This is important for any individual system so that compliance with safety and environmental legislation can be demonstrated. It can also be used at a higher level to enable a record to be made of all aircraft environmental implications, which may form part of the final qualification statement for the type.

Control function defines all the functions of a system. They may be performed by a control unit specific to that system or performed within one of the aircraft's computing networks. In the past, separate control units were used with the function performed in relay logic, semiconductor logic, semiconductor amplifiers, and switching devices. It is now more probable that the functions are implemented in software hosted in one of the computing systems such a vehicle systems management, avionic computing, or mission computing as appropriate.

Other functions may be required for the completion of the system, and they may be performed in other systems within the aircraft. These functions need to be described in detail and agreement reached with the providers to ensure compatibility.

The flight deck is where the human machine interface is addressed. The pilots will interface with the systems by a variety of means such as controls, displays, warning systems, aural warning, and speech input and output.

The ground system indicates the interfaces of the aircraft with systems, facilities, and resources available at airports, maintenance facilities, and military facilities such as airfields and carriers, some of which will be remote from any permanent home base.

2.3.2 Key Characteristics of Systems

In Chapters 3, 4, and 5, the systems will be described individually in the format shown below:

System title: The name by which the system is usually known.
Purpose of system: A brief description of the purpose of the system.
Description: A brief description of the system physical and functional characteristics.
Safety/integrity aspects: Impact on flight safety or mission availability and redundancy considerations. See notes on dispatch criteria (1, 2). For this reason, some of the definitions offered are oversimplified:
(1) For civil aircraft, some of these criteria vary greatly by vehicle type and systems, route to be flown and other operational issues, and appropriate limitations which may apply. These are defined by the aircraft Master Minimum Equipment List (MMEL) in accordance with EASA OPS.MLR.105 which specifies the requirements for producing a Master Equipment List from the MMEL.

(2) Military aircraft will have a similar MMEL equivalent for airworthiness considerations but in addition the availability of mission sensors will dictate whether the allocated mission may be prosecuted or not.

Key integration aspects: Opportunities and reasons for integration with other systems.

Key interfaces: Physical, functional, or human machine interfaces.

Key design drivers: Those design drivers having a major impact on systems engineering decisions.

Modelling: Suggestions on modelling and simulation tools are available to model the system as a part of the design process. Tools should be chosen with care and used with appropriate control of the tool and its configuration. Tools are a guide only, and the results obtained must be supplemented by testing as part of the qualification process.

References: Numbered references to further sources of information are provided as a source of information on the characteristics of individual systems. Details of the sources can be found in the References section at the end of each chapter.

For specific system designs, information should be sought in general of project-specific sources including:

- Customer requirements
- Standards
- Company design manuals
- Company codes of practice
- Supplier documents

Future considerations: Notes indicating future trends or new requirements arising from integration or environmental considerations.

Best practice and lessons learned: There have been incidents and accidents that have brought to light serious issues of design. Investigations, formal or otherwise, will usually lead to improvements in design or operations to reduce further incidents. The lessons learned from these investigations will be used in future designs to improve the product. It is important to note that there will be a cache of knowledge about errors in design within individual companies and this source must be used.

Each of the paragraphs in Chapters 3, 4, and 5 will be preceded by a system diagram to illustrate the main components of the system. The diagram will also give an indication of the complexity of individual systems and their interactions with other systems and the ground environment. This diagram format is shown in Figure 2.4. As appropriate, an example illustration of key systems will also be included to add some technical relevance.

2.4 Classification of Aircraft Roles

Many aircrafts are designed to suit particular roles and may remain in those roles for the duration of their life. Some roles, however, can be achieved by modifying aircraft designed for one purpose and re-equipping them for one or more alternative roles. It is always convenient to classify aircraft as commercial or military, but there are clear examples where use of aircraft by law enforcement agencies will blur the boundaries of the aircraft and the systems being used. In order to provide a suitable indication of the roles and the impact on the design of systems, this chapter will describe the various roles as follows:

- Commercial
- Military
- Law enforcement, emergency services and civilian agencies

There are applications in the military field for which the civil aircraft platform together with its avionic systems is well suited. It may often be economically viable to convert an existing commercial type rather than to develop a new military project. Much of the development costs of the structure and basic avionics will have been recouped from airline sales for a new platform. Alternatively, a used aircraft bought from an airline may also be an economic solution. In either case, the basic avionics fitted will have been well tried and tested and will be ideally suited for use in controlled airspace. The basic platform is often known as a 'Green' aircraft before conversion. Typical applications for reuse of commercial aircraft include the following:

- Personnel, materiel, and vehicle transport
- Air to air refuelling tanker
- Maritime patrol
- Airborne early warning (AEW)
- Ground surveillance
- Electronic warfare
- Flying classroom
- Range target/security aircraft

There are also instances where law enforcement and civilian agencies have made use of aircraft for surveillance. These include Police, Customs and Excise, Fisheries Protection, and drug enforcement agencies. These applications are often civilian helicopters or small commercial aircraft.

2.4.1 Commercial

Commercial aircraft are principally designed for paying passengers travelling on leisure or business. They can fly short haul or regional types or they can fly transcontinental very long duration flight. The pattern of use means that commercial aviation has earned a reputation for safe flying at commercial rates. This has made certain types eminently suitable for conversion to military use.

2.4.2 General Aviation

The International Civil Aviation Organization (ICAO) defines general aviation (GA) as a category representing the private transport and recreational components of aviation. It also includes activities surrounding aircraft homebuilding, flight training, flying clubs, and aerial application, as well as forms of charitable and humanitarian transportation. Private flights are made in a wide variety of aircraft: light and ultra-light aircraft, sport aircraft, business aircraft gliders, and helicopters. Flights can be carried out under both visual flight rules (VFRs) and instrument flight rules (IFRs), and can use controlled airspace with permission.

Much of the world's air traffic falls into the category of general aviation, and many of the world's airports serve GA exclusively.

Business aviation is the use of any general aviation aircraft for a business purpose. The Federal Aviation Administration defines general aviation as all flights that are not conducted by the military or the scheduled airlines. As such, business aviation is a part of general aviation that focuses on the business use of airplanes and helicopters. Some business aircraft are suitable for conversion to small-scale surveillance roles and have been used very successfully.

2.4.3 Regional

The ICAO has no definition for regional aircraft, they are merely a type of commercial air transport. A regional airliner or a feederliner is a small airliner that is designed to fly up to 100 passengers on short-haul flights, usually feeding larger carriers' airline hubs from small markets, and are also used for short trips between smaller towns or from a larger city to a smaller city.

Examples: 50-seat Canadair Regional Jet CRJ-100/200, 70–112 seat BAe 146 (or Avro Regional Jet), 34–50 seat Embraer Regional Jet ERJ-135/140/145,66–124 seat Embraer E-Jet, 98 seat Sukhoi Superjet 100, 78–98 seat Comac ARJ21, or Advanced Regional Jet; 78–92 seat Mitsubishi SpaceJet.

2.4.4 Long Haul

Long-haul and ultra-long fall into the category of Commercial Air Transport. There is no fixed definition of the term long haul, but it can be considered as any direct or non-stop flight that has a journey time of between 6 and 12 hours. An ultra-long haul flight is any flight that flies non-stop for over 12 hours.

Examples: B737, B767, B777, B787, A320, A350, and A380.

2.4.5 Military

Some military aircraft types are truly multi-role, but many other types will especially be designed for a particular role. Each of the roles designated will require a specific combination of systems and weapons to be successful. Some typical roles and example aircraft fixed wing types are described below. This is by no means an exhaustive list as new requirements arise to meet new threats, and in any event, many of the roles and types of operation will be subject to government classification.

For some roles, a large slow moving platform capable of transit over long distances and loitering for long periods of time, together with internal space for a mission crew and their workstations is ideal. Many commercial aircraft are capable of conversion to these roles, retaining the basic structure, avionics, and flight deck with the installation of additional avionics or mission systems to tailor the aircraft to a specific role. Retaining the basic avionic systems architecture makes good sense, since military aircraft make extensive use of civilian-controlled airspace during peacetime, and will often transit to theatres of operation, for training and defence purposes, using commercial routes.

The conversion to a military role, especially if the carriage of weapons is included, requires a different approach to safety and qualification, challenging the aircraft design teams to make the best use of civil and military certification rules. This often poses interesting problems in the mixing of design standards and processes.

The basic avionic systems are complemented by a set of sensors and systems to perform specific surveillance tasks. This is a situation in which the basic navigation and communication systems become part of the role-specific systems and in which there are particular issues of accuracy, integration, and security. These are especially important in instances where a commercial aircraft platform forms the basis of the military vehicle. In such instances, there may be conflicts between the characteristics of the embedded systems on the commercial vehicle, and the requirements of the military vehicle. These issues may affect the approach to design and certification of the resultant aircraft.

2.4.6 Air Superiority

The main role of the air superiority aircraft is to deny to an enemy the airspace over the battlefield to enable ground attack aircraft freedom to destroy ground targets and assist ground forces. This type is usually designed to enable the pilot to respond rapidly to a deployment call, climb to intercept or to loiter on combat air patrol, and to engage enemy forces, preferably beyond visual range. The aircraft may be directed or vectored to the engagement or to act autonomously if it has the appropriate systems and weapons. Usually, single crew, although some roles demand a second crew to operate the electronic and weapon systems.

A typical mission system would include the following:

- Radar
- Electronic support measures (ESM)
- Defensive aids sub-system (DASS)
- Mission computer
- Data loader

A typical weapon system would include the following:

- Air-to-air missile
- Internal gun

Examples: Rafale, Typhoon, F-15, F-16, F-18, F-35, F-117, Mig-21, and Mig-23.

2.4.7 Ground Attack

This type has been designed to assist the tactical situation on the battlefield. The pilot must be able to identify the right target among the ground clutter of the multiple enemy and friendly forces. This role includes close air support in which the aircraft may be directed by ground forces to deploy weapons that will be deployed in close proximity to friendly forces.

This is a role that may be assigned to air superiority vehicles as a so-called 'swing role'.

A typical mission system would include

- Radar
- Electro-optics
- Electronic support measures (ESM)
- Defensive aids sub-system (DASS)
- Laser designator
- Mission computer
- Data loader
- Cameras

A typical weapon system would include the following:

- Air-to-ground missile
- Free fall bombs
- Laser-guided bombs
- Airfield denial weapons
- Internal gun/gun pod
- Rockets

Examples: Jaguar, Harrier, A-10

2.4.8 Strategic Bombing

To penetrate deep into enemy territory and to carry out strikes to weaken defences, destroy strategic assets, destroy transport links, and to undermine the morale of the population. The strategic bomber was usually a high-altitude aircraft carrying a large bomb load. The modern aircraft is designed to fly low and fast, relying on stealth to evade enemy defences. Weapons such as cruise missiles and stand-off weapons can be used to replace mass bombing.

A typical mission system would include the following:

- Radar
- Electro-optics
- Electronic support measures (ESM)
- Defensive aids sub-system (DASS)
- Mission computer
- Data loader
- Cameras

A typical weapon system would include the following:

- Air-to-ground missile
- Free fall bombs
- Laser guided bombs
- Airfield denial weapons
- Cruise missiles

Examples: B-52, Tornado, B-2, B-18

2.4.9 Maritime Patrol

Over 60% of the earth's surface is covered by oceans – a natural resource that is exploited by many means: as a medium for transportation of cargo, as a source of food, as a mean of deploying naval assets such as capital ships and submarines and for the transport of men and materiel. It is also used for pleasure and for criminal purposes such as the smuggling of drugs, arms, alcohol, and people. All these activities are of interest to all maritime nations and their defence, police, and customs authorities.

The most practical way of carrying out surveillance is by air with close cooperation of naval, in-shore, and land forces. The information gathered by airborne sensors is combined with other forces information, with intelligence, and with satellite information. This is used to understand the situation and to determine the most suitable response. Some maritime patrol aircraft, usually military, are capable not only of gathering information but also of responding with suitable firepower.

2.4 Classification of Aircraft Roles

Typical roles performed include anti-surface unit warfare, anti-submarine warfare, search and rescue, and exclusive economic zone protection. To do this requires an aircraft with long range, long duration, large mission crew, sensors, and weapons.

A typical mission system would include

- Maritime radar
- Electro-optics
- Electronic support measures (ESM)
- Defensive aids sub-system (DASS)
- Magnetic anomaly detector (MAD)
- Acoustic system
- Mission computer
- Mission recording
- Data loader
- Cameras
- Oceanographic data base
- Mission crew workstations
- Intelligence data bases

A typical weapon system would include the following:

- Free fall bombs
- Anti-ship missiles
- Torpedoes
- Air/sea rescue (ASR) packs
- Flares
- Smoke markers
- Sonobuoys
- Mines

Examples: Nimrod MR2, P-3, P-7, Atlantic,

2.4.10 Battlefield Surveillance

Detailed knowledge of the tactical scenario on the battlefield is of importance to military commanders and planners who need real-time intelligence of enemy and friendly force disposition, size, and movement. Many commercial aircraft have been converted to this role in addition to specific military types. Suitable radar is installed on the upper or lower surface of the aircraft designed to obtain an oblique view of the ground. The aircraft will fly a fixed pattern outside the range of enemy defences to detect fixed and moving targets, the occurrence of which will be correlated to other intelligence.

A typical mission system would include the following:

- Radar
- Electro-optics
- Electronic support measures (ESM)
- Defensive aids sub-system (DASS)
- Moving target indicator
- Mission computer
- Mission recording
- Data loader
- Cameras
- Mission crew workstations
- Intelligence data bases

Examples: E-8 JSTARS, ASTOR

2.4.11 Airborne Early Warning (AEW)

Early detection and warning of airborne attack is important to give air superiority and defensive forces sufficient time to prepare a sound defence. It is also important to alert ground and naval forces of impending attack to allow for suitable defence, evasion, and countermeasures action. Operating from high altitude gives the AEW aircraft an advantage of detecting hostile aircraft at longer range than surface radar. A long-range, long-endurance aircraft allows a patrol pattern to be set up to cover a wide sector area from which attack is most likely. A radar with 360° scan and the capability to look down and look up provides detection of incoming low-level and high-altitude threats. The radar will usually be integrated with an interrogator (Identification friend or foe (IFF)/secondary surveillance) system to enable friendly aircraft to be positively identified. The aircraft will also act as an airborne command post to monitor and control all air movements in the tactical area, compiling intelligence and providing near real-time displays of the tactical situation to both local forces and remote headquarters (HQ).

A typical mission system would include

- AEW radar
- Electronic support measures (ESM)
- Defensive aids sub-system (DASS)
- Mission computer
- Mission recording
- Data loader
- Mission crew workstations
- Intelligence data bases

Examples; E-2 Hawkeye, E-3 Sentry, P-3 AEW, Tu-126 AEW, Sea King.

2.4.12 Electronic Warfare

The role of electronic warfare covers a number of tasks that required different sensors and equipment fits, including

- Electronic countermeasures to provide a mechanism for jamming or spoofing enemy defences to disrupt radar or communications. This can take the form of the emission of high-power signals at the same frequency of the threat emitter or receiver. An alternative means is the deployment of chaff and flares to create false targets.
- ESM is used to intercept, locate, analyse, and record radiated electromagnetic energy.
- Signals intelligence is used to monitor radio and radar signals to identify the type and nature of the transmitting system to aid intelligence gathering.
- Electronic countermeasures is used to try to avoid or overcome the enemy countermeasures and apply to own aircraft.

This role may be undertaken by an aircraft specifically designated and equipped for the task or some of the sub-roles can be incorporated into another role fit. In some instances, aircraft may be designed to perform only one sub-role alone.

A typical mission system would include

- Electronic support measures (ESM)
- Defensive aids sub-system (DASS)
- Mission computer
- Mission recording
- Data loader
- Mission crew workstations
- Intelligence data bases

2.4.13 Photographic Reconnaissance

Photographic imagery is used to confirm signals intelligence (SIGINT) by using high-resolution permanent or digital images using ground mapping cameras. Images can also be used to confirm targets and to assess damage after an operation. Such images are mainly obtained by satellite systems but high- or low-flying aircraft are used to overfly the battlefield or by overflying territory in peace-time

A typical mission system would include

- High-resolution cameras
- Film storage
- Film processing

- Infrared cameras
- Optional forward-looking infrared (FLIR)
- Optional synthetic aperture radar (SAR)

Examples: Canberra PR9, U-2

2.4.14 Air-to-Air Refuelling

Military aircraft of nearly all types find it necessary to extend their range or endurance. This may be the result of the global nature of conflict leading to the need to deploy over long distances to a theatre of operations, for ferry flight over long distances or to remain on combat air patrol for long duration. These aircraft will be equipped with a mechanism to allow contact with a tanker and to receive fuel at high pressure. The tanker aircraft will require large fuel storage tanks and a means of connecting with the receiving aircraft. Both types will need the necessary equipment to locate each other and to home to a suitable air space.

A typical mission system would include

- Radar
- Electronic support measures (ESM)
- Defensive aids sub-system (DASS)
- Mission computer
- Data loader
- Refuelling equipment

Examples: KC-135, P-9, Boeing KC-46 Pegasus.

2.4.15 Troop and Materiel Transport

The deployed nature of conflict and peacekeeping operations demands the movement of troops and materiel to remote theatres of operation. Whilst the bulk of the task is performed by marine transport, rapid deployment in order to establish a military position requires fast air transport. This is often seen by the military as force projection – the ability to establish a rapid presence in time of tension. The role can often include the dropping of stores, ammunition and light vehicles during a low speed, low-level transit, or by dropping by parachute. Such types are also used to assist in evacuation or to carry relief supplies for humanitarian reasons.

A typical mission system would include

- Radar
- Electro-optics turret
- Electronic support measures (ESM)
- Defensive aids sub-system (DASS)

- Data loader
- Station keeping equipment (SKE)
- Materiel deployment
- Paratroop deployment

Examples: C-130, C-17, C-5, AN124.

2.4.16 Training Aircraft

Training of aircrew is an important task, from primary training through to conversion to type, to refresher training to maintain combat readiness. Training is accomplished by a variety of means. For fast jet types, initial training (*ab initio*) is usually carried out using a dynamic simulator to gain familiarisation before transferring to a single-engine light aircraft, then on to a single-engine jet trainer. Conversion to an operational role again uses simulation before transferring to two seat version and then to solo.

Routine mission and tactics can be performed on dynamic simulators designed to have a high degree of fidelity in six degrees of motion and a computer generated outside world picture. Realism is an important aspect of training and nations have designated areas of their territory for low-level or combat training. This includes ranges where live weapons can be used in a safe environment. A typical mission system would include

- Sensors, systems, and weapons particular to the role for which the crew are being trained.
- On-board simulation of example operational scenarios.

Examples: Hawk, MB339, MB326, Alphajet, EMB-312 Tucano, Texan T-1, T-46, T-38

2.4.17 Unmanned Air Vehicles

Unmanned air vehicles are now commonplace, on the battlefield at least, where they are seen carrying out surveillance missions and deploying weapons with great accuracy. They are the solution to some of the dull, dirty, and dangerous tasks for which they were first proposed reducing risk on the battlefield and helping to perform accurate identification of assets and civilians on the ground.

Apart from military applications, there are many jobs to be performed in commercial and government applications in surveillance, monitoring and trouble-shooting in the fields of utilities, maritime rescue, customs and excise, and agriculture to name only a few. Some police forces are collaborating with industry to develop systems to replace helicopter surveillance. The main drawback

2 The Airframe and Systems Overview

Table 2.1 Classification of unmanned aerial vehicles.

Class	Weight (kg)	Size	Altitude (ft)	Range (km)	Endurance (h)
Micro	<0.25	<10 cm	<100	0.1–0.5	<1
Mini	0.25–1	10–30 cm	<500	0.5–1	<1
Very small	1–2.5	30–50 cm	<1,000	1–5	1–3
Small	2.5–25	0.5–2 m	1,000–5,000	10–100	<2
Medium	25–500	5–10 m	10,000–15,000	500–2,000	3–10
Large	5,000–15,000	20–50 m	20,000–40,000	1,000–5,000	10–20
Tactical/combat	500–10,000	10–30 m	10,000–30,000	500–2,000	5–12
Male	500–5,000	15–40 m	15,000–30,000	20,000–40,000	20–40
Hale	>2,500	20–50 m	50,000–70,000	20,000–40,000	30–50
Quadcopter	0.25–50	0.1–1 m	<500	0.1–2	<1
Helicopter	0.001–100	13 mm–2 m	<500	0.2–5	<2

Source: Austin (2010) and Sadrey (2020).

to their application has been the difficulty in obtaining certification to operate in controlled airspace.

There are many classes of unmanned vehicle in existence, and many types within each class, developed by many manufacturers. The classes range from insect-like vehicles, through hand-launched sub-scale models to full-size long endurance types. They are all capable of carrying some form of sensor and of relaying sensor information to the ground. Most are remotely piloted, and some experimental types are being operated on controlled ranges as part of a gradual progression toward full autonomy, where they will be capable of performing missions with minimum human intervention.

The classification of types of UAV has been described, and a terminology has emerged. Table 2.1 lists the types, and their classification and descriptions are available in a number of books (Austin, 2010; Sadrey, 2020).

2.4.18 Special Roles

Military aircrafts are often called upon to perform roles beyond their original design intention. This may be for research and development of sensors and systems, development of new tactics, for intelligence gathering, or for peacetime information gathering missions. Conversion to these roles may be by major modification to the type, by interim experimental modification or by adding payload externally such as in under-wing or under-fuselage pods. Some examples include the following:

- Gunship
- Casualty evacuation
- Covert troop deployments

2.4.19 Law Enforcement and Civilian Agencies

Considerable use of some aircraft types is made by civilian agencies for surveillance. These include Police, Customs and Excise, Fisheries Protection, and drug enforcement agencies for law enforcement purposes. Use by civilian agencies includes power line inspection, nuclear emission detection, atmospheric pollution monitoring, air ambulance, and aerial survey.

These applications often use civilian helicopters, small commercial aircraft, or unmanned aerial systems. The need to get quickly to a scene of an incident and the ability to search wide areas rapidly is important to achieve a speedy resolution of issues. If data is required for evidence in criminal or civilian prosecutions, then that evidence must be gathered lawfully and recorded. Suitable modifications to the basic platform are designed to incorporate suitable sensors. These modifications include the following:

- Additional communications to communicate with ground agencies for voice or data.
- Sensors to detect and track activity, often electro-optical or thermal imaging.

2.5 Classification of Systems

The total set of aircraft systems described in Chapters 3, 4, and 5 has been classified by the author into Vehicle, Avionics, and Mission systems. To a certain extent, this is a convenience, almost a traditional method of naming systems. Many organisations have structured their engineering departments in this way, and many engineers acquire their skills to suit. They elect to be educated in a way that allows them to be employed to fit into the organisation. Many supply chain organisations have carved out niches in these specialist areas.

It is the task of aircraft systems engineers to use these disparate systems and to blend them into a coherent whole in the aircraft to fulfil military and commercial roles and to satisfy the regulatory authorities of their safety and viability.

It is instructive to further classify the systems in other ways that will influence the way in which they are designed and qualified. Systems may be grouped into classification to which a common set of design procedures can be applied for the common good.

This classification of systems must not be rigidly applied, but it may help to consider systems and to define common standards and procedures for design within a project to establish a robust and safe design. As well as focussing the project toward a safe design, the existence of such standards and procedures will be useful to establish trust with the customer. Some example classifications include the following:

Safety critical systems are those in which a failure will lead to loss or severe damage to the aircraft or death or serious injury of aircraft inhabitants or the over-flown population. Such systems must be designed to strict rules to achieve a failure rate of less than 1.10^{-9} per flying hour. Multiple redundancy is often employed to achieve a fail-operational/fail-safe system (i.e. dual fault tolerance), and the process employed for software design must ensure error free results.

Safety involved systems are those that do not have full responsibility for controlling hazards such as loss of life or severe injury. The malfunction of a safety-involved system would only be hazardous in conjunction with the failure of other systems or human error.

Non-critical systems are those systems where the loss of function can be tolerated, even if it is a nuisance. An example is the galley on a commercial aircraft which may cause customer dissatisfaction but is not likely to cause any damage or injury.

Mission critical systems are those in which a failure would result in a return to base or an alternative base, whilst abandoning the original mission.

Dispatch critical is a condition in which failures can be tolerated for the next phase of flight is there is sufficient redundancy in the systems to tolerate a further failure without hazard. Thus, an aircraft may be dispatched with a known fault for a leg which returns it to a maintenance base.

Emergency use systems are those that are used to complete the mission after a failure is apparent that affects the operation of the aircraft. Examples include an arrestor hook to stop the aircraft or an emergency source of power to supplement the main engine driven source.

Dormant systems are those that cannot be tested pre-flight. Emergency use systems generally fall into this category.

2.6 Stakeholders

In order to obtain information and agreements that are essential to design systems it will be necessary to consult with all relevant parties – referred to in this book as stakeholders. Figure 2.5 is an example of the stakeholders in the aviation industry to illustrate the number and diverse needs of people and organisations. This is a

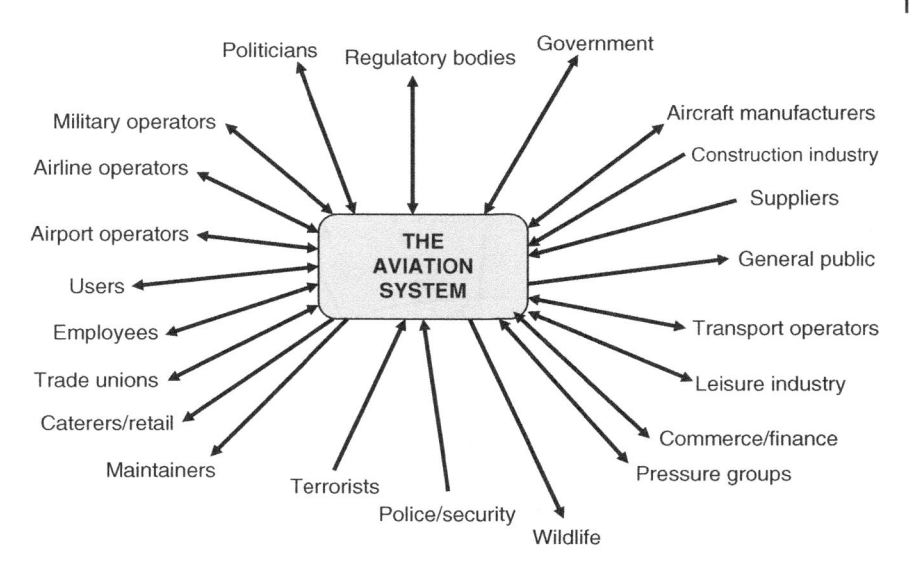

Figure 2.5 Stakeholders in the aviation system.

very generic example, and it will need to be tailored to suit a specific project to ensure that requirements are understood by all parties and agreed at significant points in the design process.

Figure 2.6 shows such a project-related stakeholder diagram. Each of these stakeholders will have a different perspective of the design and development process and each is capable of exerting an influence on the process. For those directly involved, it is vital that the design process is visible to all parties so that they can coordinate their contributions for maximum benefit to the project. A clear and well-documented process is essential to allow the stakeholders to visualise the design and development path as a framework in which to discuss their different perspectives. This can be used to establish boundaries, to air differences of opinion, and to arbitrate on differences of technical, commercial, or legal understanding.

2.7 Example Architectures

Figure 2.7 shows a top-level architecture in which blocks of systems are segregated into title groups – the titles which will be used in Chapters 3, 4, and 5. Effectively, the systems have been classified into groups which suit the concept for the project. This may fit in with the organisation work structure, and it may also fit into other classifications. It may well be that different rules will be applied to the design of vehicle systems to the rules used for avionics or mission systems.

28 | *2 The Airframe and Systems Overview*

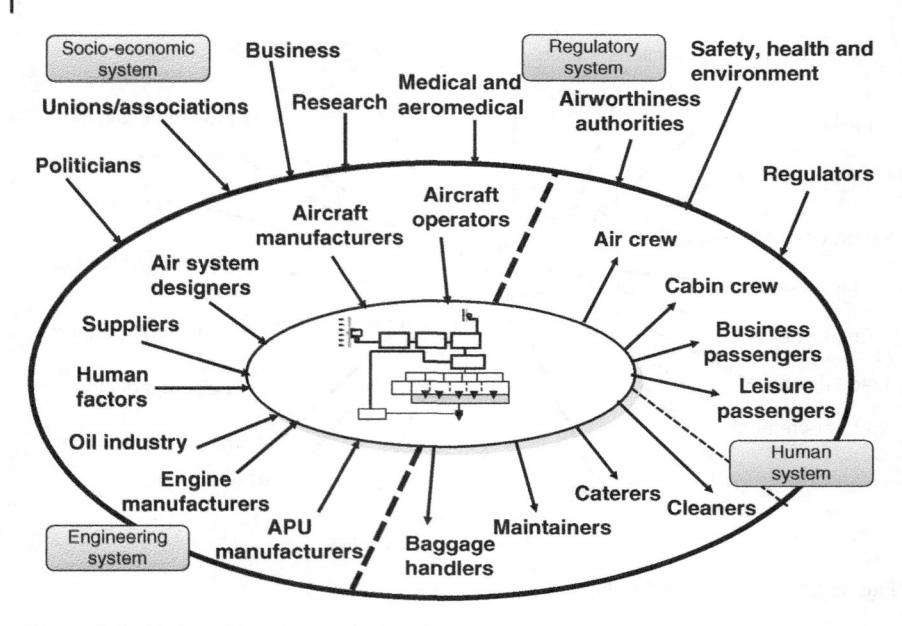

Figure 2.6 Stakeholders in a typical project.

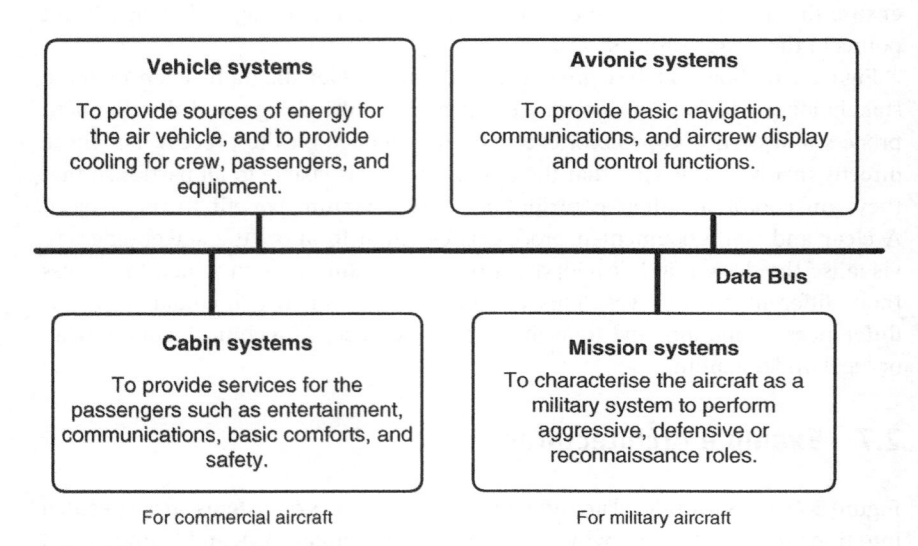

Vehicle systems	Avionic systems
To provide sources of energy for the air vehicle, and to provide cooling for crew, passengers, and equipment.	To provide basic navigation, communications, and aircrew display and control functions.

Data Bus

Cabin systems	Mission systems
To provide services for the passengers such as entertainment, communications, basic comforts, and safety.	To characterise the aircraft as a military system to perform aggressive, defensive or reconnaissance roles.

For commercial aircraft   For military aircraft

Figure 2.7 Example of top-level architecture.

2.8 Data Bus | 29

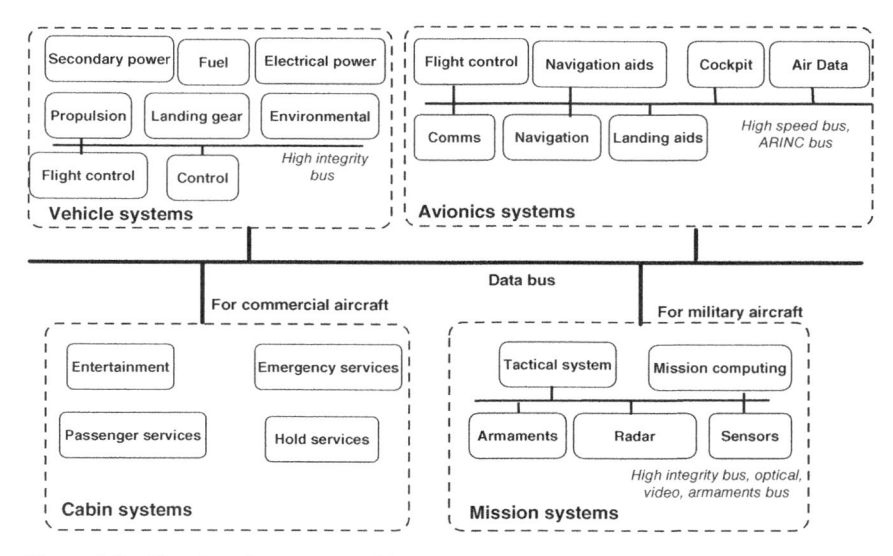

Figure 2.8 The aircraft systems architecture.

Figure 2.8 shows a more detailed architecture in which it is clear that rules are already being applied – note the description of the data bus for example. Vehicle systems will be the subject of Chapter 3, avionics the subject of Chapter 4, and mission systems the subject of Chapter 5; for convenience cabin systems are shown as a separate block because they form a major difference between commercial and military types and their functions can be incorporated into vehicle systems or avionics.

2.8 Data Bus

The data bus included in these architectures is a defining feature of the way in which some systems have evolved. The data bus can be considered a system in its own right with its own implementation and protocols. There will be no detailed description in the following chapters: it will be an assumed property of the vehicle management system, the avionics systems, and the mission system.

The development of data buses is illustrated in Figure 2.9 showing their typical transmission rates and applications. Further information on the methods of operation of the data bus types can be found in Moir and Seabridge (2013) Chapter 12 and Moir and Seabridge (2008) Chapter 3, and the standards for the bus types.

The aim of the data bus is to enable large volumes of data to be exchanged between systems at varying speeds and with a high degree of reliability. The most common data bus types in use today include the following:

2 The Airframe and Systems Overview

Data rate	Data bus	Application
1 Gbps	◄ IEEE 1394b; 800 Mbps	F-35 JSF
100 Mbps	◄ ARINC 664-P7; 100 Mbps	Airbus A380 Boeing 787
10 Mbps	◄ STANAG 3910: 20 Mbps	Typhoon, Rafale
1 Mbps	◄ MIL-STD-1553B; 1 Mbps ◄ CANbus; 1 Mbps ◄ ASCB; 670 kbps	Very widely used in military aerospace community Automotive Business jets
100 kbps	◄ ARINC 429: 100 kbps ◄ CSDB: 50 kbps	Very widely used in civil aerospace community General aviation

Figure 2.9 Commonly used data bus types.

ARINC 429 is by far the most common data bus in use on civil transport aircraft, regional jets, and executive business jets flying today. It had developed into a common standard for interfaces and functions for much avionic equipment. Since its introduction on the Boeing 757/767 aircraft and on Airbus aircraft in the early 1980s, hardly an aircraft has been produced which does not utilise this data bus. It is a single source, multi-sync, linear topology data bus. It is transmitted on a twisted pair screened cable using bipolar return-to-zero encoding. Up to 20 receiving terminals can be connected onto the bus. Connections can be made by simple splicing into the cable harness; the bit rate is comparatively low so no matched termination is required. It can still be found in many legacy aircraft and is also found in large military aircraft integrated with a military data bus type. An example structure is shown in Figure 2.10.

MIL-STD-1553B (also known as Def Stan 00-18 Part 2 or STANAG 3838) is a military standard data bus widely used on many types of military aircraft. It was originally conceived in 1973 and the 1553B standard emerged in the late 1970s.

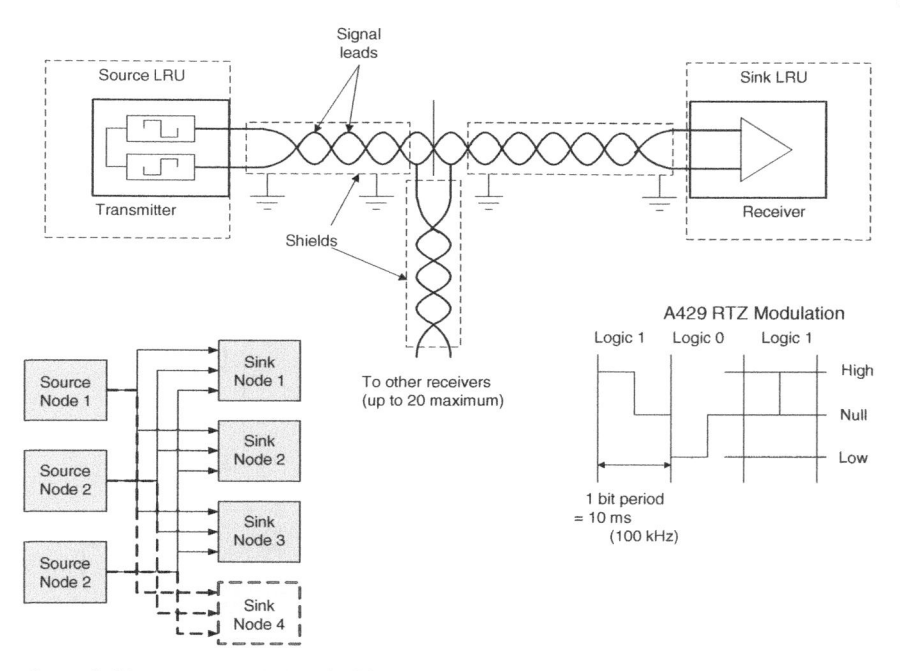

Figure 2.10 Example of ARINC 429 structure.

It is a bidirectional, linear topology, centralised control, command/response protocol bus. Data is transferred via a screened twisted wire pair in true and complement Manchester bi-phase format at 1 Mbps. Control is performed by a centralised Bus Controller (BC) which executes transactions with a Remote Terminals (RTs) embedded in each of the avionics system line replaceable units (LRU). Each transaction takes the form of a command issued by the Bus Controller, transfer of data to/from the Remote Terminal, followed by a status response from the receiving Remote Terminal. Bus communications are highly deterministic. Comprehensive message error detection and correction capability provide high-data integrity. An example structure is shown in Figure 2.11.

ARINC 629 was developed in the late 1980s to provide civil aircraft with a multi-source, multi-sync, linear topology network addressing the physical network issues of ARINC 429. It is built upon the experience of MIL-STD-1553, but specifically does not require a centralised bus controller; the civil aerospace community wanted to avoid the single point failure issues surrounding that concept and instead opted for a distributed protocol. ARINC 629 uses a 20 bit word encoded in Manchester bi-phase format operating at 2 Mbps, twice the

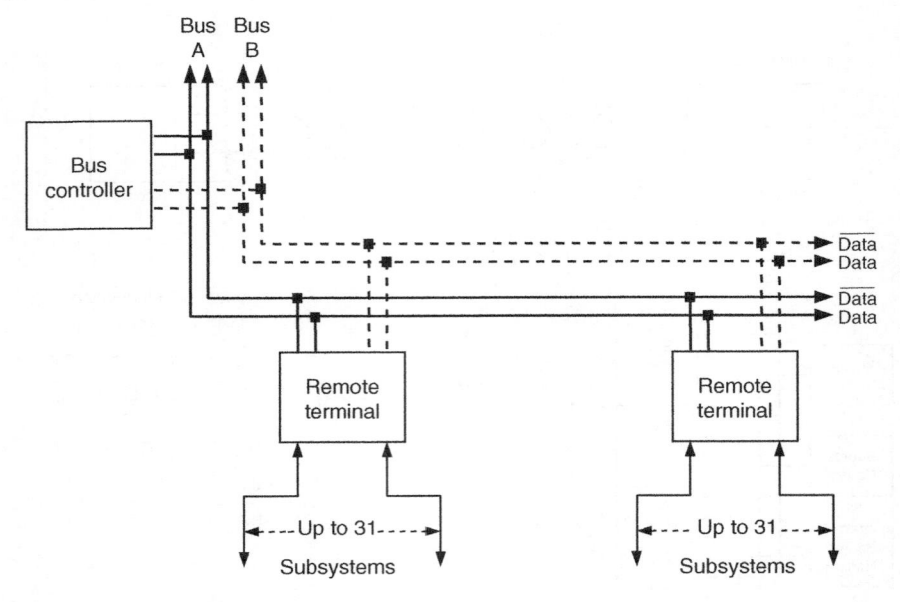

Figure 2.11 Example of MIL-STD-1553 structure.

bit rate of MIL-STD-1553. It also permits 128 terminals to be connected to the bus. Quadruple and triple redundancies are also supported in addition to dual redundancy. An example structure is shown in Figure 2.12.

The ARINC 664-P7 communications network is more than a data bus. It is currently used on the Airbus A380, A350, A400M, Boeing 787 Dreamliner, the COMAC ARJ21, and the Sukhoi Super-jet 100; and is likely to become the standard communications network on most future civil transport aircraft. The standard evolved from commercial communications packet switched origins which is why terms such as 'subscribers' are often encountered.

ARINC 664 part 7, also known as Aviation Full Duplex (AFDX, is an Airbus trademark), is based on 10/100 Mbps switched Ethernet technology (IEEE 802.3) Media Access Control (MAC) addressing, Internet Protocols (IP), User Datagram Protocol (UDP) but with special protocol extensions and traffic management to achieve the deterministic behaviour and appropriate degree of redundancy required for avionics applications. An example structure is shown in Figure 2.13.

CANbus was originally developed by Bosch in the 1980s, as a low cost data bus for automotive applications. It has been very successful and widely adopted for ground vehicles. It is now finding its way into aerospace applications. CANbus nodes are typically sensors, actuators, and other control devices with small amounts of sometimes time critical data. CANbus physical media is

2.8 Data Bus | 33

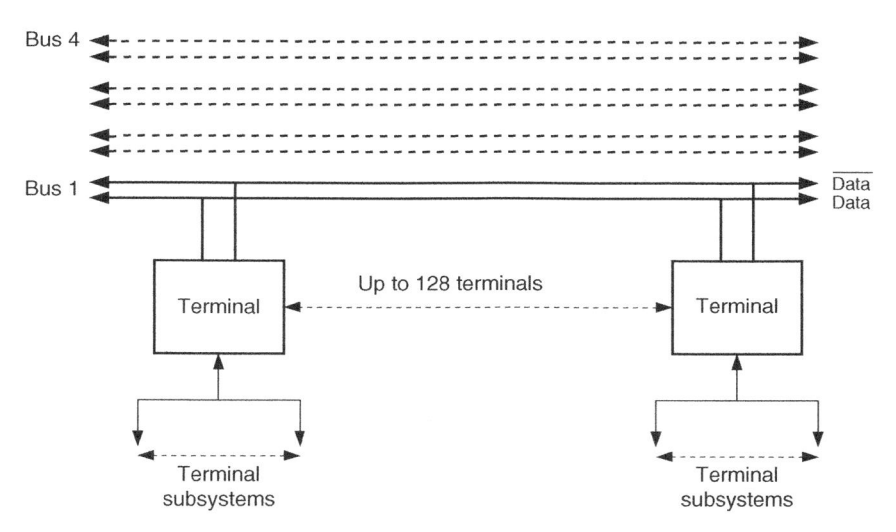

Figure 2.12 Example of ARINC 629 structure.

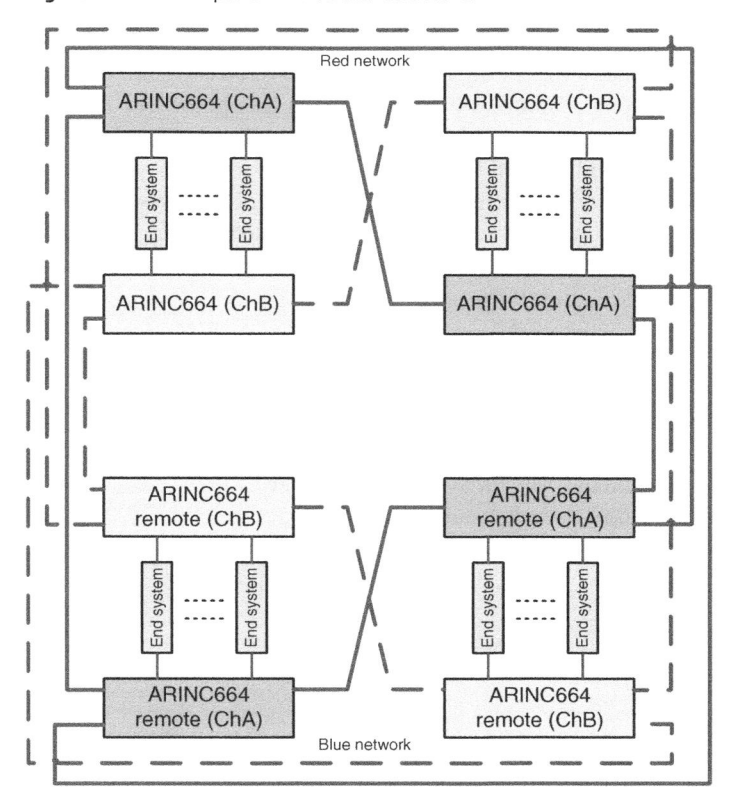

Figure 2.13 Example of ARINC 664 structure.

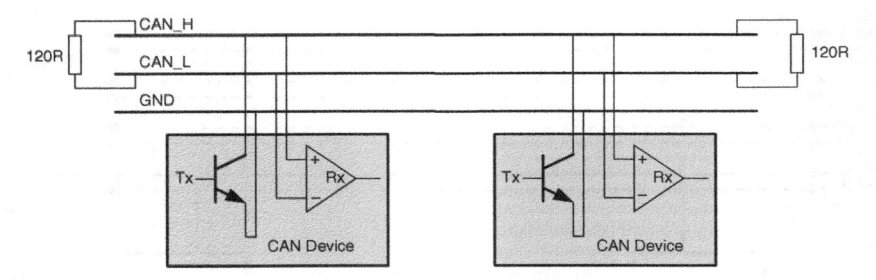

Figure 2.14 Example of CANbus structure.

most commonly a 5 V differential (true and complement) signal transmitted over a twisted pair screened cable terminated at both ends in 120 ohms. The transmitter driver is open collector. An example structure is shown in Figure 2.14.

2.9 Summary and Conclusions

This chapter has introduced some of the issues that need to be considered when analysing an aircraft system. Examples have been given as a guide to considering the characteristics of systems and are the basis for the following chapters. As examples, they are not complete, and the systems designer must carry out their own work to establish a clear and complete understanding of their system. Note also that these issues will constantly change as a result of technology advances, new customer requirements, operating experience regulatory changes, and environmental considerations.

References

Austin, R. (2010). *Unmanned Air Systems, Development and Deployment*. Wiley.
Moir, I. and Seabridge, A. (2008). *Aircraft Systems*, 3rde. Wiley.
Moir, I. and Seabridge, A. (2013). *Civil Avionics*, 2nde. Wiley.
Sadrey, M.H. (2020). *Design of Unmanned Aerial Systems*. Wiley.

Exercises

2.1 The classification of systems discussed in Section 2.4 is not complete. It may be appropriate for some projects or for some stakeholders to perform their own classifications to suit their role in the projects. For example systems may be classified according to their value – cost or investment – or according to the risk attached to their development. Consider alternative classifications according to your own role in a project or by putting yourself in the position of a representative stakeholder. How does this affect your view as an engineer?

2.2 Consider the classification systems in commercial terms, for example as high-value systems. Describe how this might lead to a conflict between the commercial and technical departments.

3

Vehicle Systems

This chapter deals with those systems of the aircraft that provide the most basic services that turn the airframe into a working vehicle. They are the source of power and energy, a function which itself requires energy sources to be loaded as consumables prior to each flight. The Vehicle Systems are known by various titles depending on the practice of the company designing and operating the aircraft and these names have changed over time. Some naming examples are the

- Power and mechanical systems
- Airframe systems
- Air vehicle systems
- Utility systems
- General systems
- Vehicle systems

In this book, the systems will be referred to as Vehicle Systems and their position in the overall aircraft system structure is illustrated in Figure 3.1.

Many of these systems are common to both civil and military aircraft; they are a mixture of systems with very different characteristics. Some are high-speed, closed-loop, high integrity control systems such as flight controls, others are real-time data gathering and processing with some process control functions such as the fuel system, and yet others are simple logical processing such as undercarriage sequencing. What they have in common is that they all affect flight safety in some way – in other words, a failure to operate correctly may seriously hazard the aircraft, crew, or passengers.

The mechanism by which many of these systems is controlled depends on the technology used in the aircraft. In many legacy types, each system will be controlled by its own specific control unit using the technology that predominated at the time of the original design – relay logic, transistors, or integrated circuits for example. Later types will use a more integrated approach with all or small groups of systems incorporated into a centralised or distributed computing system.

Aircraft Systems Classifications: A Handbook of Characteristics and Design Guidelines, First Edition.
Allan Seabridge and Mohammad Radaei.
© 2022 John Wiley & Sons, Inc. Published 2022 by John Wiley & Sons, Inc.

3 Vehicle Systems

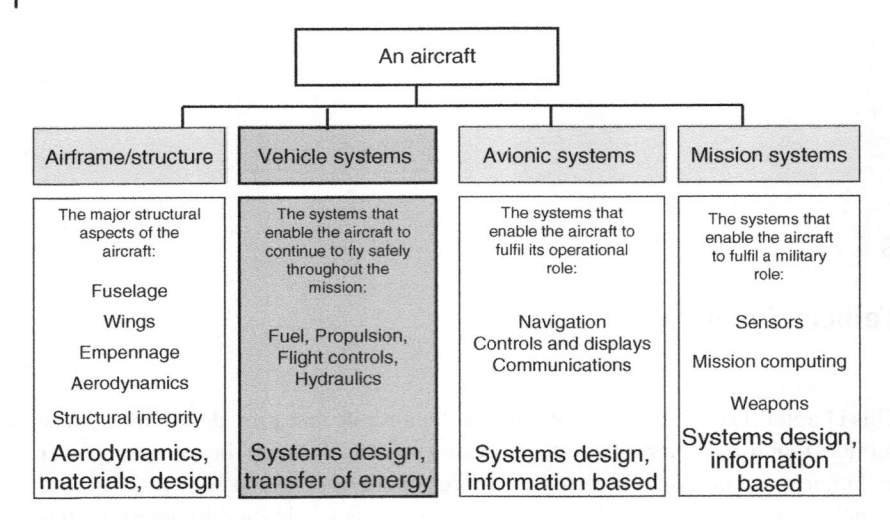

Figure 3.1 The systems described in this chapter.

An aircraft			
Airframe/structure	Vehicle systems	Avionic systems	Mission systems
The major structural aspects of the aircraft: Fuselage Wings Empennage Aerodynamics Structural integrity Aerodynamics, materials, design	The systems that enable the aircraft to continue to fly safely throughout the mission: Fuel, Propulsion, Flight controls, Hydraulics Systems design, transfer of energy	The systems that enable the aircraft to fulfil its operational role: Navigation Controls and displays Communications Systems design, information based	The systems that enable the aircraft to fulfil a military role: Sensors Mission computing Weapons Systems design, information based

The functions of many of these systems are performed by software-based control units – either individual units or an integrated processing system such as a Vehicle Management system or VMS. This means that the software must be designed to appropriate levels of robustness.

The aircraft systems will be illustrated singly below to describe the basic functions and to provide an illustration of the interconnections and interfaces to other systems.

3.1 Propulsion System

The main component of the propulsion system is the engine, but the system is so much more than this with many and varied mechanical, electrical, physical, and data interconnections and interactions with the airframe and its systems. In order to fully describe the propulsion system, it is illustrated in Figures 3.2 and 3.3.

The engine(s) at the core of this system will vary according to the type of aircraft and the technology applied. The type of engine will include the following:

- Aviation fuel engines used on general aviation types
- A turbo jet or turbo fan engine used on many long-haul airliners and military jets
- A turbo prop used on many regional commercial aircraft
- A reverse flow turbo-shaft used on helicopters
- A liquid gas fuelled engine being discussed for future 'green' solutions
- An electric engine (being discussed for future 'green' aircraft)

3.1 Propulsion System | 39

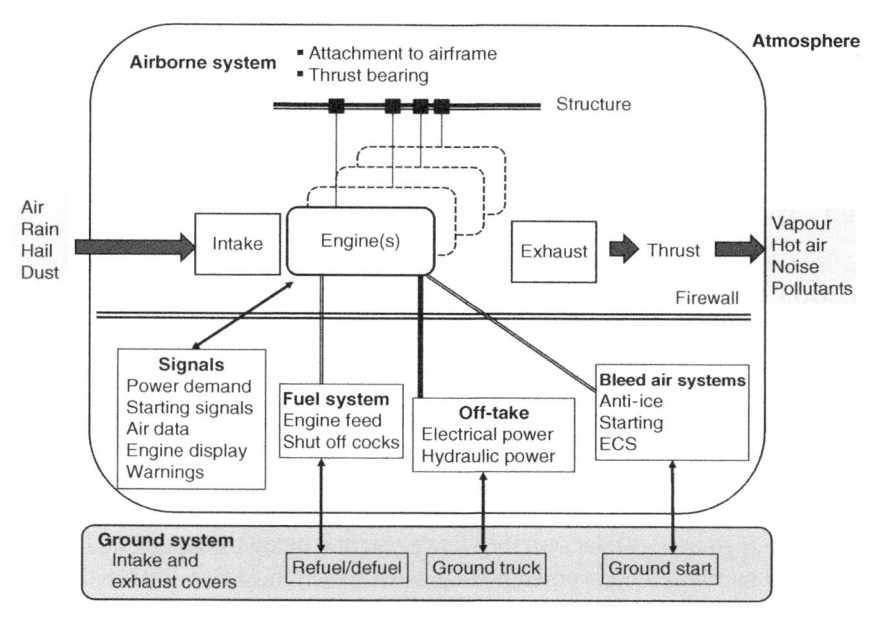

Figure 3.2 Illustration of a generic propulsion system.

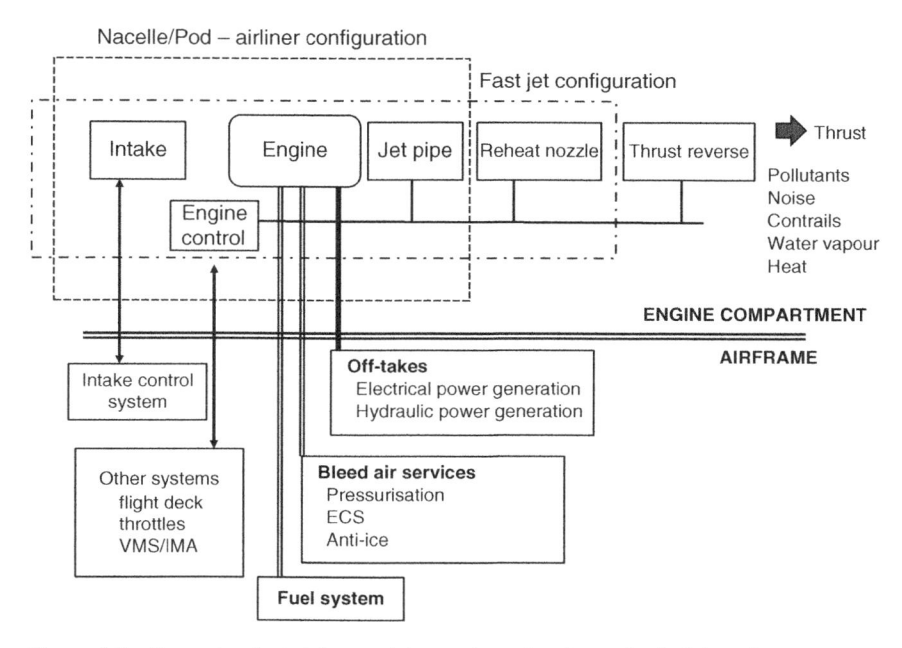

Figure 3.3 Example of a total propulsion system showing a single jet engine.

3.1.1 Purpose of System

To provide thrust for the vehicle and to provide a source of off-take power for electrical power generation, hydraulic power generation, and also to provide bleed air for pneumatic systems, pressurisation, and the environmental cooling system.

3.1.2 Description

The system includes the main propulsion units, the propulsion control system, interfaces with intake and airframe, air and mechanical power off-takes, and the provision of throttle levers, switches, and displays in the cockpit.

3.1.3 Safety/Integrity Aspects

The propulsion system is considered to be safety critical because loss of one or more engines can result in a loss of aircraft with the potential for loss of life of crew, passengers, and the over-flown population. This affects the decision on the number of engines, together with the recovery action following an engine failure. Such considerations are governed by Standards which must be considered in all aspects of the propulsion system design and will affect operational aspects such as extended twin engine operations (ETOPS) in which the aircraft must be able to continue flight on a single engine for a defined period of time.

It is possible that a crash switch or inertia switch has been included to isolate all power supplies if the specified crash case inertia has been exceeded. This must be considered as a dormant single-point failure case in the case of inadvertent operation or a defect in the crash switch mechanism. This component of the system is dormant.

3.1.4 Key Integration Aspects

The propulsion unit must be integrated with the aircraft in terms of installation, aerodynamics, and the aircraft systems. In modern commercial aircraft, the engines are usually contained in a nacelle that is attached to the airframe or wings. In this way, the propulsion unit can be provided as a self-contained unit with suitable design of the intake and exhaust which is connected to the aircraft through a thrust bearing. Military fast jet aircraft often have the engine(s) contained within the structure. In this case, the intake is contained in the fuselage, designed so that air conditions at the engine face are suitable for the engine under all flight conditions. The jet pipe and reheat nozzle, if applicable must also be designed to suit the engine and the configuration of the aircraft. In both cases, integration of off-take drives, fuel system and bleed air connections must be designed to enable the engine to be changed rapidly.

There may be integration with the flight control system in a highly agile aircraft that requires some form of thrust vectoring and may also require moving surface in the intakes to provide suitable air flow at all times. The integrity of the systems must be maintained in any such integration.

3.1.5 Key Interfaces

The interfaces of the propulsion system are many and varied, covering a range of engineering and contractual disciplines within the organisation.

These interfaces include the following:

- Electrical power
- Physical
- Fluid
- Mechanical
- Air bleed
- Exhaust
- Shaft off-take
- Thermal
- Signals to and from the FCS, air data, and flight deck
- Noise
- Contaminants
- Provision for removal and replacement of the engines
- Data for aircrew
- Data for ground crew
- Data for maintenance monitoring
- Methods of containing damaged rotating components
- Design of aircraft systems and structure to reduce uncontained component damage
- Blanks, covers, and warning pennants

To manage these interfaces effectively requires a mechanism for informing all parties involved. This is often accomplished by means of an Interface Control Document which will list all interfaces and the organisation responsible for them technically and contractually.

Many of the installation and physical interfaces are determined by the application and the type of aircraft. A typical commercial aircraft will have engines installed in a nacelle (pod) attached by a pylon to the wings, either under-wing (B737, Airbus A380, B787) or very occasionally over wing (VFW 614), or attached to the rear fuselage (BAC 1-11, Embraer 145, Fairchild A-10). The key physical interfaces can be contained in the pylon.

- **Airframe/structure**: provision of apertures and suitable fasteners for the engine nacelle, protection of the pressure shell if fuselage mounted. Most fast jet military aircraft will have the engines installed in the airframe. In this configuration, the interfaces cross the boundary between airframe and engine at a number of points. The intake and exhaust arrangements may be complicated. Airframe installation may be via a wing- or fuselage-mounted pylon or the engine may be embedded in the fuselage. In either case, there must be thrust bearings to transmit the thrust efficiently to the airframe.

- **Other systems**: exchange of information by direct wiring or data bus across the engine/airframe firewall. Provision of bleed air for pressurisation and environmental control and suitable drives for hydraulic pumps and electrical power generation.
- Flight deck for controls (throttles, reverse thrust command, and switches) and displays and human factors assessment.
- **Electrical power**: connection to appropriate bus bar with circuit protection devices, and most suitable harness runs.
- **Installation**: mounting tray for airframe-mounted electronic LRUs, and provision of suitable conditioning and cooling. Ease of access to pipe connections, off-take shafts, and wiring harnesses to facilitate simple and quick engine removal and replacement.
- Security of attachment of all components.
- **Environment**: release of exhaust, fumes, leaks, drips, and other potential environmental contamination.
- Consideration of electro-magnetic compatibility (EMC) by radiation, induction, and susceptibility.
- Ground systems for download of data, provision of services on turn around, replenishment of fluids such as fuel, oils, and greases.

3.1.6 Key Design Drivers

Aircraft performance: Commercial aircraft performance is characterised by factors such as thrust, economy, reliability, and availability. The military aircraft as a slightly different set of factors such as thrust, handling, range/endurance, and availability. For both sectors, first cost is a major consideration as well as day-to-day operating costs. Safety for both sectors is paramount.

3.1.7 Modelling

A lot of modelling of the engine performance is done using propulsion test rigs which allow the engine to be operated throughout the range of speeds and handling demands. To obtain a more realistic test at the condition that will be experienced in the aircraft, an altitude test facility is used to simulate different altitudes and intake temperatures. In some circumstances, the engine can be installed on a test aircraft that will explore the regions and rates of change that are difficult for a static test rig to achieve. Both Concorde and Tornado had their development engines attached to a Vulcan flying test bed (FTB) for in-flight testing.

3.1.8 References

Agarwal (2016), Gent (2010), Bomani and Hendricks (2016), Farokhi (2014), Farokhi (2020), Jackson (2010), Langton (2006), Langton (2010), Lawson and Judt (2022), MacIsaac and Langton (2011), Moir and Seabridge (2008), Moir and Seabridge (2013), Pornet (2016), Rolls-Royce (2005), Schutte et al. (2016), JASC Code (n.d.), ATA-100 (n.d.) Chapter 76.

3.1.9 Sizing Considerations

Typical factors to be considered include range, endurance, number of passengers, route or mission, and cost of operation. Throttle levers are included in the flight deck in accordance with human factor considerations for the project with agreed interfaces with the engine control system. The engine control unit is now typically considered as a part of the engine, although some legacy types will retain airframe mounted control equipment. Cooling will be required for engine oil, often in the form of a fuel-cooled heat exchanger which has an impact on the fuel system and on fuel temperature.

3.1.10 Future Considerations

Environmental (Green) issues are driving consideration of alternative means of propulsion such as electric, open rotor, and geared turbofan. Alternative fuels are also being considered such as hydrogen, liquid/natural gas, and bio-fuels. These will influence installation interfaces and cockpit indications. Noise reduction measures may lead to the need to design or operate to reduce noise or operational restrictions to limit noise in the vicinity of airports.

The engine control system has developed from an airframe installed system of many LRUs into an engine-mounted self-cooled computing system, usually of duo-duplex architecture.

The engine and its components have grown in robustness and reliability so that crossing oceans on two engine aircraft (ETOPS) is commonplace.

Some legacy aircraft will have the engine control system (analogue or digital) mounted in the airframe as a simple-range speed limiter. The introduction of full authority fly-by-wire with modern robust electronics has led to more modern control systems being engine-mounted.

3.2 Fuel System

Fuel is used by the main propulsion system and the auxiliary power unit and must be available in a continuous and uninterrupted flow. The fuel must be stored safely and provided to the user systems with no restriction throughout the flight. The main functions of the fuel system include the following (Figure 3.4):

- **Re-fuelling and de-fuelling**: Re-fuelling on the ground is performed by a high-pressure connection from a bowser or the airfield distribution system. For in-flight refuelling, a receptacle for the fuel hose from tanker is required – generally a fixed or retractable probe fitted to UK/European and some US aircraft with the tanker deploying a drogue, or a receptacle mating with a tanker flying boom on US aircraft.
- **Storage of fuel**: Fuel is stored in a collection of fuel tanks, together with fuel gauge probes, interconnecting pipes and couplings, together with pumps, valves, and level sensors to allow measurement and transfer of fuel. The fuel tanks are contained within the fuselage and the interconnecting pipes enable fuel to be moved from tank to tank and to the engines. Military aircraft may often

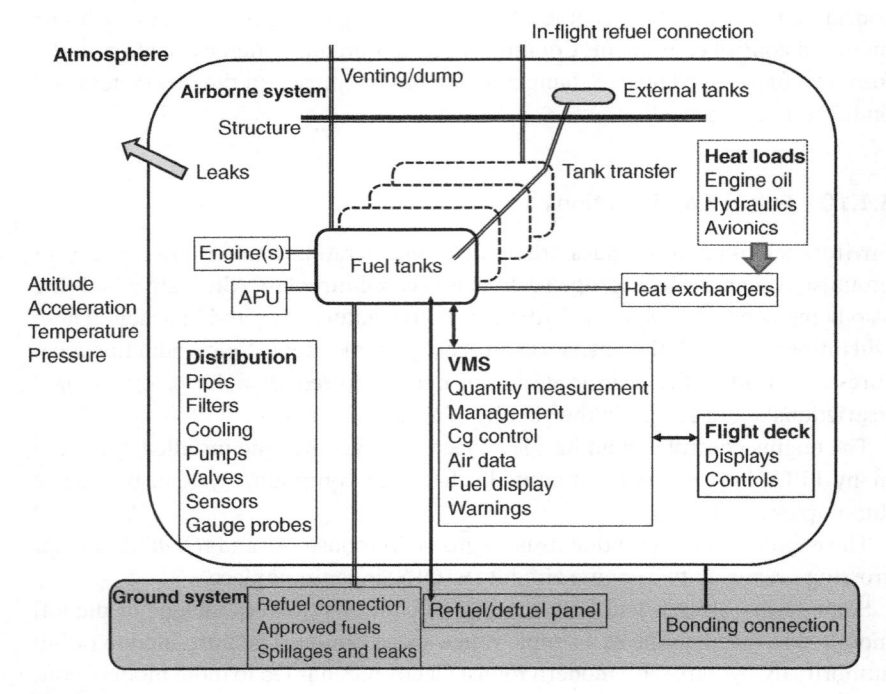

Figure 3.4 Illustration of a generic fuel system.

carry fuel in external tanks attached to the wings. The functions of determining the quantity of fuel and the procedure for moving fuel between tanks are performed by a fuel computer or by software embedded in an integrated computing function.

- **Transfer of fuel between tanks**: This is managed by a fuel computer or a VMS which recognises the quantity of fuel in the tanks and arranges the transfer to suit the longitudinal and lateral balance of the aircraft using valves and transfer pumps.
- Transfer of fuel to the engines and APU is carried out by transferring fuel into a collector tank for each engine and then via shut off valves to the engine gear pump for transfer to the combustion system.
- Measurement of fuel mass and location is performed by sensors placed in the tanks to measure the level of fuel. This level is converted to mass by the fuel control computer or VMS using tank shape and fuel density. The result can be provided to the pilot displays.
- Fuel is often used as thermal sink for aircraft heat loads, both on and off-engine, e.g. fuel cooled oil cooler, hydraulic fluid cooling, and air cooling systems.

3.2.1 Purpose of System

To enable fuel to be loaded onto the aircraft, to store fuel in tanks and to transfer fuel from tank to tank whilst measuring the quantity of fuel on board, and to provide a continuous flow of fuel to the engines. The system allows for automatic transfer of fuel between tanks to maintain feed to the engines and to play a role in control of the centre of gravity of the aircraft.

For some military aircraft types, an in-flight refuelling capability enables a receiver aircraft to obtain fuel from a tanker in flight in order to increase its range.

3.2.2 Description

The fuel system comprises the following components:

- Tanks – both integral and external
- Pipe interconnections and couplings
- Re-fuel and de-fuel points
- Quantity measurement components – gauge probes, level sensors, and density sensors
- Transfer valves and pumps
- Re-fuelling panel
- Management and quantity measurement software
- Information to displays and controls system

3.2.3 Safety/Integrity Aspects

The system is safety critical system as its main task is to provide fuel to the propulsion system. Some architectures may dictate multiple transfer paths and multiple lane control electronics. Intrinsic safety must be considered because of fire or fuel vapour explosion risk leading to the need for nitrogen or foam inerting systems in the tank ullage space, particularly on composite airframes. Foam inerting covers the surface of fuel to reduce the likelihood of a fuel/air vapour forming in the ullage space. Safety considerations apply to ground refuelling and maintenance to reduce the risk of fuel/air vapour explosion.

In-flight re-fuelling is mission-critical. There may be some safety aspects due to aircraft maintaining close formation during the re-fuelling operation.

Intrinsic safety is vital and covers all aspects of design including sound bonding of all metallic components, very low-power sensors in tanks, and design to prevent fuel surges bursting tank walls and causing structural damage.

It is possible for sources of ignition close to aircraft on the ground during maintenance to result in the risk of an explosion of air/fuel vapour. This includes electrical test equipment, hot engine components on bowsers, tractors, and delivery vans. Procedures must be in place to avoid this. Consult https://www.hse.gov.uk/fireandexplosion/atex.htm ATEX and explosive atmospheres for advice.

3.2.4 Key Integration Aspects

The fuel system is truly integrated into the airframe structure. The tanks may form part of the structure and damage to tanks may therefore lead to structural damage. The interconnecting pipes need also to be integrated into the airframe design to avoid too many holes in the structure. To do this successfully requires close collaboration between the airframe designers and the fuel system designers.

Control of the fuel system includes calculating the quantity of fuel on board, transferring the fuel from tank to tank and to the engines, and informing the crew of the status of the system. The functions required to perform these tasks can be performed by fuel management computers or they can be integrated in the VMS or into the overall integrated modular architecture (IMA).

Integration with FCS for management of aircraft centre of gravity is an important aspect in high-performance military aircraft to allow the pilot to pull high g manoeuvres without the risk of damaging the airframe (care-free handling). The system must also be integrated with the flight management systems (FMS) to allow effective route planning.

The fuel system plays a major role in managing thermal energy on the aircraft and heat exchangers are employed to make use of fuel as a heat sink for engine oil, hydraulic fluids, and for avionics cooling. The potential issue of using hot fuel in the airframe must be considered by ensuring that wiring and equipment is isolated from hot pipes or tank surfaces.

Hot fuel in the wing tanks of military aircraft will present them as a radiating 'target' to infrared sensors and missiles.

There may be an impact of equipment life if there is insufficient fuel for efficient operation of heat exchangers. This can occur at the end of a flight when fuel quantity is low and in hot day conditions.

3.2.5 Key Interfaces

The main interface with the propulsion system is a connection to the engine feed system at the pylon or through the engine fire-wall as appropriate. A connection point must be provided to allow ground refuelling from a bowser or an airfield re-fuelling point. Provision must also be made for de-fuelling the aircraft in a safe manner. For air-to-air refuelling, a probe or a connection must be provided to suit the specific requirements of the tanker and an interface to the flight deck/cockpit is required to allow information to be provided to the crew. Discussion with the cockpit designer is essential to ensure that the information is timely, intuitive, and clear under all conditions of flight from high-altitude sunlight to night.

- **Airframe/structure**: provision of apertures and suitable fasteners for ground refuelling point, in-flight refuelling probe where appropriate, protection of the pressure shell.
- **Other systems**: exchange of information by direct wiring or data bus.
- Flight deck for controls and displays and human factors assessment.
- **Electrical power**: connection to appropriate bus bar with circuit protection devices, and most suitable harness runs.
- **Installation**: mounting tray for electronic LRUs (e.g. fuel computer) and provision of suitable conditioning and cooling. Tank, wiring, and pipes to be installed to avoid hot components to reduce the risk of fire. All wiring and components that enter a tank must be designed to avoid the inclusion of electrical power above prescribed energy limitations that may lead to an explosions or ignition – intrinsic safety must be observed at all times.
- Security of attachment of all components.
- **Environment**: release of fuel vapour, leaks, drips, and other potential environmental contamination.
- Consideration of EMC by radiation, induction, susceptibility, or lightning strike.
- Ground systems for download of data, provision of services on turn around, replenishment of fuel by bowser, by hand in emergency, and suitable bonding connection for safety in explosive atmosphere conditions. (see Safety/Integrity aspects above). Availability of relevant fuel types.

- **The interface with tanker refuelling device**: drogue or boom is a key consideration that can lead to issues of non-compatibility between aircraft and tanker types.

3.2.6 Key Design Drivers

A key input to the design is the range of the aircraft or for military aircraft the range and endurance (time on station) which allow the storage capacity to be determined and from that the location and shape of fuel tanks. The gauging accuracy requirement will determine the number and location of fuel gauge probes.

3.2.7 Modelling

It used to be common to build a fuel rig to test the system at all fuel demand rates and with limited attitude changes. Modern design techniques make use of 3D modelling techniques to model tank shapes and flows between tanks. Combined with computational fluid dynamics and modelling and analysis tools, this allows confidence in the design to be established at lower cost.

In-flight refuelling will require test connections with a representative tanker first on the ground and then in flight trials.

3.2.8 References

Bomani and Hendricks (2016), Freeh (2016), Langton et al. (2009), Langton et al. (2010), Langton (2010), Lawson and Judt (2022), Purdy (2010), Moir and Seabridge (2008) Chapter 3, Rolls-Royce (2005) Chapter 2.5, Schutte et al. (2016), JASC Code (n.d.), ATA-100 (n.d.) Chapter 28.

3.2.9 Sizing Considerations

The system size in terms of fuel mass or volume is driven by the range required. This may be compounded by the considerations of cost to include re-fuelling after one journey leg, or after a return flight which can be planned to enable fuel to be bought in one currency or at a main base. Consideration must be given to the cost, mass, and complexity of an in-flight refuelling system.

3.2.10 Future Considerations

Environmental (Green) issues are driving consideration of alternative fuels such as hydrogen, liquid/natural gas, and bio-fuels. These will influence installation interfaces and cockpit indications as well as storage at airfields. Different fuel types may influence the attitude to intrinsic safety design techniques, for example handling and storage of liquid gases. In-flight refuelling has been discussed to reduce take-off mass and fuel burn for commercial types.

In 2008, a B-777 operated by British Airways crashed on approach to London, Heathrow. An investigation showed that the incident was caused by ice crystals forming in the fuel. This led to a lack of response to a demand for thrust. The aircraft was destroyed, and there was no loss of life. This happened after a long duration flight at extreme low temperatures. Care must be taken to provide adequate sensing of fuel temperatures and for the pilots to manage their fuel temperatures on long duration extreme cold conditions. It is important also to minimise the quantity of water in fuel.

There have been incidents, one fatal, in which fuel in tanks has been ignited by old wiring with poor insulation and an electrical fault and by overheating by equipment being mounted too close to the tanks. Such occurrences must be considered in the design of the airframe and equipment installation.

3.3 Electrical Power Generation and Distribution

The modern aircraft and the trend toward more electric systems mean that the demand for electrical power has grown. Many existing systems will have developed from the provision of AC power, usually 115 V 400 Hz, and DC power, usually 24 V, provided by generators, alternators, batteries, transformer rectifier units, and emergency use devices. These characteristics will still be found in many legacy aircraft and equipment, but there are trends towards different voltages and frequencies to satisfy increased loads without incurring voltage drop or using larger and heavier wiring, for example 270 VDC and 230 VAC.

Raw power is provided by generators driven by the engine, and conversion devices are used to obtain lower levels of stable voltages and constant frequencies. A distribution system of bus bars, contactors, and circuit protection devices provides a mechanism for connecting user systems throughout the aircraft. The system must be designed to reduce the risk of power interrupts, common mode failures, and damage to the generation devices whilst maintaining voltages and frequencies to meet the appropriate standards (Figure 3.5).

3.3.1 Purpose of System

In modern aircraft, the majority of systems require electrical power for control equipment and also to provide energy for loads or effectors.

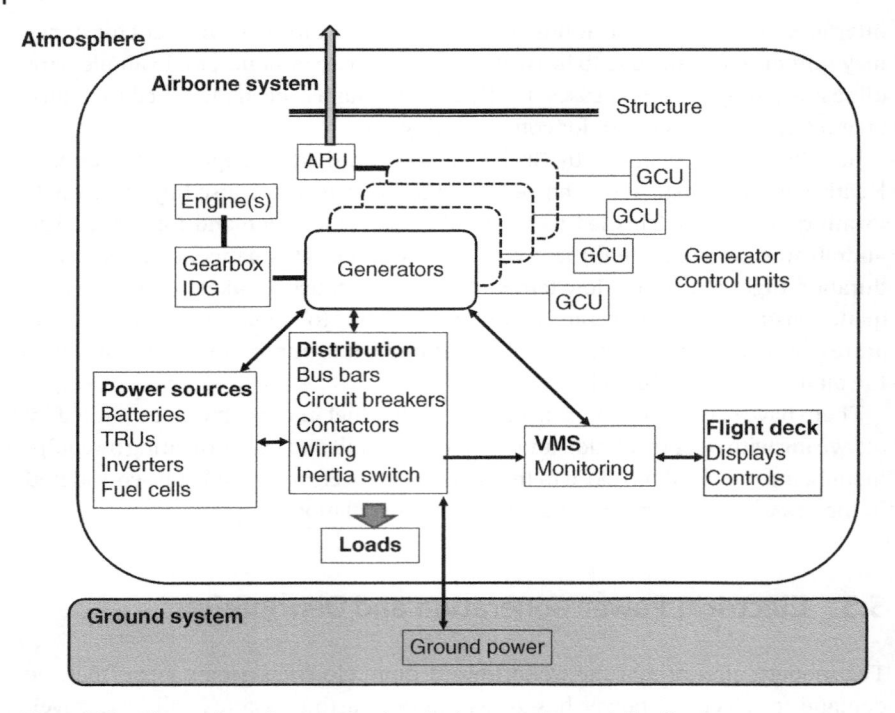

Figure 3.5 Illustration of a generic electrical generation and distribution system.

3.3.2 Description

This system provides a source of regulated AC and DC power to the aircraft systems via bus-bars and circuit protection devices. This power will be provided continuously from a combination of primary power devices such as engine-driven generators/alternators supplemented by batteries or fuel cells for emergency conditions. The power is provided to bus-bars and from there to the user equipment by wiring, with items of equipment protected by fuses, circuit breakers, or contactors in order to prevent damage to the aircraft wiring.

3.3.3 Safety/Integrity Aspects

The system is safety critical since loss of partial or total power failure could result in loss of aircraft. The sources of power are designed as a multiple redundant system with the design of bus bars and contactors providing failure propagation protection.

The use of some new battery technologies has resulted in battery fires during initial introduction into service. The Boeing 787 has suffered a number of incidents on the ground and in flight which resulted in smoke, overheat, and fire. A major fatal accident occurred in an electrical road vehicle in 2021 in which attempts to extinguish the fire led to the battery re-igniting.

3.3.4 Key Integration Aspects

Integration with the engine to provide motive power for alternators or generators, and integration with APU and ram air turbine (RAT) to provide power under emergency conditions. Engine power off-take loads must be maintained in steady conditions as far as possible.

3.3.5 Key Interfaces

- **Airframe/structure**: provision of apertures and suitable fasteners for ground power connection.
- **Other systems**: exchange of information by direct wiring or data bus.
- Flight deck for controls and displays and human factors assessment.
- **Electrical power**: connection of generators and control units to the aircraft power distribution system bus bars. Connection of ground power supply.
- **Installation**: mounting tray for electronic LRUs such as generator control unit, and provision of suitable conditioning and cooling if required.
- Consideration of EMC by radiation, induction, and susceptibility.
- Ground systems for download of data, provision of services on turn around, and replenishment of fluids or provisions.

3.3.6 Key Design Drivers

The total electrical load will drive the size of generators. A load analysis will be compiled to understand the total load and mean loads at various phases of flight and will include critical cases such as loss of one or more engines. A strategy for load shedding or overloading existing generators for very short duration will be determined to allow safe recovery of the aircraft. Electrical power quality, safety, and reliability must meet international and project standards.

3.3.7 Modelling

An electrical load analysis by phase of flight using a spreadsheet or a commercially available analysis tool will enable decisions to be made about the size of generators and the failure recovery strategy. A power generation test rig using mains electricity to run generators allows the total system to be tested using different load banks and testing failures.

3.3.8 References

Agarwal (2016), Moir (2010), Lawson and Judt (2022), Moir and Seabridge (2008) Chapter 5, Pallett (1987), Pornet (2016), Xue et al. (2016), Farokhi (2020). JASC Code (n.d.), ATA-100 (n.d.) Chapter 24.

3.3.9 Sizing Considerations

The type of aircraft and the total load combined with factors such as ETOPS has a major impact on sizing and failure strategy. The specification of the role and number of engines will provide suitable information for the specification of generators and control units, batteries, TRUs, bus bars, distribution panels, and contactors.

3.3.10 Future Considerations

Many heritage aircraft in service will be based on 115 VAC 400 Hz and 28 VDC, and most avionic equipment will be available off-the-shelf designed to those specifications. The current trend is towards higher voltages and 230 AC 400 Hz and 270 VDC have become commonplace. It is essential to provide suitable voltages such as 115 VAC and 28 VDC for off-the-shelf equipment which will require conversion systems.

Much progress has been made in the integration of generators into the engine shafts and in combining modes leading to the introduction of combined starter/generators. This reduces cost and mass, but complicated the common mode failure issue. To optimise energy consumption, consideration has been given to the use of thermal scavenging devices to produce power, and the use of hydrogen fuel cells.

It is noticeable that electrical power demand is increasing with the trend towards provision of bleed air from electrically driven compressors, the demand for more customer services and the consideration of electrical actuators as replacement for hydraulic devices.

The B787 aircraft made a determined effort to reduce bleed air off-take from the engines by introducing electrical powered compressors to provide air for ECS and to use electrical anti-icing. This has led to an increased generator capacity.

There have been advances in battery technology driven by automotive, consumer, and space markets. Battery technology and manufacturing methods to meet the large market for electric vehicles has already started to tackle the issues of mass, cost, and capacity. As a result, the cost of batteries has reduced. This trend is likely to continue as market pressures lead to increasing battery use and reduced costs, especially for electric road vehicles. Safety needs to be improved and a mechanism for extinguishing fires is needed. Note that a world shortage of Lithium is anticipated as early as 2040.

3.4 Hydraulic Power Generation and Distribution

Hydraulic power is a most effective source of power for many high-power, high-rate systems such as primary and secondary flying controls. It is also encountered in other systems such as undercarriage, brakes, and hatches or doors that need to be opened in flight. Over many years, challenges arose from proponents of electrical systems and the use of rare earth magnetic materials has led to a distinct move towards electrical actuation. Despite this, many aircraft are still dependent on hydraulic power actuation devices.

Power is provided by constant pressure displacement pumps driven by the engine or other sources of power. A distribution system of pipes, filters, valves and accumulators provides a mechanism for connecting user actuation systems throughout the aircraft. The system must be designed to reduce the risk of power interrupts, common mode failures, and damage to the generation devices whilst maintaining pressure and flow to meet the appropriate standards (Figure 3.6).

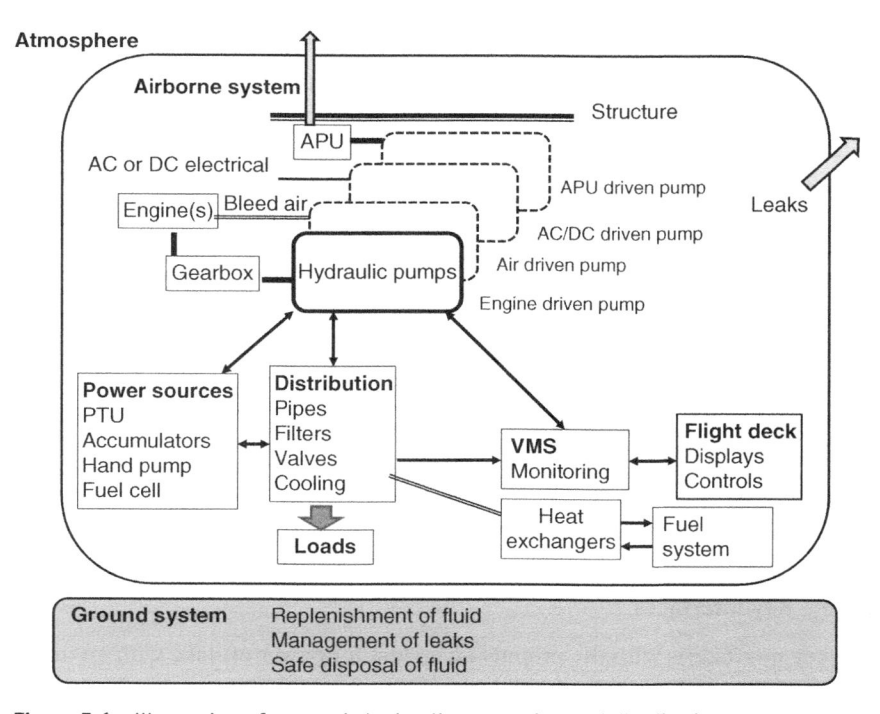

Figure 3.6 Illustration of a generic hydraulic generation and distribution system.

3.4.1 Purpose of the System

To provide a source of high-pressure motive energy for actuation mechanisms for closed loop continuous loads such as FCS and for intermittent loads such as undercarriage, brakes, air brakes, and bomb bay doors.

3.4.2 Description

A number of hydraulic pumps driven by various means including direct engine connection, electric motor, and bleed air provide a constant pressure source of hydraulic fluid. A distribution system of reservoirs, accumulators, pipes, and couplings and filters distributes the fluid to all user devices where electrical valves schedule the fluid to actuators to provide rotary or linear displacement to operate surfaces.

3.4.3 Safety/Integrity Aspects

Safety critical system. Redundancy will match that of the highest integrity system – usually the flight control system. Hydraulic system redundancy is generally three channels.

The diversity of pump power sources is important, in other words, a number of sources of pump energy or motive power will be incorporated to achieve system redundancy, each driven by a separate source, for example the engine gearbox, electrical supply either AC or DC, bleed air, and emergency sources such as APU or EPU. This mix of power sources is to avoid common mode failures.

Suitable hydraulic fluids must be used to reduce the risk of fire and mineral oil-based fluids have now been replaced by phosphate ester fluids which are less prone to fire from overheating.

Care must be exercised in the handling and disposal of hydraulic fluid to meet environmental and health and safety regulations.

3.4.4 Key Integration Aspects

Control and monitoring can be integrated in the vehicle management system (USMS or VMS) and can also be integrated into vehicle domain of IMA.

3.4.5 Key Interfaces

The key interface is with the propulsion system for power off-take with an interface to allow connection of ground hydraulic power for long periods of operation whilst parked or during maintenance. Sensors in the system provide information for flight deck for synoptic displays and warning system.

- **Other systems**: exchange of information by direct wiring or data bus. There will be an interface with the fuel system used as a heat sink for hot hydraulic oil in a fuel-cooled oil cooler (FCOC).
- Flight deck for controls and displays and human factors assessment.
- **Electrical power**: connection to appropriate bus bar with circuit protection devices, and most suitable harness runs.
- **Installation**: mounting tray for electronic LRUs and provision of suitable conditioning and cooling.
- Security of attachment of all components.
- **Environment**: release of exhaust, fumes, leaks, drips, and other potential environmental contamination.
- Consideration of EMC by radiation, induction, and susceptibility.
- Ground systems for download of data, provision of services on turn around, replenishment of fluids, or provisions.

3.4.6 Key Design Drivers

The flight control system is often the key driver depending on the number of actuators and their demand for power and rates of operation. Safety and robustness will determine the diversity of power sources and emergency or reversionary sources of supply of hydraulic fluid at the correct pressure and flow rates.

3.4.7 Modelling

Matlab/Simulink can be used to simulate system flows with flow modelling to ensure that pipes and bends are not disturbing the flow. A hydraulic test rig can be used to ensure that the main pumps operate correctly and can be tested in failure modes and normal variable power modes using realistic actuation loads. This concept can be extended to an 'Iron Bird' rig in which power at the specified system pressure can be provided to FCS actuators with flight loads exerted and landing gear actuators and in-flight refuelling probe.

3.4.8 References

Hunt and Vaughan (1996), Lawson and Judt (2022), Linden and Roddy (2002), Moir and Seabridge (2008) Chapter 4, Seabridge (2010), JASC Code (n.d.), ATA-100 (n.d.) Chapter 29.

3.4.9 Sizing Considerations

The number of surfaces, demand cycles, and rates of change of surface position can be analysed using a load analysis based on phases of flight. This can be used to determine the size of hydraulic pumps, reservoirs, valves, power transfer units, piping, accumulators, and heat exchangers. Consideration must be given to heat generated in the system in order to size the heat exchanger and to determine the impact on fuel temperature.

3.4.10 Future Considerations

The continued move towards more electric or even all electric aircraft may result in a reduced need for hydraulic power generation and more use of electro-hydrostatic and electric actuators.

Risk of fire and fluid break down in hot conditions has moved fluids away from mineral oil based towards phosphate ester fire-resistant fluid.

3.5 Bleed Air System

The bleed air system provides a form of power to systems as an alternative to electrical or hydraulic power. Bleed air systems have been long been used to make use of hot and high-pressure air readily available from the engine. Typical uses of bleed air include the following:

- Thrust reverser
- Leading edge anti-ice – wings and empennage
- Cabin pressurisation
- Cabin air conditioning
- Rain dispersal

Although seen on existing in-service aircraft, some of these systems are being superseded on modern designs by alternative sources of energy (Figure 3.7).

3.5.1 Purpose of the System

To provide a source of high-pressure bleed air for pneumatic systems contained in the aircraft, these can include, pressurisation, cabin cooling and heating, equipment cooling, thrust reverser operation, and anti-icing systems.

3.5.2 Description

A means of obtaining bleed air from various stages of the engine or from the APU and safely routing it to systems in the airframe at the appropriate temperature and pressure.

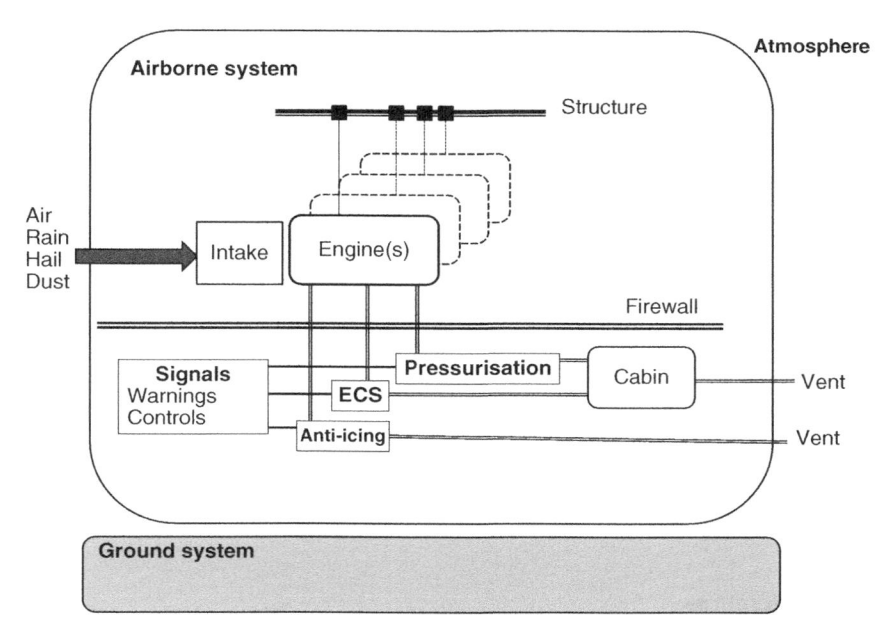

Figure 3.7 Illustration of a generic bleed air system.

3.5.3 Safety/Integrity Aspects

Safety affected system – the loss of anti-icing may lead to restricted operations or mission abort as a result of avoiding icing conditions. Hot and high-pressure air is a potential risk of overheating of structure and a maintenance personnel hazard. Integrity of seals to prevent cabin pressure loss.

3.5.4 Key Integration Aspects

Integration with the engines to prevent disturbance to normal engine operation during variable demands for air. Control and logic functions to be integrated into the VMS.

3.5.5 Key Interfaces

- **Airframe/structure**: Engine hot air pipe connections to the airframe and the impact on engine removal/replacement times. Pipes routed through the airframe will need insulating and notices to warn of hot and high-pressure air. Wing tip vent for anti-ice air.
- Flight deck for controls and displays and human factors assessment.

- **Electrical power**: connection to appropriate bus bar with circuit protection devices, and most suitable harness runs.
- Security of attachment of all components.
- **Environment**: release of exhaust air from cabin and from wing leading edge anti-icing. Possibility of contaminants in HEPA filters during cleaning and disposal.

3.5.6 Key Design Drivers

Minimum disturbance to engine air flow with varying demand to maintain optimum engine operating conditions.

3.5.7 Modelling

Matlab/Simulink for flow modelling in the distribution network, air system test rig to model a complete system with realistic loads.

3.5.8 References

Beater (2007), Gent (2010), Lawson and Judt (2022), Moir and Seabridge (2008) Chapter 6, Rolls-Royce (2005) Chapter 2.5, JASC Code (n.d.), ATA-100 (n.d.) Chapter 36.

3.5.9 Sizing Considerations

Rates of operation of thrust reversers, area of leading edges and temperature required to assure anti-icing, material of leading edge, and risk of heat damage.

3.5.10 Future Considerations

Use of alternative sources of energy to achieve optimum engine operation and to meet environmental targets. Use of efficient electrical anti-icing of the leading edges and consequent impact on electrical loads.

The use of bleed air to provide energy for a number of functions is still in use at the time of writing. There are alternative ways of doing this, and some have been tried, but the current preferred way is to use bleed air. The Boeing B787 made a radical break with this tradition by obtaining air, using ram air intake and using electric motor-driven compressors to provide air at more suitable temperatures and pressures than that obtained directly from the engine. When examining the concept for a new aircraft type, it will be worth considering alternatives, bearing in mind that there will be some impact on the total aircraft system. Some alternatives are the following:

- **Air conditioning**: from ram air and intake fans
- **Cargo compartment heating**: exhaust from cabin conditioning
- **Wing and engine anti-icing**: electric heating
- **Engine start**: Electric starter generator on the engine
- **Hydraulic reservoir pressurisation**: nitrogen from OBIGGS or ECS air
- **Rain dispersal**: using ECS air or screen coatings
- **Water tank pressurisation and toilet waste**: using ECS air
- **Air-driven Hydraulic Pump (ADP)**: use AC or DC motor-driven pump

3.6 Secondary Power Systems

There are periods during the operation of an aircraft where the main engines are not available or when starting energy is required for engines. The secondary power systems are a collection of sub-systems ancillary to the main power systems and are intended for use for short periods to support the main supply. They may not be present in all aircraft. They may provide power to support the following functions (Figure 3.8):

- **Starting of main propulsion system**: battery for starter, air starter, and chemical start.
- Provision of air and electrical power during ground operations with no engines operating to provide autonomous operation or rapid turnaround.

3.6.1 Purpose of System

To provide a form of energy in the absence of the main sources of power.

3.6.2 Description

Auxiliary power unit (APU) and the attachment of power generating devices such as electrical and hydraulic generators. Connections to airframe systems.

3.6.3 Safety/Integrity Aspects

Mission critical but are considered safety critical if the APU is to be used in flight to support the aircraft after engine failure, for example ETOPS.

3.6.4 Key Integration Aspects

Integration with air, hydraulic, and electrical ground facilities.

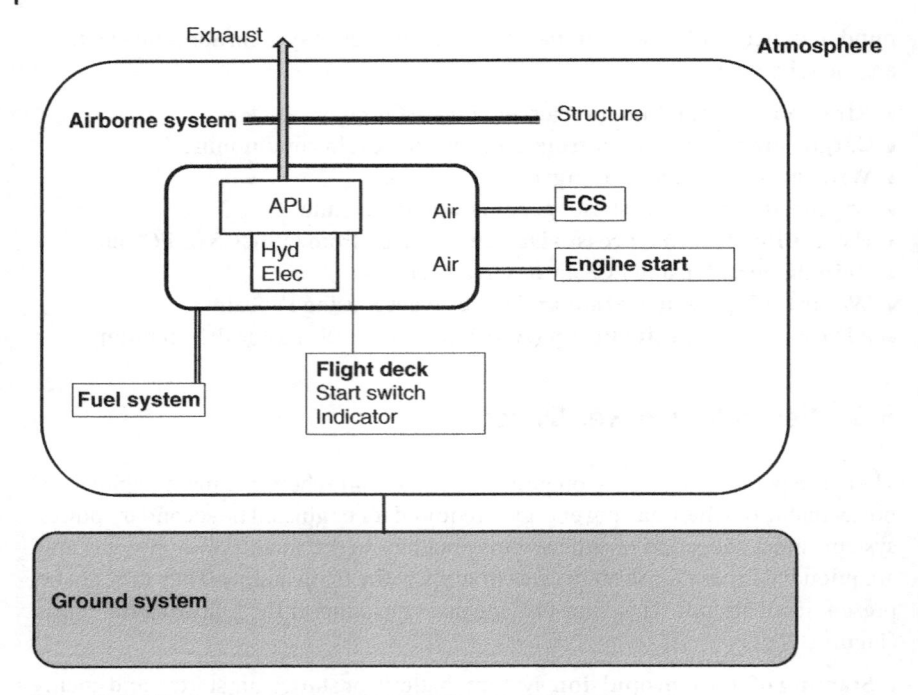

Figure 3.8 Illustration of a generic secondary power system.

3.6.5 Key Interfaces

Interface with the cockpit for starting controls and synoptic displays. Suitable connections to the bleed air system and electrical bus bars.

- **Airframe/structure**: Provision of a suitable bay for safe intake and exhaust of the APU with suitable clearances and warnings for ground personnel. Provision of doors for air intake and exhaust dust to reduce drag.
- **Other systems**: exchange of information by direct wiring or data bus.
- Flight deck for controls and displays and human factors assessment.
- **Electrical power**: connection to appropriate bus bar for APU starter motor with circuit protection devices.
- Security of attachment of all components.
- **Environment**: release of exhaust from APU, fumes, leaks, drips, and other potential environmental contamination. Noise may be an issue at airports.

3.6.6 Key Design Drivers

Mass, cost, efficiency, and noise at the airport to meet health and safety limitations.

3.6.7 Modelling

Test Rig.

3.6.8 References

Freeh (2016), Lawson and Judt (2022), Moir and Seabridge (2008). JASC Code (n.d.), ATA-100 (n.d.) Chapter 49.

3.6.9 Sizing Considerations

Loads to be connected to an APU and the length of APU operating time required will determine the size of the APU and the generator. This will have an impact on the location and the means of starting. Fire protection will be required, intake/exhaust hatches and actuation mechanism.

3.6.10 Future Considerations

More emphasis on in-flight operable APU. The ground APU will make a contribution to airport noise and pollution.

3.7 Emergency Power Systems

In certain emergency conditions, an electrical power generation system may not meet all the airworthiness authority requirements, and additional sources of energy are provided to power the aircraft systems. The aircraft battery provides a short-term power storage capability, typically up to 30 minutes. However, for longer periods of operation, the battery alone is not sufficient. The operation of twin-engine passenger aircraft on extended twin operations (ETOPS) means that aircraft must be able to operate on one engine while up to 180 minutes from an alternative or diversion airfield. This has led to the modification of some of the primary aircraft systems to ensure that sufficient integrity remains to accomplish the 180-minute diversion whilst still operating within acceptable safety margins.

There are also circumstances where hydraulic power needs to be maintained either in the event of failures to support flight trials in which one or more engines may cease to provide power.

A number of systems have evolved to provide support to electrical and hydraulic power which can be used in different combinations, including the following (Figure 3.9):

- RAT is a one-shot system with no absolute guarantee of working when required, and significant maintenance actions are required to ensure

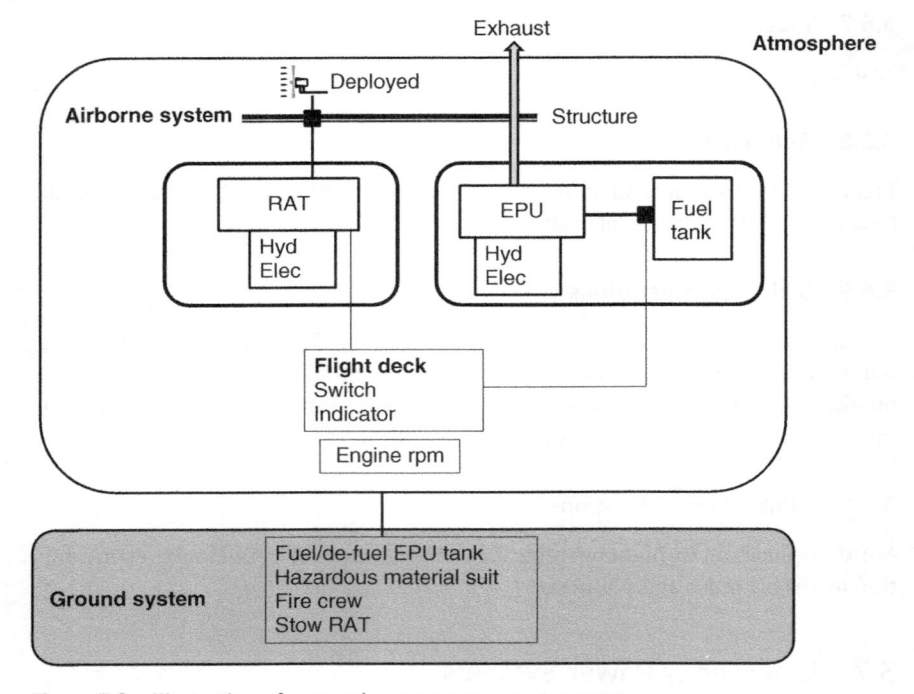

Figure 3.9 Illustration of a generic emergency power system.

condition/state-of-readiness. Again, this will provide energy for a hydraulic pump and a generator. The RAT is usually designed so that it can be deployed into the airstream using gravity.

- **Emergency power unit**: mono-fuel (such as hydrazine) powered system driving the engine gearbox. An EPU is often installed for high incidence or spinning trials on prototype aircraft. The aim is to provide hydraulic power to flight control actuators in order that the aircraft can be recovered to an attitude suitable for starting the engines. An alternative is to provide an air operable APU to provide hydraulic and electrical power for a short period of time.
- Backup power converters.
- Permanent magnet generators (PMGs) are used to provide AC or DC power for specific purposes. AC generators include a PMG to boot-strap the excitation system. PMGs producing an AC supply are used to power independent lanes of the engine control units from an engine or APU.
- Fuel cells using a chemical reaction to provide electrical energy for a short period of time.
- One-shot battery to provide a one off very short time source of energy for emergency use – such as electro-hydraulic pump.

- Electro-hydraulic pumps can be used to provide short-term hydraulic power for lowering of the undercarriage, either powered from the main battery or from a one-shot battery.
- Hydraulic accumulators are contained within the hydraulic power generation system and charged with nitrogen to maintain fluid pressure for a short time for specific systems only.

3.7.1 Purpose of System

To provide power for a short period of time in order to recover an aircraft in the event of a departure and engine failure or a major system failure that results in the loss of main power. This can be achieved by the use of one of a number of different means.

3.7.2 Description

A choice must be made of the sources of emergency power required and a selection made of the most appropriate devices available. Control and activation of the devices must be by automatic operation or by pilot selection.

3.7.3 Safety/Integrity Aspects

If the RAT is to be deployed into the air stream by gravity, then the system will be dormant and will need to be designed accordingly and provision made for periodic testing. The mechanism MUST operate when required, and MUST NOT operate inadvertently and careful design is required to ensure that this is the case. Testing, including flight-testing, will be required to validate this. It must form a part of a safety critical analysis.

Mono-fuels used in EPUs have a strong oxidising function and will be a source of fire or explosion. They are also a personnel health and safety hazard and the design must ensure all precautions are taken to allow safe fuelling and de-fuelling. Some fuels, e.g. hydrazine may be carcinogenic.

Most of these systems are dormant and cannot be tested pre-flight.

3.7.4 Key Integration Aspects

The power units must be installed and integrated with the airframe for optimum intake performance to ensure start up. The RAT must be deployed to achieve optimum energy extraction from airflow.

3.7.5 Key Interfaces

This will need provision of suitable isolating contactors in the electrical distribution system and maybe an emergency bus bar.

- **Airframe/structure**: provision of apertures and suitable fasteners for APU, EPU, and RAT. Automatic doors to be provided with ground crew warning notices.
- **Other systems**: interfaces with secondary sources of electrical and hydraulic power and connection to the aircraft power distribution system in order to provide power to a limited number of devices essential for the safe recovery of the aircraft.
- Flight deck for controls and displays and human factors assessment.
- **Electrical power**: connection to appropriate bus bar with circuit protection devices, and most suitable harness runs.
- Security of attachment of all components.
- **Environment**: release of exhaust, fumes, leaks, drips, and other potential environmental contamination. EPU mono-fuel is hazardous and may be carcinogenic.
- Consideration of EMC by radiation, induction, and susceptibility.
- Ground systems for download of data, provision of services on turn around, and replenishment of fluids or provisions. Suitable clothing to be available for re-fuelling EPU.

3.7.6 Key Design Drivers

Availability, effective operation.

3.7.7 Modelling

Three-dimensional modelling (Catia) of the installation and clearances for deployment, virtual mock up, test, and rig.

3.7.8 References

Giguere (2010), Lawson and Judt (2022), Moir and Seabridge (2008) Chapter 8.

3.7.9 Sizing Considerations

Electrical and hydraulic demand and duration of demand to determine the capacity of pumps and generators.

3.7.10 Future Considerations

Although the emergency power unit system is only used for a short part of the testing of aircraft, it is an important system. An alternative to hazardous mono-fuel systems would be an advantage, and fuel cells may be a suitable alternative.

3.8 Flight Control System

The flight control system is required to ensure that the aircraft is always stable and responsive to pilot demands from the control column or system demands from the auto-pilot or flight management system. The system will isolate the control surface forces from the pilot. In the case of unstable aircraft, the system operates continuously to ensure that the aircraft is stable and safe (Figure 3.10).

3.8.1 Purpose of System

To translate the pilots' commands into a demand for power to drive primary and secondary control surfaces and to respond to auto-pilot demands for automatic

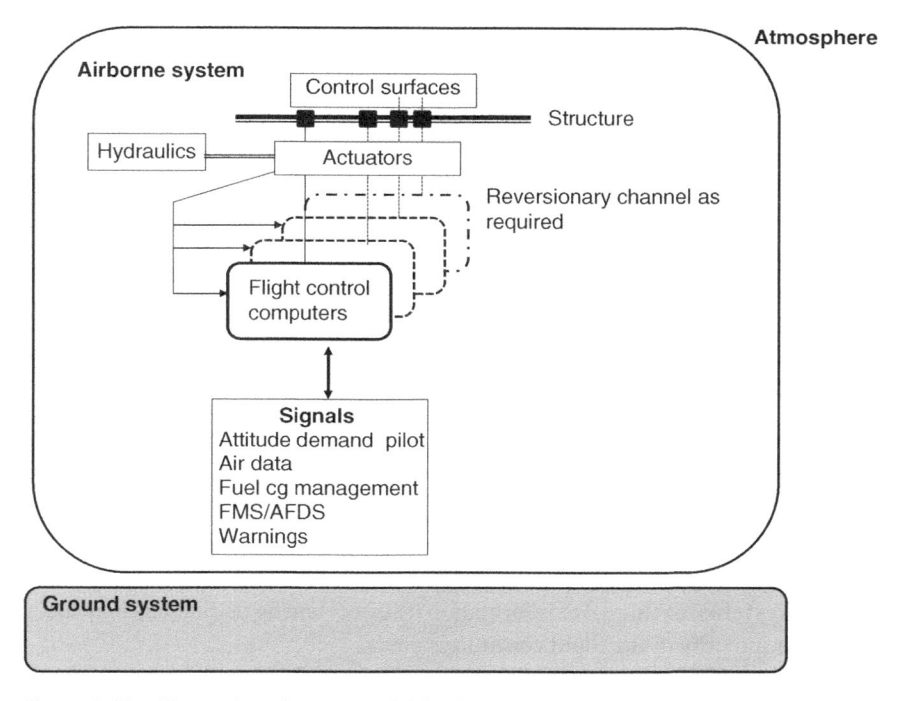

Figure 3.10 Illustration of a generic Flight Control System.

control and stability. For unstable military aircraft to ensure that demands are acted upon rapidly, to limit demands to a safe operating envelope, and constantly react to external aerodynamic and atmospheric conditions.

3.8.2 Description

A control system designed to translate the pilot's control column inputs and aircraft attitude changes into continuous demands for primary control surface actuators and also to respond to pilot's demands for operation of the secondary flight controls. To do this requires demand input sensors, a robust computing system, actuators, surface position and rate feedback sensors, and knowledge of the aircraft attitude.

3.8.3 Safety/Integrity Aspects

The system is safety-critical and must be designed to the suitable integrity and safety standards. A suitable standard of redundancy must be provided to withstand at least two failures without degradation in performance. Decisions need to be made with respect to the use of dissimilar redundancy of hardware and software or alternatively to prove that a safe design can be achieved without it. Manual or alternate electric reversion can also be used.

Software must be designed as safety-critical using a suitable and validated high-order language in accordance with a validated design process.

3.8.4 Key Integration Aspects

Integration with air data system, auto-pilot, flight management, propulsion, landing aids to complete guidance, and control integration. Integration with fuel system for centre of gravity control in unstable aircraft can be employed to achieve better manoeuvre performance (high g) and carefree handling. Integration with the propulsion system for thrust vectoring.

3.8.5 Key Interfaces

- **Airframe/structure**: provision of apertures and suitable fasteners for control surface actuators on wings and empennage. Location and attachment of foreplanes as required.
- **Other systems**: exchange of information by direct wiring or data bus to respect the high integrity of the flight control system.
- Flight deck for control column (stick) and displays and human factors assessment.

- **Electrical power**: connection to appropriate bus bar with circuit protection devices, and most suitable harness runs. Will require main and emergency supplies to respect integrity.
- **Installation**: mounting tray for electronic LRUs and provision of suitable conditioning and cooling. Suitable location and spatial separation of flight control computers and wiring harness to respect control lanes.
- Security of attachment of all components.
- **Environment**: may be leaks and drips of hydraulic fluid at actuators.
- Consideration of EMC by radiation, induction, and susceptibility.

3.8.6 Key Design Drivers

Safety, structural limitations, flight envelope, and performance.

3.8.7 Modelling

Matlab/Simulink can be used for preliminary testing and feasibility. More exacting modelling must be performed throughout the design process which will include control loop modelling, control law modelling, flight simulation for pilot in the loop testing, and pilot familiarisation. Testing of the hardware and some failure philosophies can be performed by iron bird testing in which the entire system will be produced and made to work with a hydraulic system and against realistic flight loads applied to control actuators. Depending on the degree of novelty of the control system and the aircraft performance, it may be necessary to use a prototype aircraft. This approach was used in the early development of unstable aircraft and full authority fly by wire systems.

3.8.8 References

Bryson (1994), Lloyd and Tye (1982), Langton (2006), Lawson and Judt (2022), Moir and Seabridge (2008) Chapter 1, Moir and Seabridge (2013), Raymond and Chenoweth (1993), Weller (2018). JASC Code (n.d.), ATA-100 (n.d.) Chapter 27, Pratt (2000).

3.8.9 Sizing Considerations

The degree of redundancy will determine the number of flight control computers required and the nature of redundancy will determine whether or not they need to be dissimilar in build standard or software. The same is largely true for actuators and sensors. Human factors and overall aircraft design philosophy will have an input into the type and location of the control column on flight deck resulting in centre or side location and displacement of force type. If a force control is selected consideration must be given to the way in which both pilots are informed of its position to prevent conflicting stick inputs.

3.8.10 Future Considerations

Full electric actuation will remove some of the weight and complexity of multiple channel hydraulics distribution. On many aircraft today, the FCS is a stand-alone system, in accordance with a design philosophy that segregates high-integrity systems from all others. Consideration must be given to advances in system reliability which will justify a move towards integration of controls into VMS or IMA for a more efficient computing philosophy.

3.9 Landing Gear

An aircraft will be provided with a number of wheels depending on the weight of the aircraft, usually in a configuration of steerable nose wheel and a number of main wheels descending from the wings or the centre fuselage. The wheels are attached to pneumatic oleo struts that absorb the main landing loads and will be designed with the appropriate mechanism to retract and stow in appropriate bays with doors. The mechanisms also contain locks to prevent inadvertent deployment, and a number of position sensors to inform the control system and warning/status indications (Figure 3.11).

The gear must be designed to respect the speed and weight of an aircraft from a fully laden emergency landing to a rejected take-off and full braking. The wheels must deal with a wide variety of landing surfaces and the turning radius of taxy ways. Very large passenger or transport aircraft will have steerable wheels in their main gear.

3.9.1 Purpose of System

To enable the aircraft to be mobile on the ground and to allow take-off and landing on different types of surface from rough field to fully prepared runways according to the type of aircraft.

3.9.2 Description

Installation of main landing gear forgings for attachment to the airframe and the inclusion of oleos and wheel attachments. The design will include the mechanism for deployment and retraction of the entire assembly with an appropriate folding

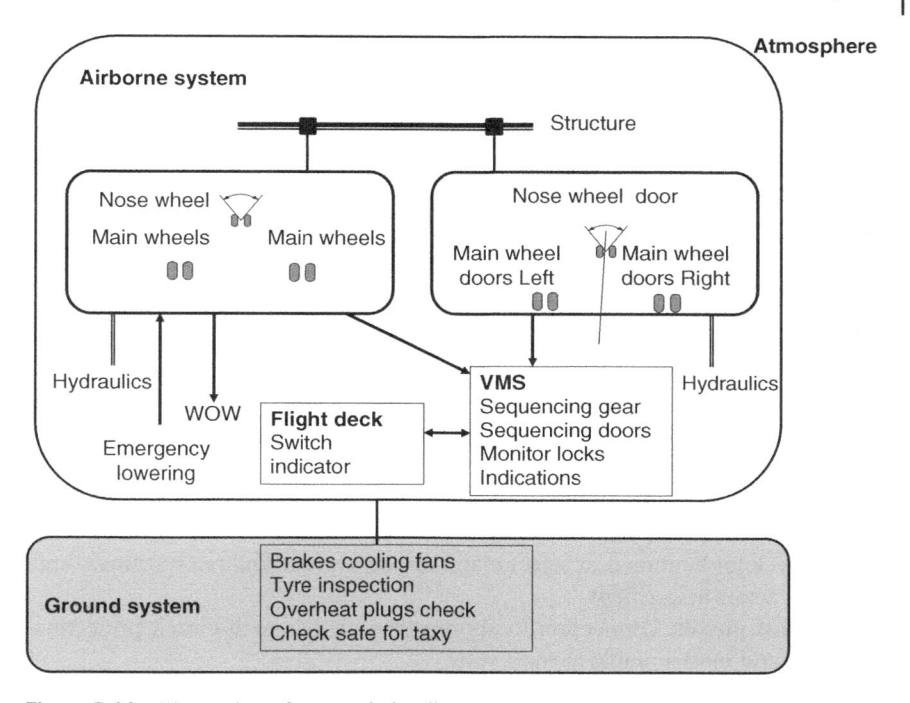

Figure 3.11 Illustration of a generic landing gear system.

sequence to fit into the wheel bays without risk of fouling. Locks and proximity switched will be included so that the sequence can be monitored to enable warnings to be generated if the sequence fails or the locks are not engaged.

3.9.3 Safety/Integrity Aspects

Safety critical: usually provided with a mechanism for manual lowering of the gear if the normal means fails. This may be gravity, electro-hydraulic pump, or crew-operated hand pump.

Clearances to be maintained to prevent fouling, and provision must be made for the wheels spinning during retraction and shedding moisture and grit. No equipment shall be placed in the bay that could be damaged.

Hot brakes and risk of fire during long taxy, short landings, or rejected take-off are a potential risk from the thermal energy generated in the brakes. Brake cooling fans may be required, and may be available at the aircraft stand as a matter of course to achieve the required turnaround time.

Plugs are provided in tyres to deflate the tyre in hot conditions to avoid a burst.

3.9.4 Key Integration Aspects

The legs, wheels, and oleos must be designed so that they can be integrated into airframe to provide for efficient stowage of gear. Weight on wheels signals will be required by other systems, and a cockpit warning system is required for unambiguous indication of safe gear positions. Control and management functions to be integrated with the VMSs.

3.9.5 Key Interfaces

- **Airframe/structure**: provision of bays and suitable attachment points for landing gear structure and inclusion of doors. Volume of bays to include gear retraction and folding envelopes. Stone guards may be required on short field aircraft to protect the fuselage from stone damage. Precautions to accommodate spinning wheels on retraction.
- **Other systems**: hydraulic or electric power supplies, electrical signalling of gear and doors positions.
- Flight deck for landing gear selector and displays (green and red warnings) and human factors assessment.
- **Electrical power**: connection to appropriate bus bar with circuit protection devices, and most suitable harness runs.
- Security of attachment of all components.
- Ground systems for download of data, provision of services on turn around such as tyre inspection, replacement tyres, and ground cooling fans.

3.9.6 Key Design Drivers

Mass, aircraft all up weight, aborted take-off mass, and airfield condition (Runway LCN and braking conditions).

3.9.7 Modelling

Three-dimensional (Catia) modelling of extension and retraction of gear using animation to check the deployment, retraction, and stowage sequence is correct with no fouls with the airframe. An iron bird test rig will provide more realistic tests. Supplier test rigs will be used to demonstrate wheel drop onto a rolling road under dynamic conditions.

3.9.8 References

Conway (1957), Currey (1984), Lawson and Judt (2022), Moir and Seabridge (2008) Chapter 4, JASC Code (n.d.), ATA-100 (n.d.) Chapter 32.

3.9.9 Sizing Considerations

Aircraft all up weight, type of runways and weather conditions, braking and rejected take-off loads, and typical perimeter track radii. Landing loads and catapult loads for carrier borne aircraft. Consideration is to be given to rough field and carrier landings.

3.9.10 Future Considerations

All electric actuation is a serious consideration with the move towards all-electric aircraft.

3.10 Brakes and Anti-skid

Wheel brakes are included in the main wheels to decelerate the aircraft, and anti-skid is used to avoid any skid developing under adverse conditions. The brakes are usually operated by depressing the pedals. A handbrake function is also incorporated for parking the aircraft (Figure 3.12).

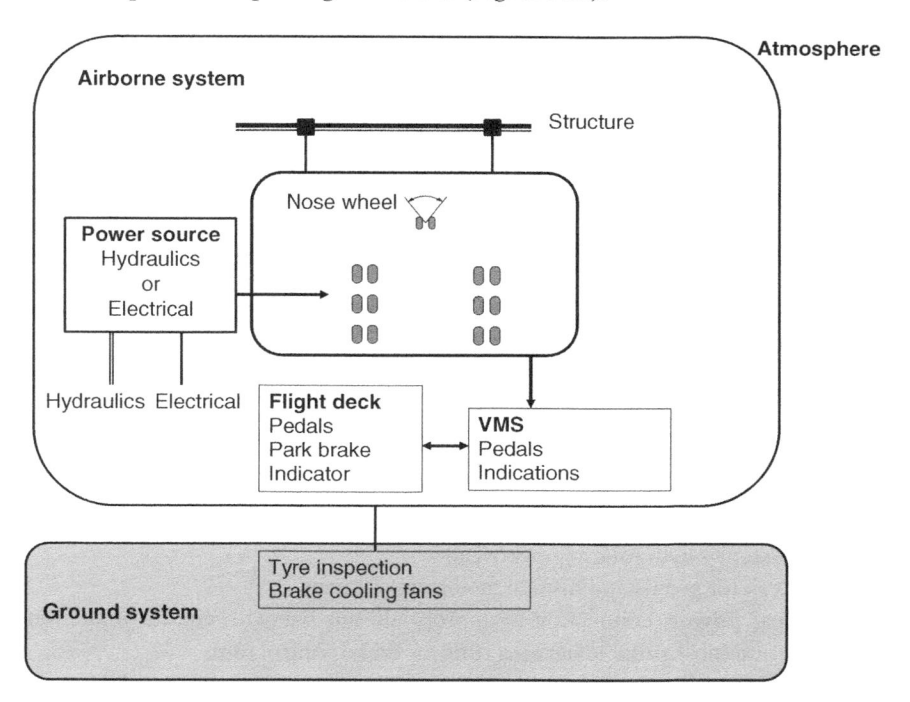

Figure 3.12 Illustration of a generic braking system.

3.10.1 Purpose of System

To allow the aircraft to be decelerated on the ground, to absorb braking energy, and to prevent loss of wheel traction during braking. A separate function is provided for parking.

3.10.2 Description

The pedals have a function that allows the crewmembers to operate the brakes at the appropriate parts of the take-off, landing, and taxying cycle. The pedals allow variable braking force to be applied and operate the wheel pads against the brake discs using hydraulic or electric actuation. In most modern systems, the function is contained in a brake control unit or integrated into VMS. This function senses differential wheel rotation and acts to prevent a skid. Brake cooling fans may be used to prevent brake overheating, especially during long taxy operations with many stop and hold points.

3.10.3 Safety/Integrity Aspects

Safety critical to prevent a skid developing and to absorb the energy of a maximum weight maximum thrust aborted take-off.

3.10.4 Key Integration Aspects

High-speed closed-loop servo system for the high bandwidth/brake and antiskid control functions. Control of the systems can be performed by separate control units or integrated into the VMS.

3.10.5 Key Interfaces

- **Airframe/structure**: recognition of bay temperatures on retraction if brakes have been used during taxy.
- **Other systems**: hydraulics or electrical power depending on technology of braking actuation. Sensors to detect brake temperatures, wheel rotation, and differential rotation. Connection to the pedal sensors and to the actuation power system – hydraulics or electrical.
- Flight deck for pedals and human factors assessment.
- **Electrical power**: connection to appropriate bus bar with circuit protection devices, and most suitable harness runs for brake control unit.
- **Installation**: mounting tray for electronic LRUs such as brakes control unit and provision of suitable conditioning and cooling.

- Security of attachment of all components.
- **Environment**: release of brake pad particles and vapour into the atmosphere.
- Consideration of EMC by radiation, induction, and susceptibility.
- Ground systems for download of data, provision of services on turn around such as brakes cooling fans.

3.10.6 Key Design Drivers

Aircraft all-up weight, maximum rejected take-off clearances, landing characteristics, dissipation of brake energy for ramp departure – cooling fans. Runway surface type.

3.10.7 Modelling

The braking system supplier or the landing gear supplier will have a dynamic landing test rig to prove the system at different speeds and loads.

3.10.8 References

Lawson and Judt (2022), Moir and Seabridge (2008) Chapter 4. JASC Code (n.d.), ATA-100 (n.d.) Chapter 32.

3.10.9 Sizing Considerations

Braking system and energy source, energy during braking, energy dissipation, and cooling mechanism. Consideration is to be given to rough field and carrier landings.

3.10.10 Future Considerations

Hydraulic power is common today, but there will be a trend towards all electric actuation. Brakes heat can be used by for scavenging devices to re-use energy on future vehicles.

3.11 Steering System

When the aircraft is on the ground, it needs to be manoeuvred when under its own power and when being towed. To provide suitable steering of the aircraft during the transition of rudder effect on landing to steering, to steer the aircraft around the perimeter track without over running the track edges. The aircraft can be steered

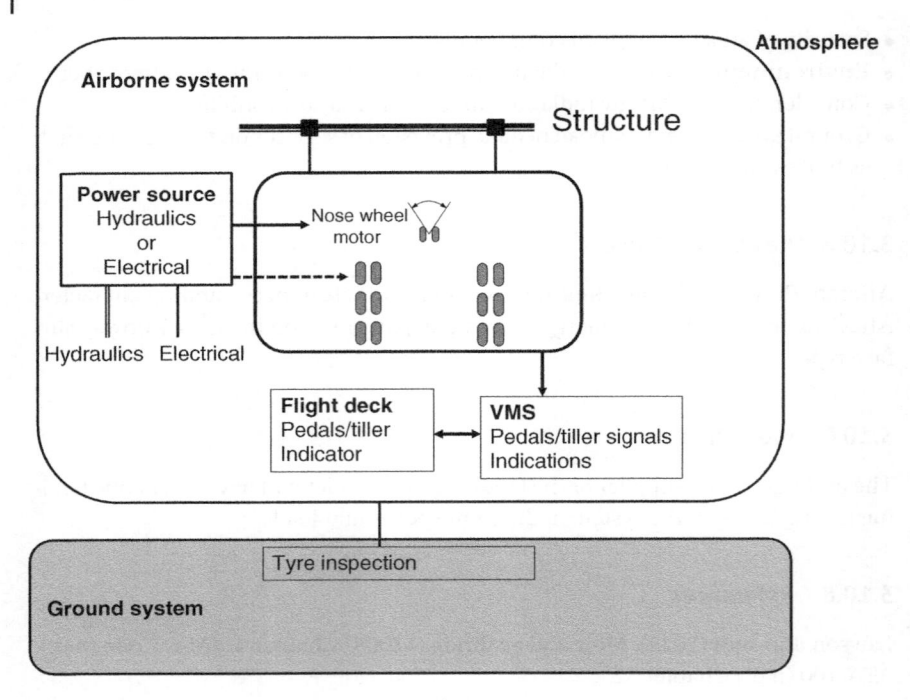

Figure 3.13 Illustration of a generic steering system.

by use of a steering tiller or the pedals. Predominantly a steering actuator or motor acts on the nose wheel, but in a large aircraft, there may be steerable wheels in large bogies (Figure 3.13).

3.11.1 Purpose of System

To provide a means of steering the aircraft under its own power or whilst being towed with consideration being given to the radius of turns on the perimeter track and aerodynamic effects on the runway.

3.11.2 Description

Provision of a steering mechanism to relevant wheels using the most appropriate power source – electrical or hydraulic. Steering can be performed using the pedals or by means of a steering tiller mechanism.

3.11.3 Safety/Integrity Aspects

Safety affected – failure to steer correctly at high speeds can lead to a departure from runway or taxiway or scrubbing of the tyres. In the worst case, poor steering performance can damage the undercarriage legs.

3.11.4 Key Integration Aspects

Human factors and the pilot's ability to control the aircraft. Hydraulic system, pilot's displays, including video and wheel-monitoring cameras (some models).

3.11.5 Key Interfaces

Integration with flight control to ensure correct hand-over from rudder steering during landing run:

- **Airframe/structure**: clearance of steered wheel on full deflection.
- Other systems: hydraulics or electrical power depending on technology of steering actuation.
- Flight deck for pedals or steering tiller and human factors assessment.
- **Electrical power**: connection to appropriate bus bar with circuit protection devices, and most suitable harness runs for steering control unit.
- **Installation**: mounting tray for electronic LRUs such as steering control unit and provision of suitable conditioning and cooling.
- Security of attachment of all components.
- Consideration of EMC by radiation, induction, and susceptibility.

3.11.6 Key Design Drivers

The nature of common airport taxi ways and radius of turns combined with landing and taxi speeds.

3.11.7 Modelling

CAD, Matlab/simulink.

3.11.8 References

Lawson and Judt (2022), Moir and Seabridge (2008) Chapter 4. JASC Code (n.d.), ATA-100 (n.d.) Chapter 32.

3.11.9 Sizing Considerations

Steering mechanism and source of energy number of steerable wheels and bogies.

3.11.10 Future Considerations

Auto steering from runway to terminal.

3.12 Environmental Control System

The passengers and crew must be provided with cabin conditions that are comfortable, clean, and with appropriate humidity. The cabin must be maintained at a suitable 'altitude' condition to provide optimal conditions for absorption of oxygen. The air is to be filtered to remove solid and moisture particles and some bacterial or virus contamination. Above a certain altitude, ozone is removed from the air. Air must also be provided to avionic equipment for cooling purposes and to cargo compartments to reduce condensation and to protect live cargo. Some military aircraft will require air to maintain bomb bay carried weapons at a suitable temperature (Figure 3.14).

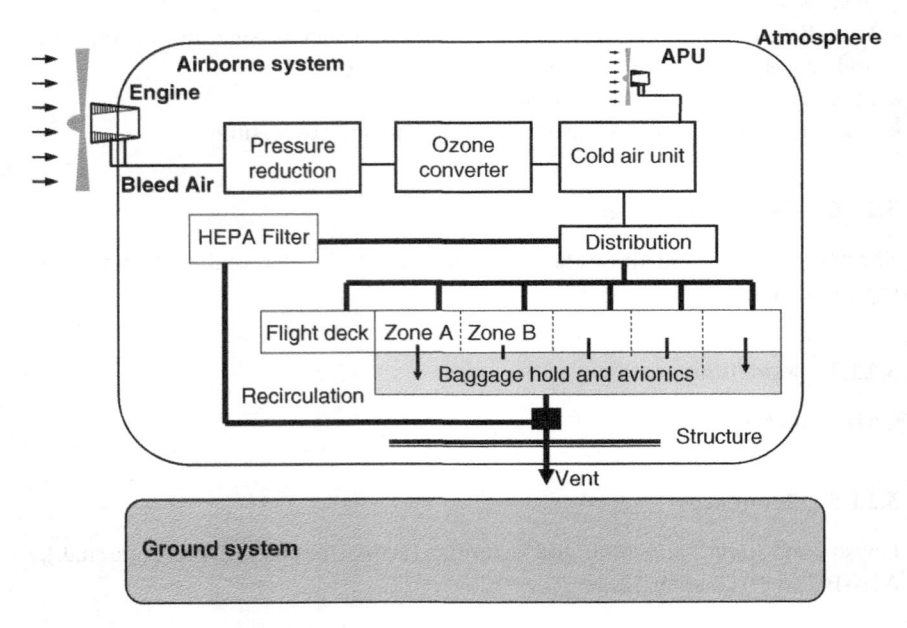

Figure 3.14 Illustration of a generic ECS.

3.12.1 Purpose of System

To provide heating and/or cooling air for passengers, crew, and avionics equipment using air available as bleed from the engines or from compressed ram air, to provide a safe and comfortable environment with appropriate temperature, humidity, and cleanliness, and to allow passengers and crew to breathe throughout the aircraft flight envelope.

3.12.2 Description

Provision of pressurised air to the cabin with appropriate circuits for air filtration, humidity control, temperature control, and ozone reduction. Air is generally obtained from the compressor stages of the engines and fed to one or more cold air units after pressure reduction and initial cooling before entering the cabin at close to 0 °C. The very cold air is mixed with warm air to obtain the correct set temperature and may be re-circulated through filters. Exhaust air is commonly directed to avionics bays and cargo compartments before being exhausted overboard to atmosphere.

3.12.3 Safety/Integrity Aspects

Safety affected – loss of all cooling and pressurisation can lead to crew's loss of consciousness, passenger discomfort, and equipment malfunction or reduction in operating life.

3.12.4 Key Integration Aspects

Air must be extracted from the engines without affecting engine performance. Insulation of bleed air off-take pipes in the airframe is required to reduce heat loss and to protect maintenance crew and other aircraft components. Location of air ducts to cabin and sound insulation. Location of air vents to provide optimum circulation for passengers.

3.12.5 Key Interfaces

- **Airframe/structure**: provision of apertures for air intake and cabin exhaust or connection to engine for bleed air.
- **Other systems**: exchange of information by direct wiring or data bus.
- Flight deck and cabin crew panels for controls (setting of temperature and cabin pressure) and displays and human factors assessment.
- **Electrical power**: connection to appropriate bus bar with circuit protection devices, and most suitable harness runs.

- **Installation**: mounting tray for electronic LRUs and provision of suitable conditioning and cooling. Providing paths for the routing of high-pressure air pipes and air ducts for high volume air to the cabin.
- Security of attachment of all components.
- **Environment**: potential contamination risk to maintenance crew during replacement and servicing of HEPA filters.

3.12.6 Key Design Drivers

The number of passengers and the electronic equipment in their seats add to the general heating of the cabin (up to 200 W per person). Crew and passenger comfort and safety are the principle concerns. Knowledge of ambient operating conditions – regional or world-wide is essential to design for hot and cold soak and high humidity conditions. Good filtration is required to reduce particle and biological contamination risk and to produce clean cabin air. Standards must be consulted for temperature range, number of changes of air, and for filtration parameters.

The fast jet military cockpit demands protection of the pilot wearing survival clothing and subject to solar radiation via the cabin transparency, electronic equipment in the cockpit also contributes to the cockpit heat load. It is important to note that the pilot in a combat situation may be physiologically stressed, and a high volume flow of cool air is required. For fast jet aircraft, a source of pressurised air will be required for the anti-g suits.

3.12.7 Modelling

MatLab/Simulink models of flows and temperatures. Modelling of airflow in ducting using CFD.

3.12.8 References

Lawson (2010), Lawson and Judt (2022), Moir and Seabridge (2008) Chapter 7. JASC Code (n.d.), ATA-100 (n.d.) Chapter 21.

3.12.9 Sizing Considerations

Volume of cabin, number of occupants, ambient temperature ranges, pressurisation, air intakes (drag), air distribution system, cold air units, filters, redundancy, and emergency air supply.

3.12.10 Future Considerations

The need to reduce intermittent demands for engine bleed has started a trend towards electrical compressors to provide pressurised air. This provides a more stable set of condition for the engine but results in an increased AC power demand to power the air compressors. The B-787 is the first example of this type of system.

There will need to be a continuous assessment of contamination risk for local and pandemic-type infection risks in order to refine the air filtration mechanism, HEPA filter, or otherwise. Examples are severe acute respiratory syndrome (SARS), avian bird flu, influenza, and the Covid-19 virus in 2020 and 2021. The need for distancing measures will affect seating arrangements, passenger density, and may also lead to different designs for air distribution.

3.13 Fire Protection System

There are areas of the aircraft that are at risk of overheating or combustion. Particular areas include the APU and the engine compartments, where there is a source of high temperature and also the risk of a fuel leak. These areas must be protected by detectors to recognise an overheat situation and by extinguishers to extinguish a fire (Figure 3.15).

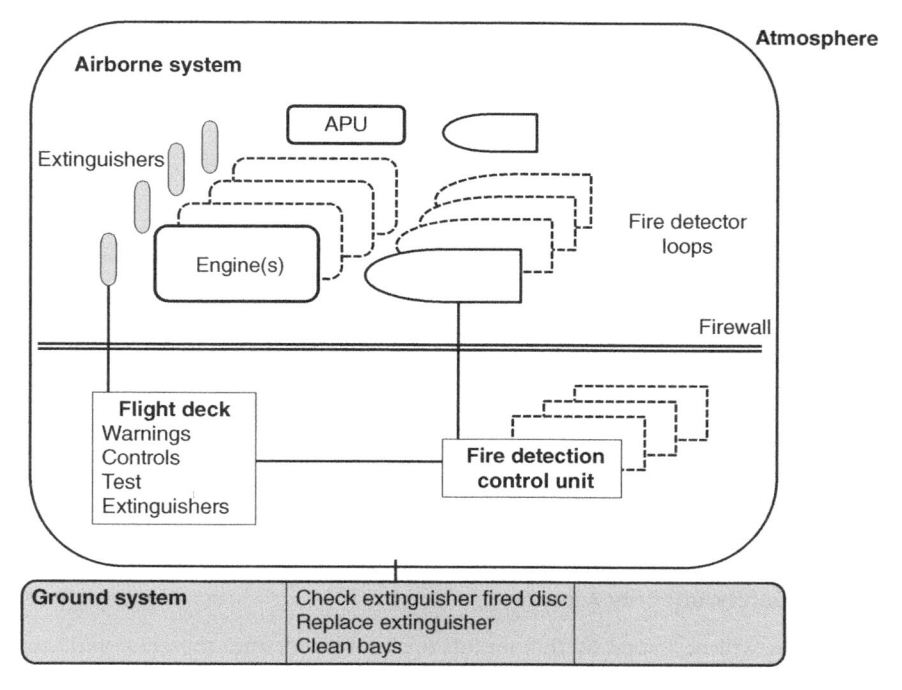

Figure 3.15 Illustration of a generic fire protection system.

3.13.1 Purpose of System

To sense, annunciate, and provide protection for fire, torching flame, hot gas leaks, and overheat situations and to provide a source of extinguishing fluid. The system is used in the following locations:

- Engine bays to sense fuel fires or hot gas leaks.
- Wing leading edges using hot air anti-icing to detect overheat to minimise damage to internal surfaces.
- Cargo bay and bomb bays to detect fires and/or overheat.

3.13.2 Description

Overheat or fire and flame detectors installed in a bay to provide wide area coverage. All equipment bays at risk must be protected with firewalls and fitted with fire extinguishers. Protection may be provided by fire extinguishers using liquid or powder extinguishing fluids.

Fire detection is undertaken by specialised temperature sensors whereby high temperatures cause a partial breakdown between inner and outer elements. A control unit measures this characteristic and signals a warning to the pilot when an appropriate threshold is exceeded. In many civil applications, the sensor loop is duplicated.

In the event of a fire warning, the pilot action will be specified in pilot's notes and may include such a procedure as follows:

- Shut down hazardous engine
- Isolate fuel at firewall
- Initiate emptying of chemical extinguisher(s) into bay
- All initiated with minimum number of actions preferably at the pressing of a single button

Fire bottles that may be discharged into the engine bay by the flight crew upon receipt of a warning. Some installations may use a double-shot system.

Fire extinguishers are provided in the cabin for small fires within the cabin or electrical equipment bay. Clearly, the active components have to be chosen carefully due to release in a confined space.

All extinguisher fluids must comply with current standards and regulations for performance and to meet environmental standards.

3.13.3 Safety/Integrity Aspects

Procedures will be issued by the aircraft operator to instruct the crew with an appropriate means of dealing with detected fire. This may result in deliberate engine shut down.

This is a dormant system and cannot be pre-flight tested, it is essential that it operates when required and must be designed accordingly. There may be a test available for the fire-detector loop operation to ensure circuit operation and warning lamp operation.

3.13.4 Key Integration Aspects

Human factors analysis and design for appropriate design of warning and extinguisher switches in the flight deck/cockpit. Location and direction of extinguisher spray nozzles. Location of the extinguisher loop to cover most potential hazard locations.

3.13.5 Key Interfaces

Cockpit warning system

- **Airframe/structure**: routing and clipping of detector loops in engine, APU bays and anti-ice system in wing leading edge. Provision of visual inspection of extinguishers.
- **Other systems**: exchange of information by direct wiring or data bus.
- Flight deck for controls and displays and human factors assessment. Provision of suitable warning lights and attention getters, with in-built switch for extinguisher operation.
- **Electrical power**: connection to appropriate bus bar with circuit protection devices and most suitable harness runs.
- **Installation**: mounting tray for electronic LRUs and provision of suitable conditioning and cooling.
- Security of attachment of all components.
- **Environment**: potential environmental contamination by fire extinguishing fluid voided into engine bays.
- Ground systems for check of fire extinguisher serviceability (ruptured disc inspection).

3.13.6 Key Design Drivers

Rapid and unambiguous detection mechanism. Human factors design. Minimisation of false warnings.

3.13.7 Modelling

Simple simulation, fire rig testing, and past experience.

3.13.8 References

Giguere (2010), Lawson and Judt (2022), Moir and Seabridge (2008) Chapter 8, Rolls-Royce (2005) Chapter 3.2. JASC Code (n.d.), ATA-100 (n.d.) Chapter 26.

3.13.9 Sizing Considerations

The number of engines and APUs and the size of their bays must be assessed. Other situations where fire or overheat may be a hazard include battery bays, fuel cell bays, and cargo compartments. The detection loop, sensor type, and control unit will designed to suit and the number of extinguishers located for maximum effect.

3.13.10 Future Considerations

All extinguishing fluids must conform to standards to meet contemporary environmental standards and (e.g. Montreal Protocol) and these standards must be continuously assessed.

3.14 Ice Detection

The aircraft will frequently enter airspace in which the conditions are appropriate for the formation of ice. This usually happens on leading surface of wings and empennage and will seriously impair the performance of the aircraft. An ice detection system is designed to provide advance warning of the formation of ice to allow ice protection measures to be used (Figure 3.16).

3.14.1 Purpose of System

To detect entry into icing conditions that may lead to the accretion of ice on leading edges of wing, empennage, or intake lips.

3.14.2 Description

Ice detector probes will be located on the airframe, and a detection unit will provide the necessary warning that the aircraft is entering icing conditions to inform the aircrew. The ant-icing system may be manually switched on by the crew or preferably automatically engaged.

3.14.3 Safety/Integrity Aspects

Safety involved.

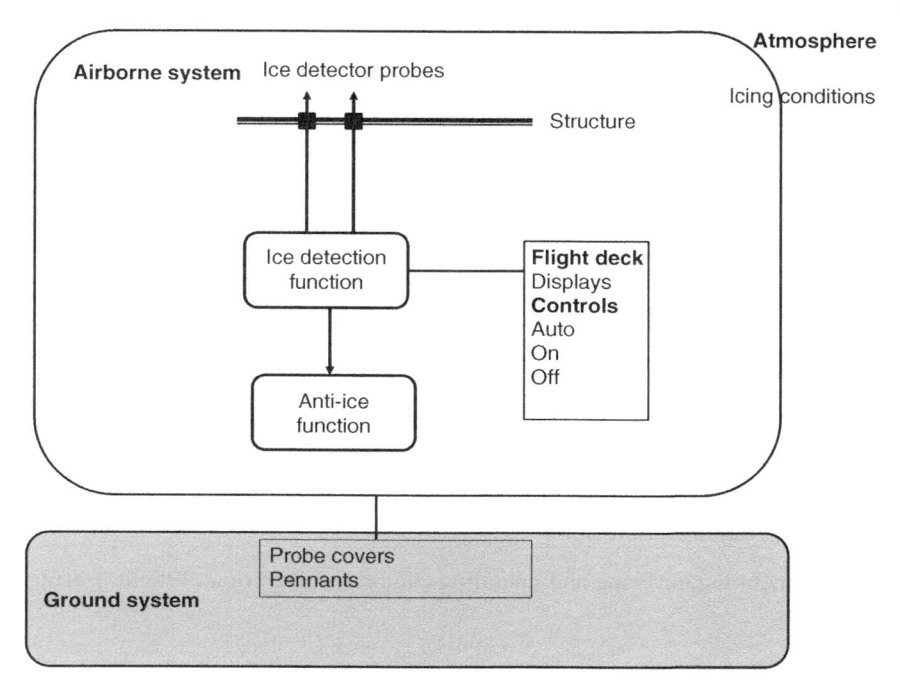

Figure 3.16 Illustration of a generic ice detection system.

3.14.4 Key Integration Aspects

Integration with ice protection system and flight deck display and controls.

3.14.5 Key Interfaces

Cockpit warnings

- **Airframe/structure**: provision of apertures and suitable fasteners for ice detectors in the most suitable location and protection of the pressure shell.
- **Other systems**: exchange of information by direct wiring or data bus.
- Flight deck for controls and displays and human factors assessment.
- **Electrical power**: connection to appropriate bus bar with circuit protection devices and most suitable harness runs.
- **Installation**: mounting tray for electronic LRUs such as ice detector unit and provision of suitable conditioning and cooling.
- Security of attachment of all components.
- Consideration of EMC by radiation, induction, and susceptibility.

3.14.6 Key Design Drivers

Aircraft operating envelope and operating conditions. Duration of flight in icing conditions.

3.14.7 Modelling

Simple simulation, specialised wind tunnel for controlled ice accretion tests, in flight testing.

3.14.8 References

Bedard (2016), Gent (2010). JASC Code (n.d.), ATA-100 (n.d.) Chapter 30, Lawson and Judt (2022).

3.14.9 Sizing Considerations

Aircraft operating envelope and operating conditions. Duration of flight in icing conditions.

3.14.10 Future Considerations

New methods of prediction and measurement of the approach to icing conditions, use of artificial intelligence (AI) in control unit to analyse meteorological data in flight.

3.15 Ice Protection

The aircraft will frequently enter airspace in which the conditions are appropriate for the formation of ice. This usually happens on leading surfaces of wings and empennage. It is usual for the aircraft systems to prevent ice formation (anti-icing) by heating the surfaces at risk. This may be applied automatically and to do so requires a device to detect the presence of conditions that may lead to ice formation (Figure 3.17).

3.15.1 Purpose of System

To prevent the build-up of ice and/or to remove ice already formed.

3.15 Ice Protection | 85

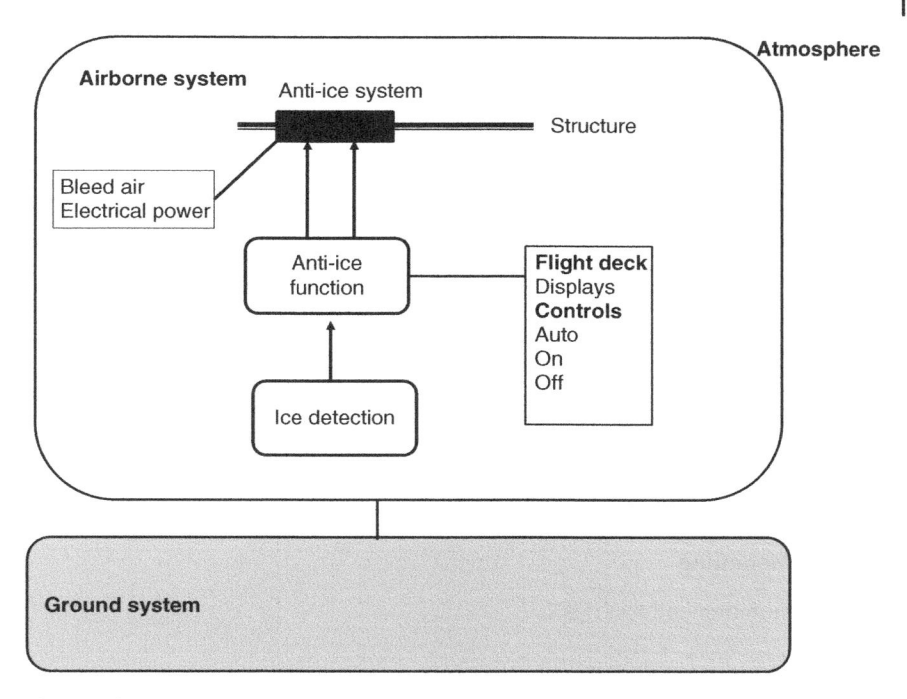

Figure 3.17 Illustration of a generic ice protection system.

3.15.2 Description

For large aircraft and military types, the anti-icing system is usually electrically operated with the aim of preventing ice formation. For some general aviation types pneumatic flexible boots are used to dislodge ice after it has formed (de-icing) or by vibrating to prevent ice formation.

3.15.3 Safety/Integrity Aspects

Safety involved – must work when required or aircraft must rapidly leave icing conditions.

This is a dormant system and cannot be pre-flight tested, and it is essential that it operates when required and must be designed accordingly. There may be a test of the ice detector and the warning system operation.

3.15.4 Key Integration Aspects

Integration with ice detection system for automatic operation and display of information on the flight deck displays.

3.15.5 Key Interfaces

Avionics for air data and external temperature calculations:

- **Airframe/structure**: provision of apertures and suitable fasteners for anti-icing system.
- **Other systems**: bleed air for leading edge heating or rubber boots.
- Flight deck for controls and displays and human factors assessment.
- **Electrical power**: connection to appropriate bus bar with circuit protection devices and most suitable harness runs.
- Security of attachment of all components.
- **Environment**: release of fluids if an anti-ice fluid system is used.

3.15.6 Key Design Drivers

Mass, electrical load, drag, and successful ice prevention.

3.15.7 Modelling

Simple simulation and test rigs.

3.15.8 References

Gent (2010), Lawson and Judt (2022), Moir and Seabridge (2008) Chapter 6. JASC Code (n.d.), ATA-100 (n.d.) Chapter 30.

3.15.9 Sizing Considerations

Type of ice protection mechanism and potential electrical load. Area of surface to be protected and its thermal inertia.

3.15.10 Future Considerations

Reduced dependence on engine bleed air, there may be a trend towards electric anti-icing (re B-787). Alternative, low power ice prevention mechanisms may be available.

3.16 External Lighting

The aircraft must be visible at all times to other aircraft either in the air or on the ground, and also to air traffic controllers. A recognisable pattern of lights has evolved that is a standard for most aircraft.

3.16.1 Purpose of System

To ensure that the aircraft is visible to other airspace users and to provide lighting for landing and taxying, and for identification by air traffic control towers. Provision of lighting of company logos on the fin.

3.16.2 Description

Lights are located at various points on the airframe to fulfil a specific purpose of identification, warning or illumination during taxy and take-off. Lights include wing tip navigation lights, high-intensity strobe lights, fuselage strobe lights or anti-collision beacons, and logo lights. Military users will include formation lights and air-to-air refuelling probe light (Figure 3.18).

3.16.3 Safety/Integrity Aspects

Safety involved. Optimum location of external lights to ensure visibility to other aircraft and control tower is essential.

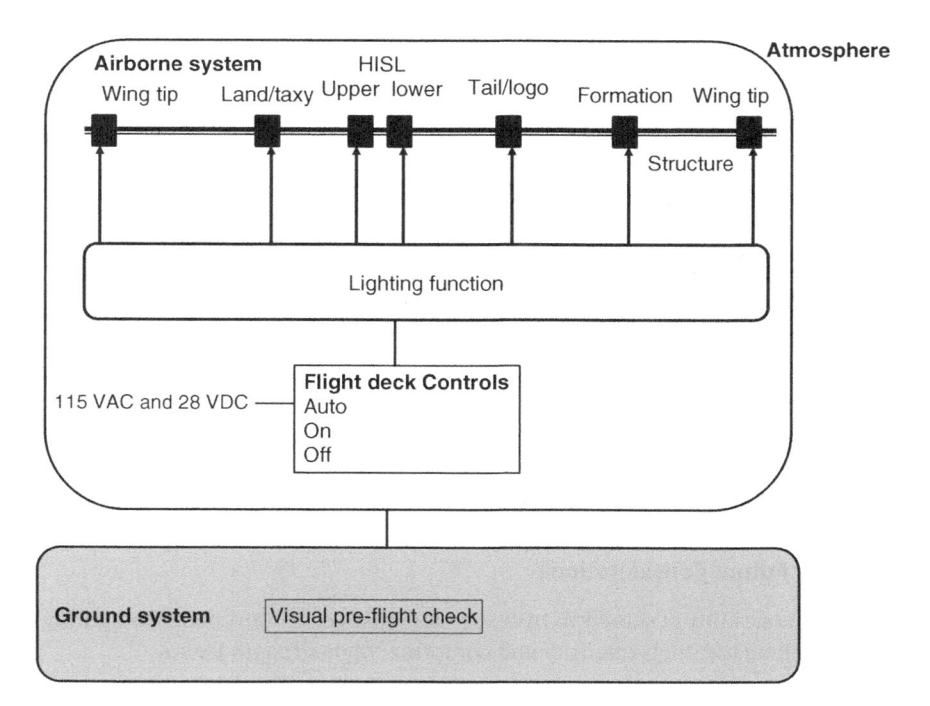

Figure 3.18 Illustration of a generic external lighting system.

3.16.4 Key Integration Aspects

All lights will be mounted on the aircraft structure – wing tips, fuselage, or undercarriage bay doors. A low drag installation is preferred to suit aircraft performance.

3.16.5 Key Interfaces

- **Airframe/structure**: provision of apertures and suitable fasteners for lights and lenses, protection of the pressure shell.
- Flight deck for controls and displays and human factors assessment.
- **Electrical power**: connection to appropriate bus bar with circuit protection devices, and most suitable harness runs.
- Security of attachment of all components.
- Ground systems for download of check and replacement of lights.

3.16.6 Key Design Drivers

Regulations governing purpose, colour, angle of viewing, and intensity. All lights must be visible to other aircraft in all weather conditions. The angle of coverage of lights must be suitable for landing and taxying.

3.16.7 Modelling

Simple simulation or models to specify beam angles and dispersion, long-range visibility in different weather conditions.

3.16.8 References

JASC Code (n.d.), ATA-100 (n.d.) Chapter 33.

3.16.9 Sizing Considerations

Size of aircraft, flying conditions, and current regulations.

3.16.10 Future Considerations

Continuous scrutiny of standards to ensure compliance. Improvements in lighting design for long life, high-intensity and conformal high-strength lenses.

3.17 Probe Heating

The air data probes provide information about air conditions that is used by safety critical systems such as flight control, engine, and the pilot's displays. This information must be available and correct throughout the flight and in all weather conditions. All probes are provided with suitable heating to prevent the formation of ice which will lead to misleading information (Figure 3.19).

3.17.1 Purpose of System

To provide a means of heating the pitot-static and temperature probes on the external skin of the aircraft to ensure that they are kept free of ice.

3.17.2 Description

An electrical heater is built into the air data probes to prevent ice forming and blocking the orifices. The heater is often a low power heater for ground and taxy

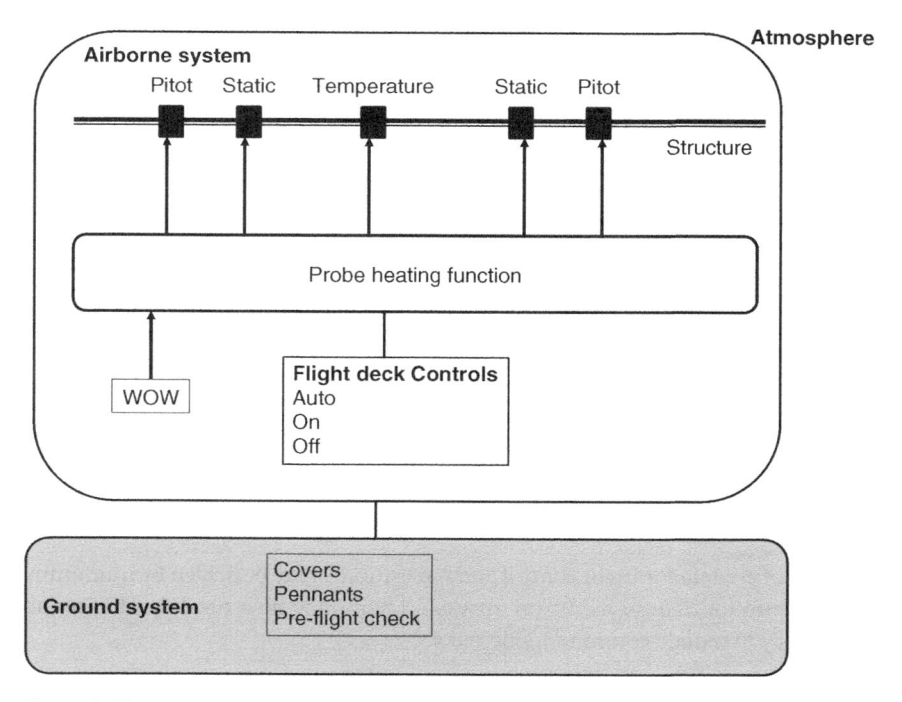

Figure 3.19 Illustration of a generic probe heating system.

operations, with a higher power heater operated by the oleo weight on wheels (WOW) switch. The probes are usually protected on the ground by a cover and a safety pennant.

3.17.3 Safety/Integrity Aspects

Safety critical. Failure of heaters will affect accuracy of air data sensing and will affect cockpit indications and flight and propulsion control system input data. This is a dormant system. On small aircraft, the ground crew (with care) may test the heating action by hand if the probes are within reach (health and safety implications).

3.17.4 Key Integration Aspects

Flight control system, cockpit displays, and controls.

3.17.5 Key Interfaces

- **Airframe/structure**: provision of apertures and suitable fasteners for pitot-static probes, protection of the pressure shell.
- **Other systems**: exchange of information by direct wiring or data bus. Connections to respect the integrity of flight control system, air data computer, and altitude displays. Air – ground/weight-on-wheel signals from landing gear via VMS or by direct wiring from the undercarriage oleo switches.
- Flight deck for controls and displays and human factors assessment.
- **Electrical power**: connection to appropriate bus bar with circuit protection devices, and most suitable harness runs.
- Security of attachment of all components.
- Ground systems for provision of covers and warning pennants and for removal before flight.

3.17.6 Key Design Drivers

Accuracy of air data for flight control and navigation – may be driven by minimum height separation requirements on airways. There may be a need for dissimilar redundancy to reduce common mode risks.

3.17.7 Modelling

Simple simulation.

3.17.8 References

Moir and Seabridge (2013) Chapter 8, JASC Code (n.d.), ATA-100 (n.d.) Chapter 22.

3.17.9 Sizing Considerations

Electrical load, compatibility with FCS.

3.17.10 Future Considerations

Use of redundant heating elements and/or use of a current monitoring system to provide warning of element failure. It may be necessary to make of dissimilar types of probe to reduce the probability of common mode failure.

Air France AF 447 (A330 aircraft) suffered a fatal accident in June 2009 whilst flying in the inter-tropical convergence zone. The incident started with inconsistent airspeed data on the flight deck. A number of subsequent crew actions resulted in the loss of the aircraft. Further investigation showed that air data probes were freezing on other aircraft and recommendations were made on the type and redundancy of the probes.

3.18 Vehicle Management System (VMS)

All the vehicle systems need to be connected to a computing system to be controlled and monitored to enable them to perform their control functions and to provide some form of action to operate devices. This requires an interface to be provided to all relevant sensors and all relevant power conversion devices. Connection to other systems and to the flight deck is provided by an interface to a data bus network. These functions are performed by a collection of control units or by a central computing function known as VMS (Figure 3.20).

3.18.1 Purpose of System

To provide an integrated processing and communication system for interfacing with vehicle system components, performing built in test, performing control functions, providing power demands to actuators and effectors, and communicating with cockpit display and other systems.

3.18.2 Description

A number of interfacing and processing units geographically dispersed in the airframe to reduce wiring lengths. A standard data bus is used to interconnect the units and to communicate with other systems. Standard data buses in general

3 Vehicle Systems

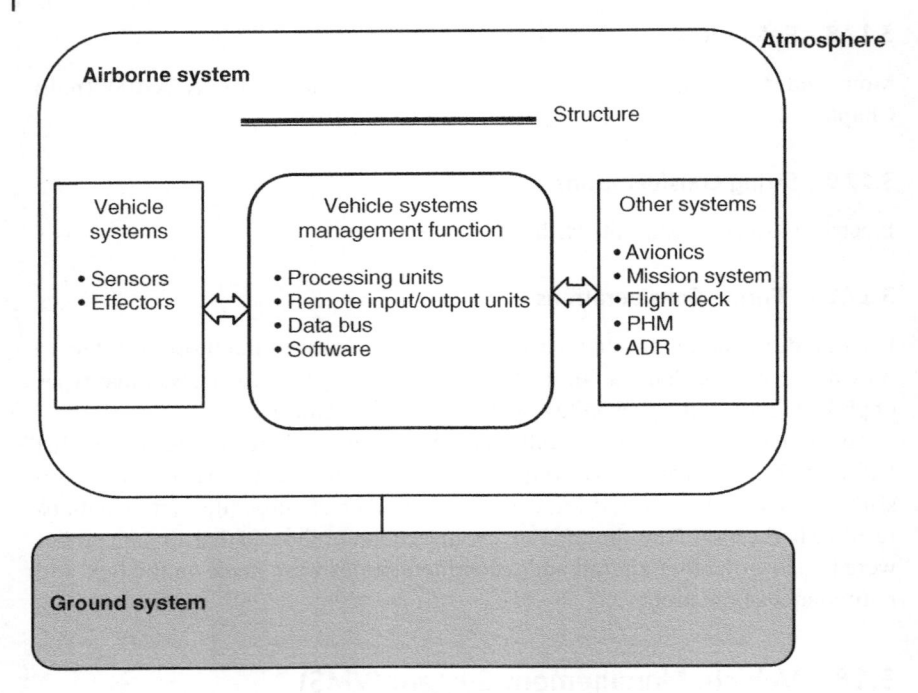

Figure 3.20 Illustration of a generic vehicle management system.

use include ARINC 429, ARINC 629, and ARINC 664 for commercial aircraft; MIL-STD-1553 and IEEE 1398 for military use; although these types may be combined to suit particular needs.

3.18.3 Safety/Integrity Aspects

Integrity depends on control functions – generally safety involved or safety critical.

3.18.4 Key Integration Aspects

Integration with avionics systems, flight deck displays, and controls and with VMS.

3.18.5 Key Interfaces

- **Other systems**: exchange of information by direct wiring or data bus. Interfaces to vehicle system components such as sensors and effectors of very different interface characteristics such as impedance, rate of change, slewing rate, capacitance, inductance, voltage levels, current drives.

- Flight deck for controls and displays and human factors assessment.
- **Electrical power**: connection to appropriate bus bar with circuit protection devices, and most suitable harness runs.
- **Installation**: location in the airframe to provide separation to avoid common mode damage such as engine disc burst or munitions damage, mounting tray for electronic LRUs, and provision of suitable conditioning and cooling.
- Security of attachment of all components.
- Consideration of EMC by radiation, induction, and susceptibility.
- Ground systems for download of data, provision of services on turn around.

3.18.6 Key Design Drivers

Safety, availability.

3.18.7 Modelling

Integrated modelling across the systems using computer-aided tools, individual system test rigs, and integrated test rigs.

3.18.8 References

Avionics Communications Inc (1995), Lawson and Judt (2022), Lloyd and Tye (1982), Moir and Seabridge (2008) Chapter 5, Moir and Seabridge (2010), Spitzer (1993).

3.18.9 Sizing Considerations

Number and complexity of component interfaces influences the number of control and interface units. Number and complexity of functions to be performed in software influences redundancy, memory capacity, and computer throughput.

3.18.10 Future Considerations

This function may be integrated into the aircraft IMA as part of the utilities domain. Integration of safety critical functions such as FCS.

3.19 Crew Escape

There may come a time when the crew of a fast jet aircraft need a means of escape from the aircraft either at high-speed flight or on the ground. An escape system

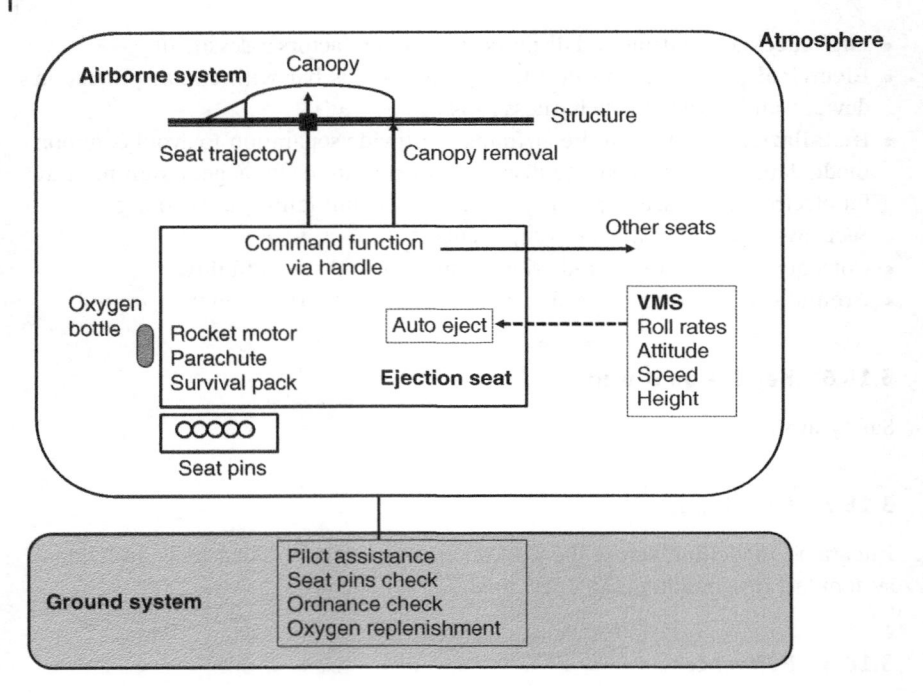

Figure 3.21 Illustration of a generic crew escape system.

is required to deal with that circumstance. The most usual system is an ejection seat, though escape modules have been used. The seat must be equipped with a means of rapidly leaving the aircraft with the pilot protected during the ejection and to provide life support and supplies for over-sea escape and survival. The seat mechanism ejection lines must be designed to prevent the pilot being in contact with any piece of internal or external structure or equipment during the ejection process (Figure 3.21).

3.19.1 Purpose of System

To enable fast jet aircrew to escape from the aircraft under a wide variety of conditions with minimum risk of injury or death – range from high altitude to zero speed, zero altitude, (zero/zero).

3.19.2 Description

Rocket-assisted seat equipped with parachute, life support and survival pack, and emergency oxygen. The seat contains an oxygen supply for the pilot and a survival pack. The seat is operated by a handle attached to the seat and travels up an inclined rail to maintain clearances for the pilot's limbs. The canopy is ejected before the seat or alternatively the transparency is shattered to allow free exit. The seat is equipped with a parachute and a drogue to deploy the main parachute.

An external selector housed in a clear panelled hatch is required to allow the canopy to be released on the ground after an accident with appropriate hazard notices.

3.19.3 Safety/Integrity Aspects

This is a dormant system and cannot be pre-flight tested, it is essential that it operates when required and must be designed accordingly.

There are explosive devices and propellants that will require specialised skills. The seat is made safe by insertion of pins into specific parts of the mechanism. To prevent inadvertent firing and injury to ground crew, the pins are stowed in a visible stowage device and it is the responsibility of the crew and ground crew to ensure that the pins are in place while the aircraft is being worked on and are clearly stowed before flight.

Ejection hazard notices must be situated on the external fuselage as a ground crew warning, and clear instructions included in the aircraft operating manuals. There have been serious and fatal accidents during maintenance and all personnel must be aware of the hazard to life.

3.19.4 Key Integration Aspects

Integration with canopy jettison or shattering mechanism.

3.19.5 Key Interfaces

- **Airframe/structure**: provision of seat rails at correct angle, clearance, and exit lines from airframe and tail fin on ejection to prevent injury to aircrew. Provision of warning notices.
- **Other systems**: connection to canopy jettison system. Provision of aperture for ground rescue panel.
- Flight deck for pin stowage, ejection clearances, and human factors assessment.
- **Electrical power**: connection to appropriate bus bar with circuit protection devices and most suitable harness runs. Essential supply required.
- Security of attachment of all components.
- **Environment**: release of exhaust from seat motors, shattered plastic from canopy.
- Consideration of EMC by radiation, induction, and susceptibility.

- Ground systems for seat security, pilot strapping in procedure and safety pins, and replenishment of emergency oxygen cylinder. Periodic inspection of the parachute by the safety equipment teams.

3.19.6 Key Design Drivers

Clear ejection lines, crew physiology, and safety.

The use of an ejection system may cause severe injury to aircrew and the system must be designed to prevent or minimise injury. Careful design of the escape trajectory and clearance from all items of structure and equipment shall be ensured and demonstrated by design, by modelling, and by test.

3.19.7 Modelling

Calculation – 3D modelling ejection clearance lines. A pull-out test with a man in the seat to demonstrate clearances. Rig test including high-speed dynamic sled testing on a range with instrumented mannequins and high-speed photography or video.

3.19.8 References

Giguere (2010), Lawson and Judt (2022), Moir and Seabridge (2008) Chapter 5. JASC Code (n.d.), ATA-100 (n.d.) Chapter 95.

3.19.9 Sizing Considerations

Size and percentile range of typical crew which may need to be varied to suit some ethnic characteristics and gender-determined sizes and injury modes.

3.19.10 Future Considerations

VTOL and rapid rolling – auto ejection, steering to compensate for rapid change of attitude at low height to avoid ejection into the ground.

The system is normally not connected to any other system to avoid inadvertent operation. With the advent of VTOL activities, there has been a move to providing some form of automatic ejection and limited steering in the event of a low height, rapidly rolling situation. This may require some form of interconnection with the aircraft systems to obtain suitable information. Such a move demands a robust system design or the use of devices on the seat.

Some serious and fatal accidents have occurred on the ground during servicing and repair. The system and ground-operating procedures shall be designed to prevent or minimise accidental operation of the system.

3.20 Canopy Jettison

There are times when it will be necessary to remove the canopy or the transparency, certainly during ejection, in the unlikely event that the seat does not eject when commanded and also to allow an alternative means of leaving the aircraft or to allow rescue of the crew from a crashed aircraft (Figure 3.22).

3.20.1 Purpose of System

To provide a means of removing or fragmenting the canopy material on ejection command or to provide a means of access to the pilot by rescue crews.

3.20.2 Description

Rocket-assisted jettison mechanism of the entire canopy or miniature detonating cord embedded in canopy material, the pattern and charge are determined by the type of canopy material. This mechanism is normally connected to the seat ejection handle – an external mechanism is provided to enable rescue crews to assist

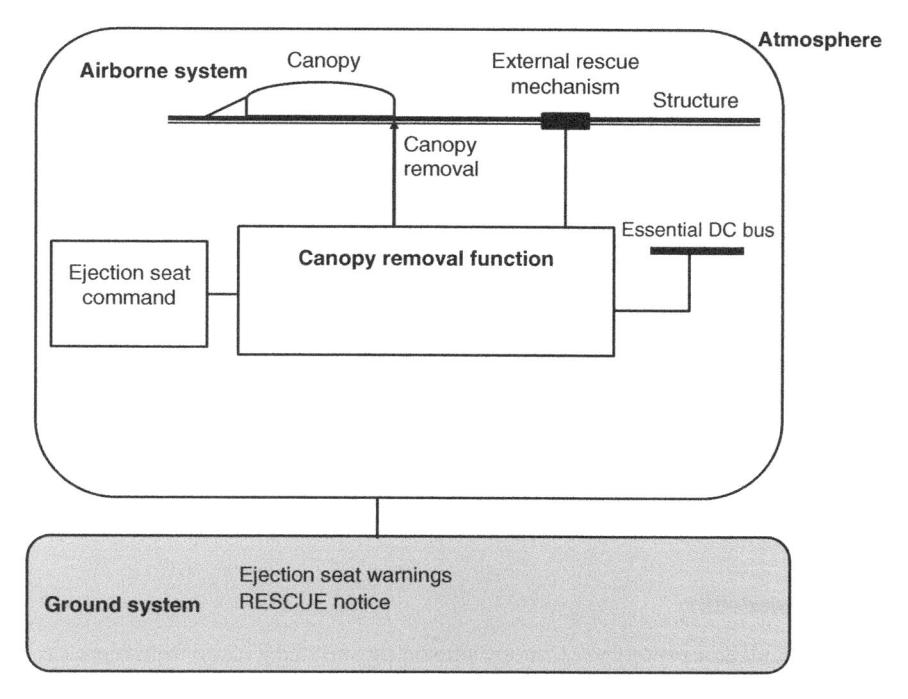

Figure 3.22 Illustration of a generic canopy jettison system.

the pilot if the aircraft had crashed and the canopy is still in place. This mechanism is usually a frangible plastic cover which is broken to reveal a D-handle attached to a long cable. The rescue crew can retire to a safe distance and pull the handle to jettison the canopy or initiate the explosive devices in the transparency.

3.20.3 Safety/Integrity Aspects

Safety critical: danger to ground crew if not isolated on the ground.

A means of commanding jettison from the exterior of the aircraft during rescue from a crashed aircraft. Suitable warning notices must be provided.

This is a dormant system and cannot be pre-flight tested, it is essential that it operates when required and must be designed accordingly.

3.20.4 Key Integration Aspects

This function must be closely integrated with crew escape initiation.

3.20.5 Key Interfaces

- **Airframe/structure**: provision of hinges and canopy-locking mechanism. Provision of warning notices.
- **Other systems**: connection to ejection seat system. Provision of aperture for ground rescue panel and notices.
- Flight deck for operation of locking mechanism, ejection clearances, and human factors assessment.
- **Electrical power**: connection to appropriate bus bar with circuit protection devices, and most suitable harness runs. Essential supply required.
- Security of attachment of all components.
- **Environment**: release of exhaust from canopy jettison devices, shattered plastic from canopy.
- Consideration of EMC by radiation, induction, and susceptibility.

3.20.6 Key Design Drivers

Must allow the crew to exit the aircraft without injury.

3.20.7 Modelling

Physical models or prototypes. Canopy jettison rig and sled jettison from representative fuselage model at high speed.

3.20.8 References

Moir and Seabridge (2008) Chapter 8. JASC Code (n.d.), ATA-100 (n.d.) Chapter 95.

3.20.9 Sizing Considerations

Canopy, jettison type, and mechanism.

3.20.10 Future Considerations

The use of new materials in the canopy may need a new mechanism for fracturing the transparency to allow unimpeded exit, for example the different methods required for acrylic or polycarbonate materials because of their differing fracture behaviour.

3.21 Oxygen

Oxygen must be provided to passengers in the event of a depressurization by means of individual oxygen masks on commercial aircraft. Oxygen for this purpose is usually provided by a simple chemical reaction. The fast jet pilot breathes pressurized oxygen at all times via a facemask. This may be from liquid oxygen or a cylinder for pure oxygen, or by a catalytic reaction process designed to produce oxygen-enriched air (Figure 3.23).

3.21.1 Purpose of System

Civil Airworthiness Requirements state that operations above 25 000 ft must make provision for emergency oxygen. This system provides a source of breathable oxygen for crewmembers and passengers. The system is required to cover the following operational scenarios. Oxygen must be available immediately so that the aircraft can descend to an altitude at which breathing can be maintained, generally around 15 000 ft.

- Cabin depressurisation
- Failure of the environmental control system
- Contamination of the ECS

3.21.2 Description

Commercial: to cover descent to safe altitude in the event of pressurisation loss: bottled gaseous oxygen is provided for pilots with quick-don masks. Oxygen masks for passengers descend automatically for each passenger on

100 | *3 Vehicle Systems*

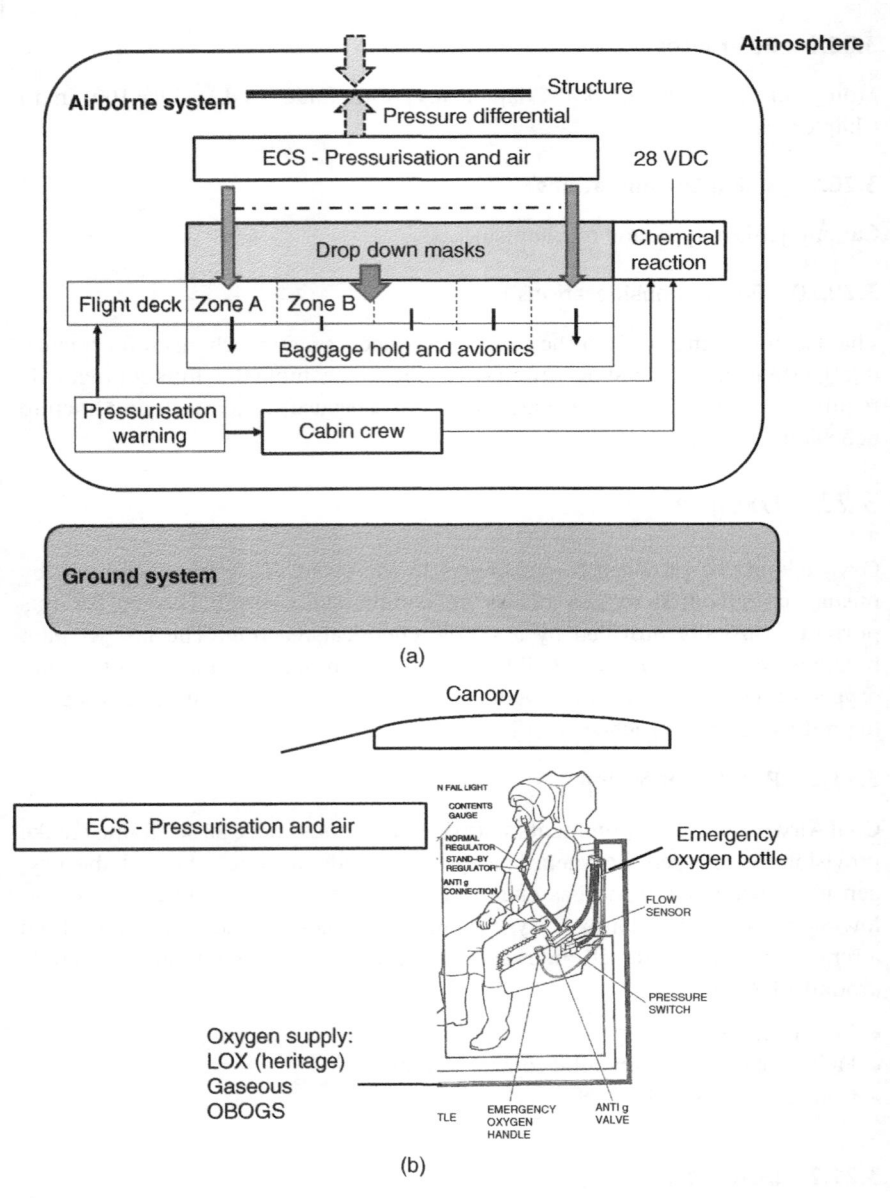

Figure 3.23 Illustration of a generic oxygen system (a) commercial and (b) fast jet.

de-pressurisation, providing oxygen derived from a chemical reaction, generally by heating sodium-chlorate for a period of 15 minutes.

Military: continuous pressure breathing from liquid oxygen or oxygen-enriched air from on board oxygen generation system (OBOGS). A robust gaseous oxygen cylinder is provided on the ejection seat.

3.21.3 Safety/Integrity Aspects

Commercial: must be available on demand to enable pilots to fly the aircraft to a safe altitude, must be available for passenger safety and comfort. Instructions must be given to passengers and a safety card provided to explain the use of the system. Flight deck crew are provided with a quick don mask and bottled gaseous oxygen.

Military: pressure air/oxygen mixture must be available at all times in combat aircraft. Supply also available on ejection seat.

3.21.4 Key Integration Aspect

Commercial: integration with emergency system.
Military: integration with ECS, human factors, and crew escape systems.

3.21.5 Key Interfaces

- **Other systems**: exchange of information by direct wiring or data bus. Cabin crew alerts and cabin crew control of the system. Automatic drop-down masks, quick don mask, and gaseous oxygen for aircrew and cabin crew.
- Flight deck for controls and displays and human factors assessment.
- **Electrical power**: connection to appropriate bus bar with circuit protection devices, and most suitable harness runs.
- Ground systems for download of data, provision of services on turn around, and replenishment of fluids or provisions. Provision of notices and seat back information leaflets.
- Human factors and aero-medical advice to be sought.

3.21.6 Key Design Drivers

Rapid and automatic deployment for passenger and crew safety. Autonomous operation or availability of LOX or gaseous oxygen at remote sites.

3.21.7 Modelling

Simple simulation.

3.21.8 References

Giguere (2010), Lawson and Judt (2022), Moir and Seabridge (2008) Chapter 5. JASC Code (n.d.), ATA-100 (n.d.) Chapter 36.

3.21.9 Sizing Considerations

Number of passengers, type of breathing air supply, and emergency sources.

3.21.10 Future Considerations

Alternative methods for providing oxygen or enriched air for short durations, potential for the use of on-board oxygen generation system for passenger cabins.

3.22 Biological and Chemical Protection

For aircraft operating in a hostile environment, there is a threat to ground crew and aircrew of attack by chemical or biological weapons. Protection is required to enable the aircraft to continue operations. The ground crew and aircrew need to be protected as far as possible from the immediate threat of fluid, liquid, or aerosol contamination until they can be washed down and subjected to appropriate de-contamination measures (Figure 3.24).

3.22.1 Purpose of System

To protect the crew from the toxic effects of chemical or biological contamination. Washing of aircraft to continue operations. Design of control panel in cockpit or external panels to allow operation with hazard protection suits and gloves.

3.22.2 Description

Provision of suitable gloves, suits, masks, and goggles to allow operation of the aircraft after a chemical or biological incident. Provision of filtered air and oxygen supply, protective clothing and respirators, and a wash-down facility. Consideration to be given to allow cockpit controls to be used using suitable gloves.

3.22.3 Safety/Integrity Aspects

Mission-critical.

3.22 Biological and Chemical Protection | 103

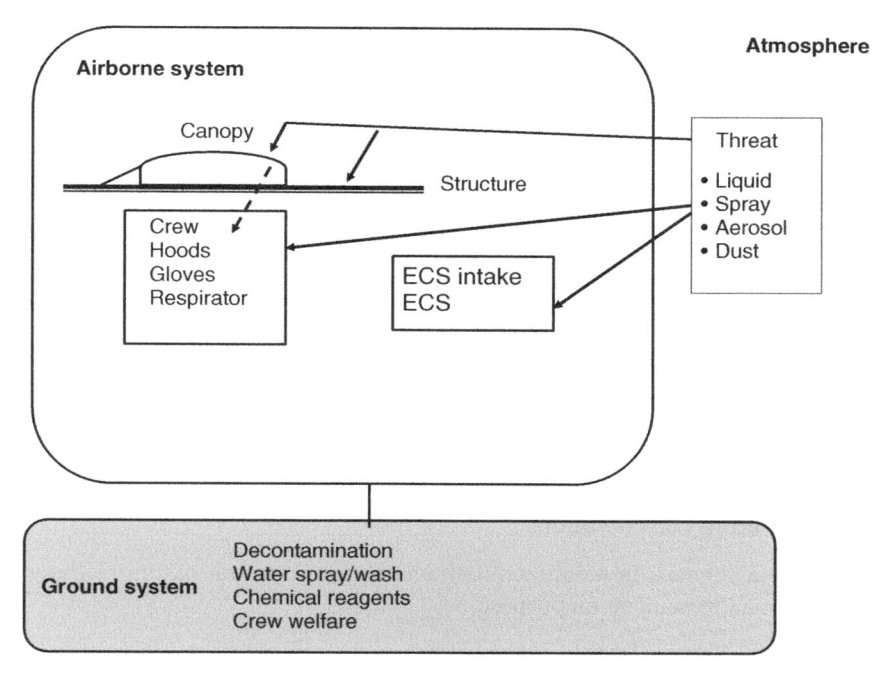

Figure 3.24 Illustration of a generic biological & chemical protection system.

3.22.4 Key Integration Aspects

Human factors, compatibility with control panels.

3.22.5 Key Interfaces

Cabin air system, cockpit controls must be compatible with protective gloves to ensure control operation and airframe/structure – sealing of all hatches.

- Flight deck for controls and displays for use and visibility wearing BC clothing and gloves, to avoid fouling adjacent keys, switches or controls, and human factors assessment.
- **Other systems**: cabin air system.
- **Environment**: wash down procedure will require safe operation by ground crew and safe disposal of residues.
- Ground systems for wash down procedures and disposal of residues.

3.22.6 Key Design Drivers

Operator safety.

3.22.7 Modelling

Simulation of operability of controls with gloves, simulation of controls separation using VAPS, full-scale panel mock-ups, or cockpit mock-ups.

3.22.8 References

This subject is usually classified, and there is very little published information available, searches on the Internet will reveal some information.

Eldridge (2006), Hart and Garrett (2007).

3.22.9 Sizing Considerations

Threat substances, filters, air crew clothing, and respirator mechanism.

3.22.10 Future Considerations

New threat materials to be monitored continuously to ensure that existing clothing and respiration remains fit for purpose.

3.23 Arrestor Hook

A failure of the brakes or hydraulic system can lead to an aircraft requiring some assistance with deceleration on landing, especially fast jet types. The arrestor hook system consists of two parts: the hook attached to the aircraft and the ground system with which the hook engages. The system is usually in three parts – an arrestor wire, a net, and a sand pit. The total system is for occasional use, but for aircraft designed to land on carrier decks, it is a normal part of landing (Figures 3.25 and 3.26).

3.23.1 Purpose of System

To stop the aircraft by engaging a runway arrestor gear wire if the brakes should fail. This is normally used for military aircraft, but may also be used for prototype commercial aircraft. The on-board system is used in conjunction with a ground-based system to engage the aircraft and absorb the landing energy. A backup to this may also be used.

For naval aircraft, the use of an arrestor hook is normal for all landings, the hook engaging with one of a number of wires laid across a carrier deck. The hook is normally deployed before landing.

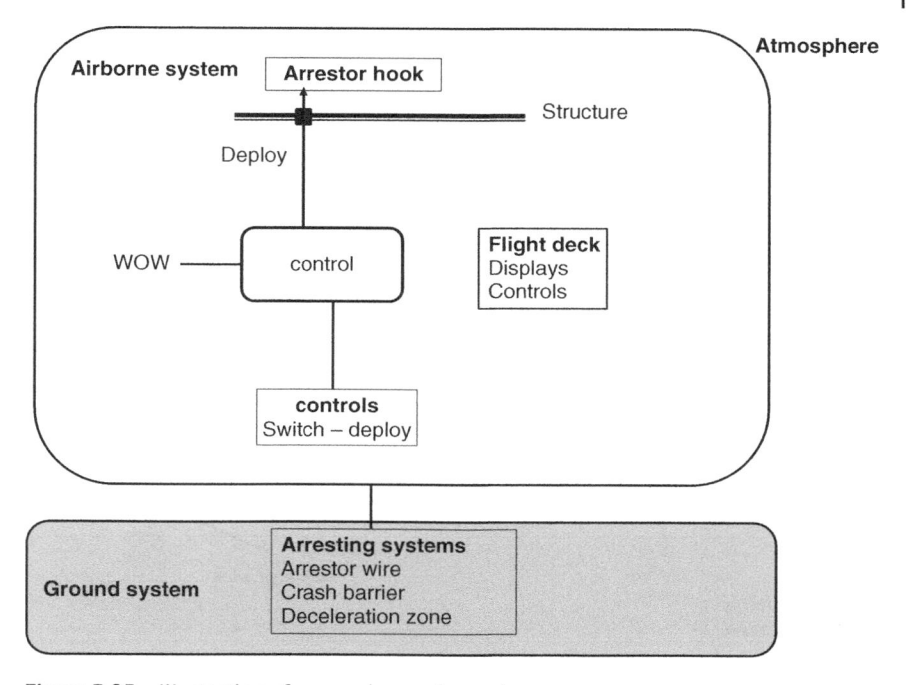

Figure 3.25 Illustration of a generic arrestor system.

3.23.2 Description

An arrestor hook stowed at rear of an aircraft with a mechanical locking mechanism which can be unlocked using a solenoid connected to a switch in the cockpit. The hook falls to the ground by gravity and should not bounce and miss the wire.

3.23.3 Safety/Integrity Aspects

This is a dormant system and cannot be pre-flight tested, it is essential that it operates when required and must be designed accordingly. Care must be taken to ensure that the mechanical unlock mechanism does not jam and cannot operate inadvertently. Wiring to release solenoids in the unlock mechanism must be protected from any form of damage that could lead to a short circuit that prevents operation of initiated inadvertent operation. The designer may consider that the wiring may need screening or mechanical protection and separation from other wiring.

3.23.4 Key Integration Aspects

Integration with the airframe to reduce drag.

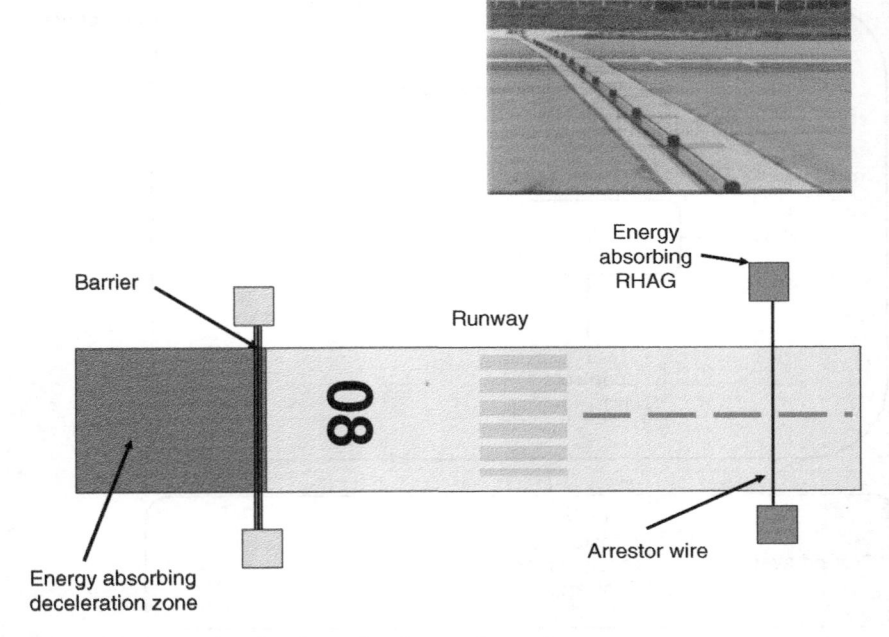

Figure 3.26 An example of associated ground arresting system.

3.23.5 Key Interfaces

- **Airframe/structure**: provision of apertures and suitable fasteners for arrestor hook hinge and locking/release mechanism. Stress for strength of attachment of hook to aircraft and to check engagement with the ground system wires.
- Interface with in-service arrestor gear at military airfields and carriers.
- Flight deck for controls and displays and human factors assessment.
- **Electrical power**: connection to essential bus bar with circuit protection devices and most suitable harness runs.
- Ground systems for maintenance of ground element, rescue of aircraft, and crew.

3.23.6 Key Design Drivers

Safety, emergency operations, airframe mass, and speed (energy) requirements and their impact on the hook and its attachment.

3.23.7 Modelling

Stress calculation, 3D (e.g. Catia) modelling.

3.23.8 References

Moir and Seabridge (2008) Chapter 8.

3.23.9 Sizing Considerations

Energy requirement (speed of landing, all up mass of aircraft) determines the design of the hook and its attachment to the airframe.

3.23.10 Future Considerations

Fast military aircraft will always need an emergency arresting system. The mass of the hook must be sufficient to allow the hook to fall using gravity and not to suffer from ground effect and miss the wire; therefore, new lighter materials may not always be suitable.

3.24 Brake Parachute

The brake parachute is deployed to decelerate aircraft with high landing speeds and short runway capability. It is usually jettisoned by the pilot in a safe place off the runway for collection by ground crew. This is to avoid issues to do with FOD, fouling with runway or taxiway furniture, and to avoid any damage to the aircraft and other aircraft. The system can also be used following loss of brakes (Figure 3.27).

3.24.1 Purpose of System

Used on military types and some commercial prototypes to decelerate the aircraft for ultra-short stopping distances or on short runways.

3.24.2 Description

The brake parachute is normally stowed in a canister in the aircraft rear fuselage. On selection, the parachute will deploy into the airstream in the direction of the aircraft. When the aircraft has decelerated, the parachute or canister can be jettisoned if required, preferably on the taxy-way and not the runway for collection by ground services.

3.24.3 Safety/Integrity Aspects

This is a dormant system and cannot be pre-flight tested, it is essential that it operates when required and must be designed accordingly. Care must be taken to

108 | *3 Vehicle Systems*

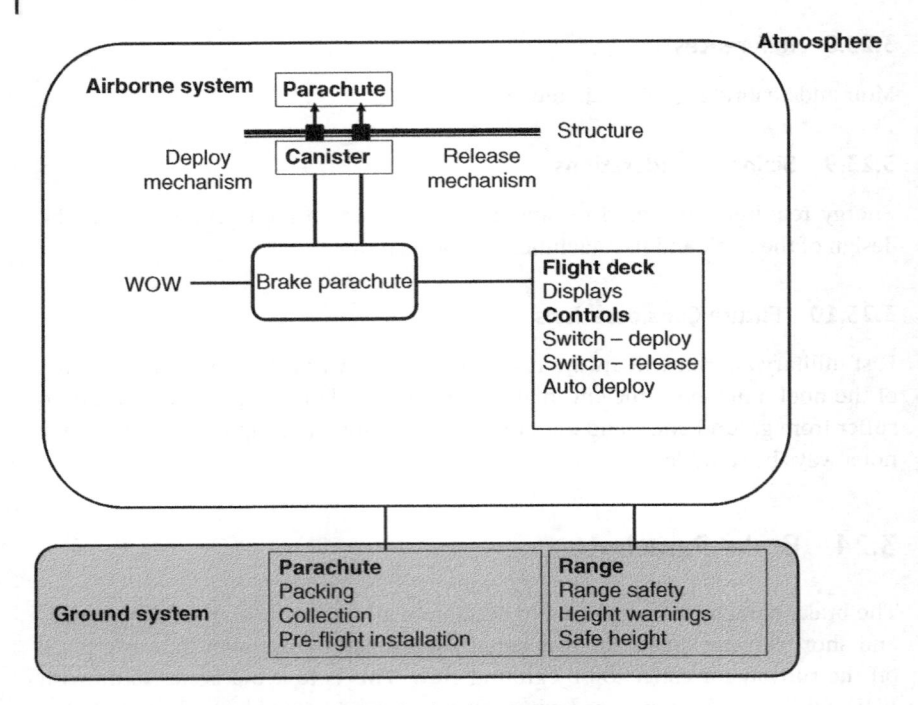

Figure 3.27 Illustration of a generic brake parachute system.

ensure that the mechanical unlock mechanism does not jam and cannot operate inadvertently. Wiring to release solenoids in the unlock mechanism must be protected from any form of damage that could lead to a short circuit that prevents operation of initiated inadvertent operation. The cutting mechanism must also be carefully designed to prevent inadvertent jettison. The designer may consider that the wiring may need screening or mechanical protection and separation from other wiring.

3.24.4 Key Integration Aspects

This is a single system with no opportunity for redundancy. The system must be integrated with the airframe with appropriate security of attachments and with no risk of fouling any aircraft structure.

3.24.5 Key Interfaces

- **Airframe/structure**: provision of apertures and suitable fasteners for parachute canister and locking/release mechanism. Stress for strength of attachment of parachute cable to aircraft and suitable material.

- Flight deck for controls and displays and human factors assessment.
- **Other systems**: simple manual operation by the pilot with a guarded and clearly labelled switch with a position for automatic operation when weight on wheels has been detected. For this automatic mode, an interface with the landing gear weight on wheels system is required.
- **Electrical power**: connection to essential bus bar with circuit protection devices, and most suitable harness runs.
- Ground systems for maintenance of parachute, collection of discarded chute after landing.

3.24.6 Key Design Drivers

The aircraft landing speed, typical mass on landing, and the required stopping distance are key aspects of the design requirement. This will determine the strength of the attachment and the size of parachute harness. Aircraft support is required to permit parachute collection by ground support after jettison, and for repackaging and replacement.

3.24.7 Modelling

A simple simulation of the system using CAD may be sufficient. Ground test during taxy trials will be required to verify the design.

3.24.8 References

Moir and Seabridge (2008) Chapter 8.

3.24.9 Sizing Considerations

Calculation of energy requirement based on aircraft mass, typical landing speed, and runway surfaces.

3.24.10 Future Considerations

Lighter parachute and packaging materials to reduce weight. Some modern aircraft make use of aerodynamic-braking effects to slow the aircraft before using brakes. Consideration should be given to this if possible to eliminate the need for a parachute.

3.24.11 Best Practice and Lessons Learned

There have been instances where the entire canister and parachute have been ejected on the runway after landing. Although there have been no reported accidents as a result of such occurrences, all efforts must be made to prevent it. Despite no accidents, there will be a cost incurred as a result of investigation of the incident and any subsequent re-design. Careful design and stringent scrutiny of the design is essential.

3.25 Anti-spin Parachute

The anti-spin parachute is used during flight trials may depart and enter a spin from which it is possible that recovery cannot be guaranteed. The parachute will act on the spinning aircraft so as to stabilise it in a vertical dive during which the pilot can relight the engines and regain normal flight. The parachute would then be jettisoned using a mechanical release, and if that fails to work, then a small explosive charge can be detonated to separate the aircraft from the parachute (Figure 3.28).

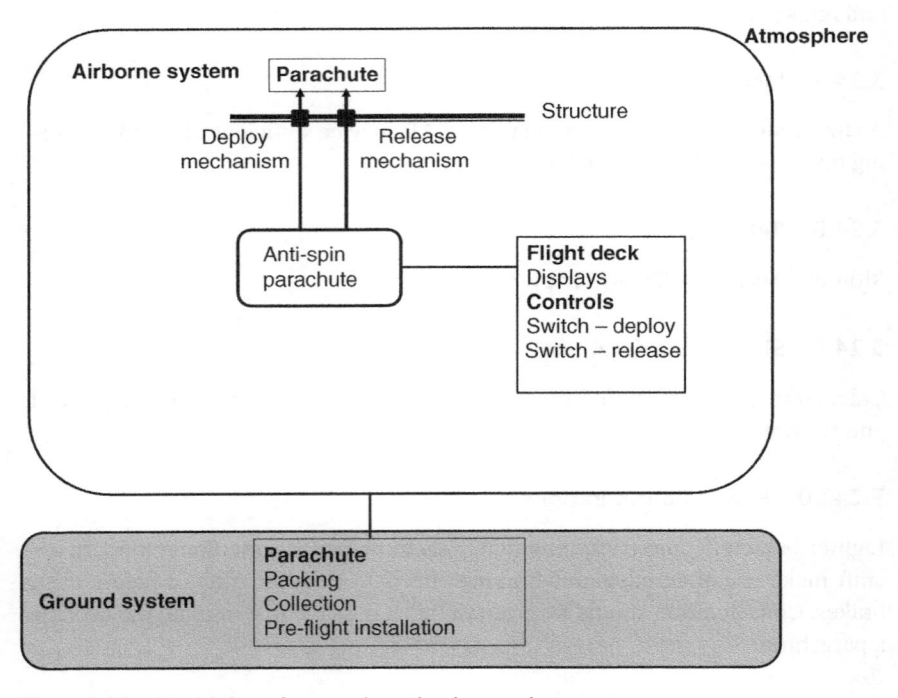

Figure 3.28 Illustration of a generic anti-spin parachute system.

3.25.1 Purpose of System

Used on military types and some commercial prototypes during high-altitude, high-attitude, low-speed flight trials to enable an aircraft to be recovered from a spin.

3.25.2 Description

The parachute is normally stowed in a canister in the aircraft. On selection, the parachute will deploy into the airstream at an angle to the direction of the aircraft. This angle is calculated from aerodynamic data or from models to be the optimum for recovering the aircraft from the spin. When the aircraft has recovered, the parachute or the canister can be jettisoned using a guillotine or explosive device in a safe place. Normally, such trials would be conducted on a test range which will limit the risk of damage to property, animals, and people.

3.25.3 Safety/Integrity Aspects

This is a dormant system and cannot be pre-flight tested, it is essential that it operates when required and must be designed accordingly. Care must be taken to ensure that the mechanical unlock mechanism does not jam and cannot operate inadvertently. Wiring to release solenoids in the unlock mechanism must be protected from any form of damage that could lead to a short circuit that prevents operation of initiated inadvertent operation. The designer may consider that the wiring may need screening or mechanical protection and separation from other wiring.

Trial test procedures must cater for engine flameout and must provide safe height warnings.

3.25.4 Key Integration Aspects

This is a single system with no opportunity for redundancy. The system must be integrated with the airframe with appropriate security of attachments and with no risk of fouling any aircraft structure.

3.25.5 Key Interfaces

- **Airframe/structure**: provision of apertures and suitable fasteners for parachute canister and locking/release mechanism. Stress for strength of attachment of parachute cable to aircraft and suitable material.

- Flight deck for controls (guarded and clearly labelled switch) and displays and human factors assessment.
- **Electrical power**: connection to essential bus bar with circuit protection devices and most suitable harness runs.
- Ground systems for collection of parachute after use.

3.25.6 Key Design Drivers

The aircraft typical mass and spin rates are key aspects of the design requirement. This will determine the strength of the attachment and the size of parachute harness. Aircraft support is required to permit parachute collection by ground support after jettison, and for repackaging and replacement for the next trial. The system must be capable of operation when the pilot is disorientated and during violent manoeuvres.

3.25.7 Modelling

Mathematical modelling using aerodynamic data, wind tunnel models may provide information about the attitude of the aircraft during departure and spin to allow suitable positioning of the parachute. A simple simulation of the system using CAD may be sufficient.

3.25.8 References

Moir and Seabridge (2008) Chapter 8.

3.25.9 Sizing Considerations

Energy requirement based on speed and rates of spin.

3.25.10 Future Considerations

It is most likely that any aircraft entering that regime of the flight envelope that encourages departure will need a mechanism for recovery. As a short-term method for flight trials, the parachute is a simple and well-tried solution. An alternative is for the flight control system to be commanded to provide a combination of control surfaces that will prevent a spin. Some unstable aircraft control systems provide a 'care-free handing' system.

3.26 Galley

The passengers on a commercial aircraft need to be provided with food and drinks. The number and size of galleys provided depends on the nature of the

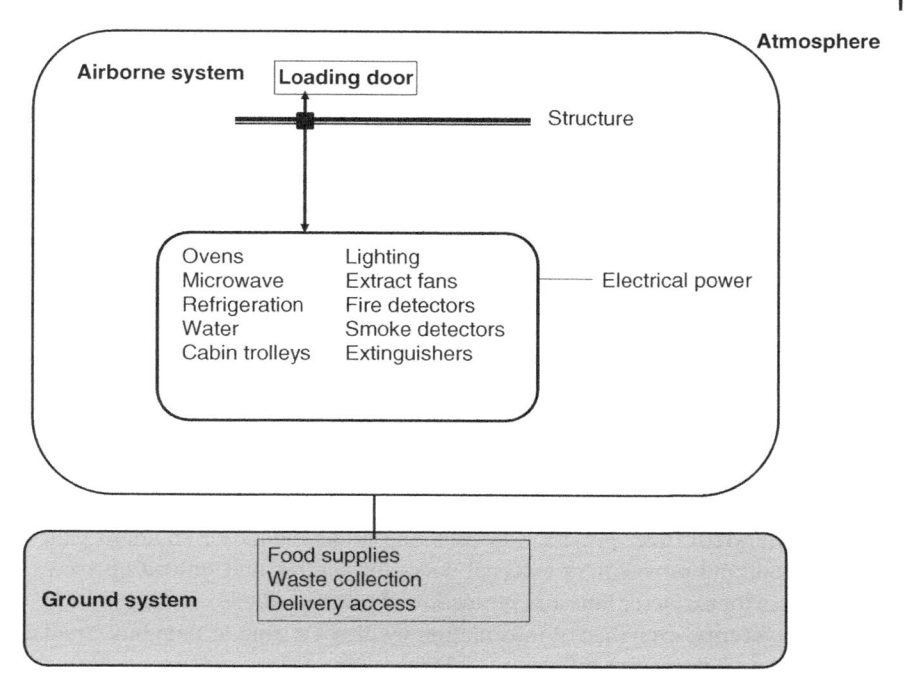

```
┌─────────────────────────────────────────────────┐
│                                      Atmosphere │
│  Airborne system  [Loading door]                │
│  ─────────────────────■──────────── Structure   │
│                       ↓                         │
│         ┌─────────────────────────────┐         │
│         │ Ovens      Lighting         │         │
│         │ Microwave  Extract fans     │──── Electrical power
│         │ Refrigeration Fire detectors│         │
│         │ Water      Smoke detectors  │         │
│         │ Cabin trolleys Extinguishers│         │
│         └─────────────────────────────┘         │
└─────────────────────────────────────────────────┘
                         │
┌────────────────────────┴────────────────────────┐
│                  Food supplies                  │
│   Ground system  Waste collection               │
│                  Delivery access                │
└─────────────────────────────────────────────────┘
```

Figure 3.29 Illustration of a generic galley.

flight. Business jets and short-haul aircraft will provide hot and cold drinks and sandwiches or simple meals, whereas a long-haul aircraft will provide full meals to different classes several times in any one flight. Generally, provision is made for safe storage of food, a means of cooking, refrigeration, and the provision of alcoholic drinks (Figure 3.29).

3.26.1 Purpose of System

To provide a safe and hygienic method of food preparation and cooking for passengers and crew. For refrigerated products, very precise Health and Safety requirements must be applied.

3.26.2 Description

Storage must be provided for food and pre-packed meals, refrigeration and cooking (heating and microwave) appliances. Containers for alcoholic beverages must meet customs requirements.

3.26.3 Safety/Integrity Aspects

May be mission-critical for long-range flights. Health and safety regulations, crew electrical shock, and fire risks to be minimised.

3.26.4 Key Integration Aspects

Interface with primary electrical system which includes precise fault protection schemes (the galley is an airline furnished item). The galley/passenger provision power requirements for a long-range passenger aircraft may equate to ~40% to 50% of overall connected load.

Protection from spills of fluids to avoid corrosion of structure at floor level with risk of explosive decompression.

3.26.5 Key Interfaces

- **Airframe/structure**: provision of rails and attachments for standard galley equipment and provision of external doors for loading and unloading access. Apertures for extractor fans and protection of pressure shell.
- **Other systems**: exchange of information by direct wiring or data bus. Smoke detectors and extinguishers.
- Flight deck for controls and displays and human factors assessment.
- **Electrical power**: connection to appropriate bus bar with circuit protection devices and most suitable harness runs.
- Mounting tray for electronic LRUs and provision of suitable conditioning and cooling.
- Security of attachment of all components.
- **Environment**: potential risk from waste food disposal.
- Consideration of EMC by radiation, induction, and susceptibility.
- Ground systems for provision of comestibles and materials, loading and unloading, and waste disposal.
- Interfaces with standard airline provisions supplier for roll on, roll off modules and food packaging.

3.26.6 Key Design Drivers

Health and safety and customer comfort/preference.

3.26.7 Modelling

Load analyses performed by Galley supplier.

3.26.8 References

JASC Code (n.d.), ATA-100 (n.d.) Chapter 25, Lawson and Judt (2022).

3.26.9 Sizing Considerations

Number of passengers and cabin areas, number of galleys, galley equipment and trolleys, and electrical loads. Consideration of special diets.

3.26.10 Future Considerations

Reduced need for galley in short-haul budget flights.

3.27 Passenger Evacuation

Passengers in commercial aircraft must be safely evacuated in the event of an accident on land or water. Seats will be provided with fully equipped life vests and aisles will be of suitable size for multiple, quick, and safe evacuation without leading to jams of passengers. The doors will be suitably signed and opened by the crew to reveal escape chutes. The cabin crew are an essential part of the system as they are well trained, provide the safety demonstration, and will guide and assure passengers (Figure 3.30).

3.27.1 Purpose of System

To allow safe evacuation of passengers from the cabin when the aircraft is on the ground or has ditched in water.

3.27.2 Description

Safe evacuation is accomplished by a robust combination of procedures, equipment, and the active participation of the cabin crew. The aircraft aisles are lit by emergency lighting to provide a pathway to the exits. Passengers are informed by the cabin crew of the exit locations and safety cards in the seatback also provide this information. The cabin crew are trained to assist passengers in the escape process, and they will also unlock the doors and deploy the chutes. Seats are provided with life vests, and rafts are available with survival packs.

Equipment is provided in the cabin for crew and passenger use including Smoke hoods, Axes, and Torches/flashlights.

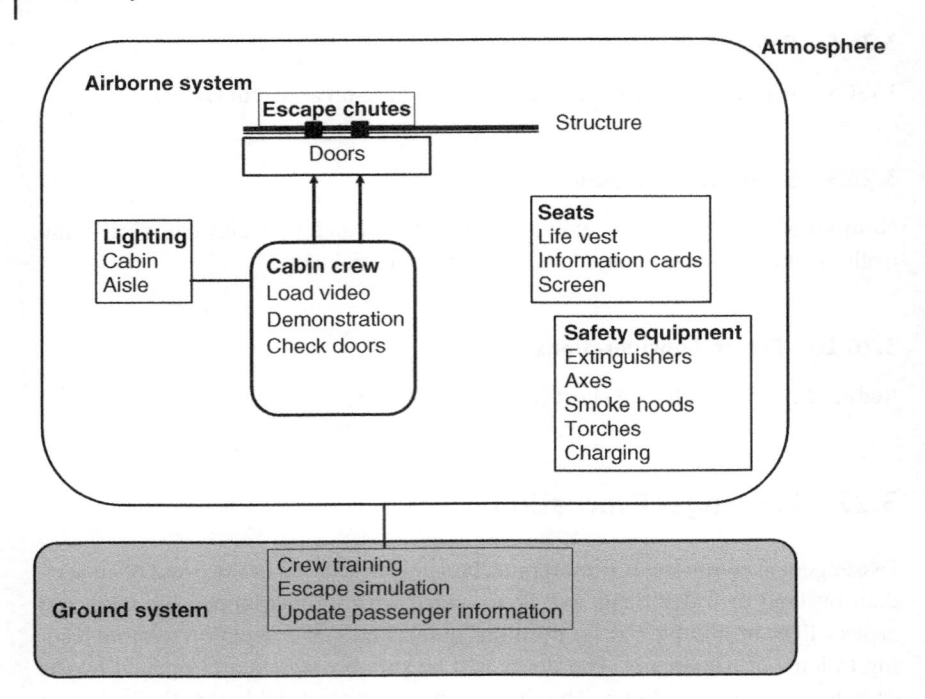

Figure 3.30 Illustration of a generic passenger evacuation system.

3.27.3 Safety/Integrity Aspects

Much of the system is dormant but must be available when required.

3.27.4 Key Integration Aspects

Door and slide operation. Flight deck awareness, passenger awareness via seat pocket notes.

3.27.5 Key Interfaces

Passenger and evacuation requirements and demonstration. Interface with airports for very large aircraft and high passenger numbers.

- **Airframe/structure**: provision of locking of doors, clearance for escape chutes.
- **Other systems**: exchange of information by direct wiring or data bus. Linked to emergency lighting.
- Flight deck for controls and displays and human factors assessment communications with cabin crew.

- **Electrical power**: connection to appropriate bus bar with circuit protection devices and most suitable harness runs. Essential supplies to be used.
- Ground systems for emergency and rescue services.

3.27.6 Key Design Drivers

Availability, passenger safety.

3.27.7 Modelling

Mock ups and evacuation test facilities, evacuation of a prototype aircraft in realistic conditions of smoke, visibility, and cabin crew assistance.

3.27.8 References

Giguere (2010), Lawson and Judt (2022).

3.27.9 Sizing Considerations

Number of passengers, number of exits and escape equipment, and configuration of aircraft.

3.27.10 Future Considerations

Blended wing body configurations may prove to be a challenge for doors and escape chutes to provide escape routes over wing.

3.28 In-Flight Entertainment

Passengers expect to be provided with entertainment in the form of music and video and on modern aircraft each seat is equipped with suitable equipment, usually connected to a central broadcast system operated by the cabin crew. The nature and sophistication of the service may vary with class.

3.28.1 Purpose of System

To provide audio and video entertainment for passengers at their seats and also to provide routine and emergency announcements (Figure 3.31).

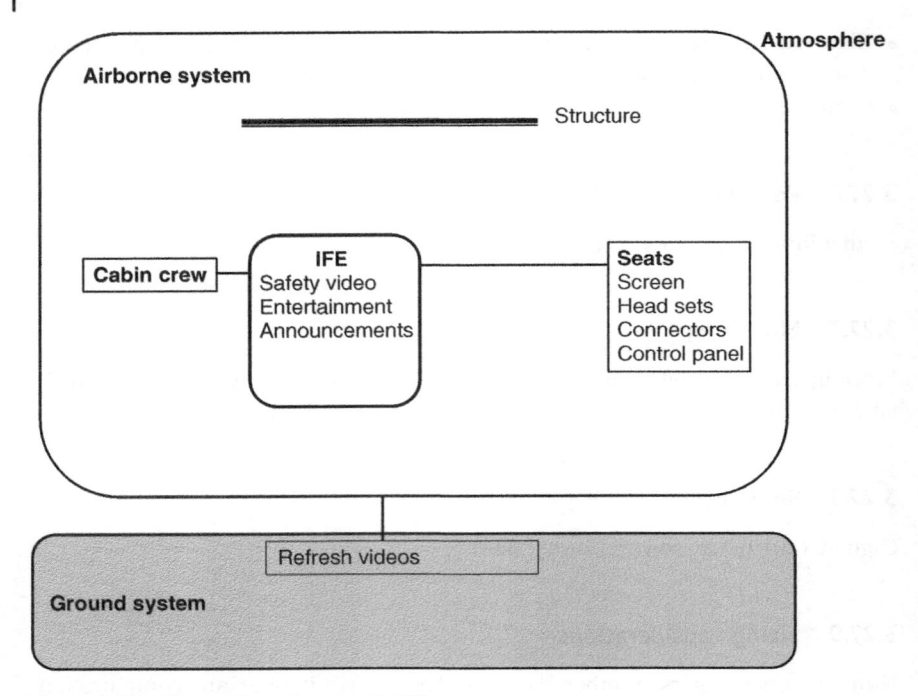

Figure 3.31 Illustration of a generic IFE.

3.28.2 Description

All seats to be equipped with screens, telephones, and remote control as appropriate to route and seating class. Networked audio and video signals to cabin screens or seat-located devices. Connection to the cabin PA system to enable announcements by flight crew and cabin crew so that passengers are aware of the correct route of exit. IFE will vary according to route and class with long-haul first class having the highest standard of equipment.

3.28.3 Safety/Integrity Aspects

Dispatch critical for passenger preference reasons.

3.28.4 Key Integration Aspects

Integration with cabin PA.

3.28.5 Key Interfaces

- Flight deck for controls and displays and human factors assessment.
- Cabin crew control of IFE services.
- **Electrical power**: connection to appropriate bus bar with circuit protection devices, and most suitable harness runs.
- Mounting tray for electronic LRUs and provision of suitable conditioning and cooling.
- Security of attachment of all components.
- Ground systems for provision of recorded material and cabin seat back information.

3.28.6 Key Design Drivers

Passenger satisfaction, marketing appeal, and compliance with safety regulations.

3.28.7 Modelling

Simulation and integration off-aircraft.

3.28.8 References

Lawson and Judt (2022), Moir and Seabridge (2013).

3.28.9 Sizing Considerations

Number of seats, cabin class variations, electrical loads, and impact on cabin heat load.

3.28.10 Future Considerations

Provision of e-mail and text for all passengers. Provision of a service to allow passengers to make telephone calls and access the Internet in flight. Possible streaming video/TV. In-seat telephone handsets and personal computer/portable electronic device charging capability.

3.29 Toilet and Water Waste

Basic provision for toilets and baby changing must be incorporated into the cabin design with regard to the number of passenger and the duration of the flight.

120 | *3 Vehicle Systems*

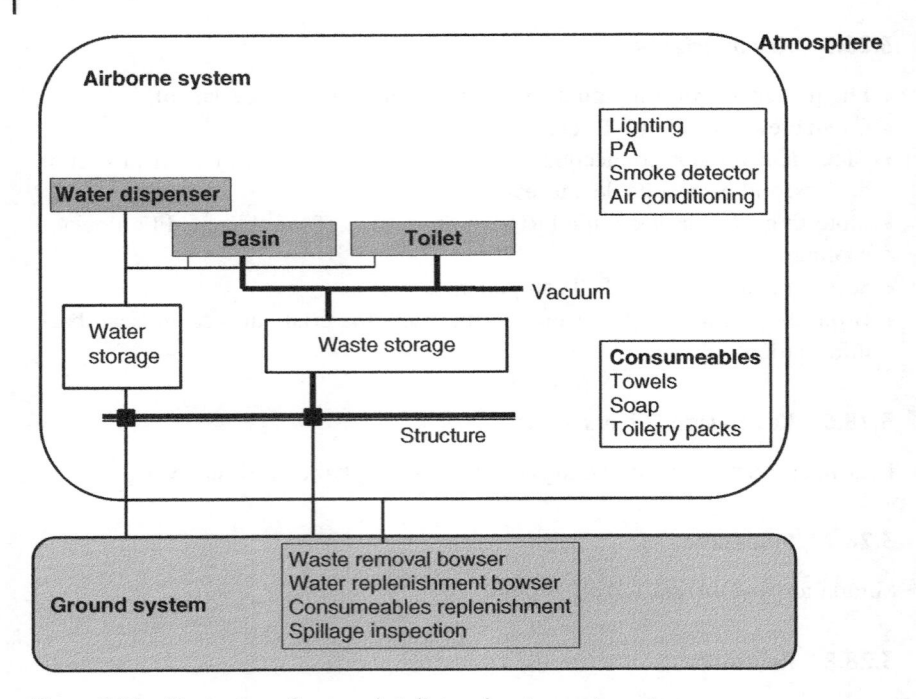

Figure 3.32 Illustration of a generic toilet and water waste system.

The water flush system must be sanitary and closed so that no effluent is discharged into the atmosphere (Figure 3.32).

3.29.1 Purpose of System

To provide hygienic management of toilets and water waste with consideration to the number of passengers at strategic times of day.

3.29.2 Description

Provision of flushing toilets, hot and cold water and disposal. Wash basins, showers, baby nursing, and changing provision. Storage of waste products throughout the flight and hygienic transfer to ground disposal services.

3.29.3 Safety/Integrity Aspects

Dispatch critical due to the implications of the inability of passengers to use toilet facilities.

3.29.4 Key Integration Aspects

Human factors, cabin furnishings, and safety. Location and number of toilets to avoid passenger obstructing the aisles.

3.29.5 Key Interfaces

Ground waste disposal and water replenishment systems

- **Airframe/structure**: provision of apertures and suitable connections for water, drinking water, sewage waste disposal, and protection of the pressure shell. Protection of cabin floor to reduce risk of urine corrosion and threat to pressurisation seals.
- **Other systems**: smoke detectors and sprinklers.
- **Electrical power**: connection to appropriate bus bar with circuit protection devices, and most suitable harness runs.
- Security of attachment of all components.
- **Environment**: potential for contamination from spillage during waste removal.
- Ground systems for waste disposal, replenishment of materials, and cleaning.

3.29.6 Key Design Drivers

Passenger satisfaction, hygiene, health, safety, and HSE regulations.

3.29.7 Modelling

Simple simulation.

3.29.8 References

JASC Code (n.d.), ATA-100 (n.d.) Chapter 38, Lawson, and Judt (2022).

3.29.9 Sizing Considerations

Number of passengers, cabin class variations, compliance with health and safety regulations, and duration of flight.

3.29.10 Future Considerations

System for military pilots on long missions or combat air patrol.

3.30 Cabin and Emergency Lighting

Lighting must be provided for all areas of the cabin for comfort, safety, and information reasons as follows (Figure 3.33):

- General lighting
- Seat spot and reading lights
- Evacuation path lights
- Statutory notices – exits, toilets, and call lights
- Galley lighting

3.30.1 Purpose of System

To provide general lighting for the cabin and galley, reading lights, exit lighting, and emergency lights to provide a visual path to the exits.

3.30.2 Description

General light in the cabin ceiling/overhead, reading lights with personal controls above each seat, statutory exit lighting, and emergency lighting.

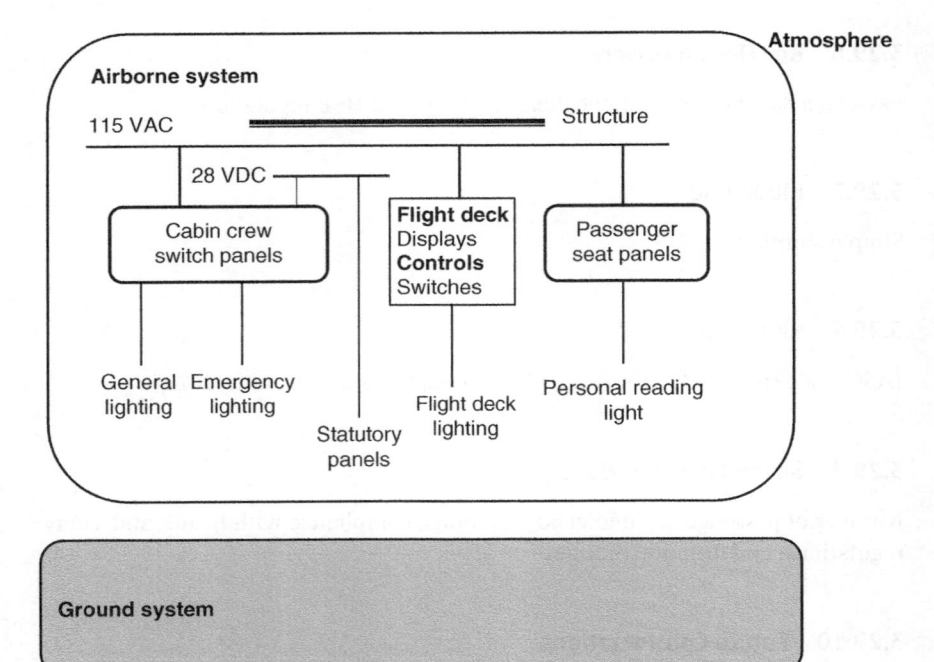

Figure 3.33 Illustration of a generic cabin and emergency lighting system.

3.30.3 Safety/Integrity Aspects

Must be available for emergency evacuation – dispatch critical.

3.30.4 Key Integration Aspects

Integration with other emergency systems.

3.30.5 Key Interfaces

- Flight deck for controls and displays and human factors assessment.
- **Electrical power**: connection to appropriate bus bar with circuit protection devices, and most suitable harness runs. Normal and emergency power generation system and batteries.
- Security of attachment of all components.

3.30.6 Key Design Drivers

Human factors for lighting, safety, passenger satisfaction, health, and safety regulations.

3.30.7 Modelling

Evacuation mock up.

3.30.8 References

Giguere (2010), Lawson and Judt (2022), JASC Code (n.d.), ATA-100 (n.d.) Chapter 33.

3.30.9 Sizing Considerations

Size of cabin and number of exits.

3.30.10 Future Considerations

Long life lighting based on commercial technology to reduce failure and maintenance.

References

Agarwal, R. (2016). Energy optimization for solar powered aircraft. In: *Encyclopedia of Aircraft Engineering, Green Aviation* (ed. R. Argawal, F. Collier, et al.). Wiley.
Air Transport Association of America (ATA), Specification 100 code (n.d.).

Avionics Communications Inc (1995). *Principles of Avionics Data Buses*. ACI.

Beater, P. (2007). *Pneumatic Drives: System Design, Modelling and Control*. Springer.

Bedard, A.J. Jr., (2016). Meteorology. In: *Encyclopedia of Aircraft Engineering, Green Aviation*, Chap. 18, Sec. 4 (ed. R. Argawal, F. Collier, et al.), 459–475. Wiley.

Bomani, B.M.M. and Hendricks, R.C. (2016). Biofuels for green aviation. In: *Encyclopedia of Aircraft Engineering, Green Aviation* (ed. R. Argawal, F. Collier, et al.), 179–191 Wiley.

Bryson, R.E. Jr., (1994). *Control of Spacecraft and Aircraft*. Princeton University Press.

Conway, H.G. (1957). *Landing Gear Design*. Chapman & Hall.

Currey, N.S. (1984). *Landing Gear Design Handbook*. Lockheed Martin.

Eldridge, J. (ed.) (2006). *Jane's Nuclear, Biological and Chemical Defense 2006–2007*, 19the. Jane's Information Group. ISBN 0-7106-2763-7.

Farokhi, S. (2014). *Aircraft Propulsion*. Wiley.

Farokhi, S. (2020). *Future Propulsion Systems and Energy Sources in Sustainable Aviation*. Wiley.

Freeh, J.E. (2016). Hydrogen fuel cells for auxiliary power units. In: *Encyclopedia of Aircraft Engineering, Green Aviation* (ed. R. Argawal, F. Collier, et al.), 193–199. Wiley.

Gent, R.W. (2010). Ice detection and protection. In: *Encyclopedia of Aerospace Engineering*, Chap. 408, vol. 8 (ed. R.H. Blockley and W. Shyy), 5005–5015. Wiley.

Giguere, D.'.A. (2010). Aircraft emergency systems. In: *Encyclopedia of Aerospace Engineering*, Chap. 407, vol. 8 (ed. R.H. Blockley and W. Shyy), 4995–5003. Wiley.

Hart, B.C. and Garrett, J. (2007). *Historical Dictionary of Nuclear, Biological, and Chemical Warfare*. Lanham, MD: Scarecrow Press. ISBN 978-0-8108-5484-0.

Hunt, T. and Vaughan, N. (1996). *Hydraulic Handbook*, 9the. Elsevier.

Jackson, A.J.B. (2010). Choice and sizing of engines for aircraft. In: *Encyclopedia of Aerospace Engineering*, Chap. 401, vol. 8 (ed. R.H. Blockley and W. Shyy), 5123–5134. Wiley.

JASC Code. Joint Aircraft System/Component. FAA and Joint Aviation Authority (European Civil Aviation Conference) code table for printed and electronic manuals.

Langton, R. (2006). *Stability and Control of Aircraft Systems*. Wiley.

Langton, R. (2010). Gas turbine fuel control systems. In: *Encyclopedia of Aerospace Engineering*, Chap. 405, vol. 8 (ed. R.H. Blockley and W. Shyy), 4973–4984. Wiley.

Langton, R., Clark, C., Hewitt, M., and Richards, L. (2009). *Aircraft Fuel Systems*. Wiley.

Langton, R., Clark, C., Hewitt, M., and Richards, L. (2010). Aircraft fuel systems. In: *Encyclopedia of Aerospace Engineering*, Chap. 402, vol. 8 (ed. R.H. Blockley and W. Shyy), 4919–4938. Wiley.

Lawson, C.P. (2010). *Environmental control systems*. In: *Encyclopedia of Aerospace Engineering*, Chap. 406, vol. 8 (ed. R.H. Blockley and W. Shyy), 4985–4994. Wiley.

Lawson, C.P. and Judt, D.M. (2022). *Aircraft Systems - A Design and Development Guide*. Wiley.

Linden, D. and Roddy, T.B. (2002). *Handbook of Batteries*, 3rde. McGraw Hill.

Lloyd, E. and Tye, W. (1982). *Systematic Safety*. Taylor Young Ltd.

MacIsaac, B. and Langton, R. (2011). *Gas Turbine Propulsion Systems*. Wiley.

Moir, I. (2010). *Electrical power generation and distribution*. In: *Encyclopedia of Aerospace Engineering*, Chap. 404, vol. 8 (ed. R.H. Blockley and W. Shyy), 4955–4972. Wiley.

Moir, I. and Seabridge, A. (2008). *Aircraft Systems*, 3rde. Wiley.

Moir, I. and Seabridge, A.G. (2010). *Vehicle management systems*. In: *Encyclopedia of Aerospace Engineering*, Chap. 401, vol. 8 (ed. R.H. Blockley and W. Shyy), 4903–4917. Wiley.

Moir, I. and Seabridge, A. (2013). *Civil Avionics*, 2nde. Wiley.

Pallett, E.H.J. (1987). *Aircraft Electrical Systems*. Longmans Group Limited.

Pornet, C. (2016). Electric drives for propulsion of transport aircraft. In: *Encyclopedia of Aircraft Engineering, Green Aviation* (ed. R. Argawal, F. Collier, et al.), 201–219. Wiley.

Pratt, R. (2000). *Flight Control Systems: Practical Issues in Design & Implementation*. IEE Publishing.

Purdy, S.I. (2010). *Probe and drogue aerial refuelling systems*. In: *Encyclopedia of Aerospace Engineering*, Chap. 409, vol. 8 (ed. R.H. Blockley and W. Shyy), 5018–5027. Wiley.

Raymond, E.T. and Chenoweth, C.C. (1993). *Aircraft Flight Control Actuation System Design*. Society of Automotive Engineers.

Rolls-Royce (2005). *The Jet Engine*. Wiley.

Schutte, J.S., Payan, A.P., Briceno Simon, I., and Marvis, D.N. (2016). *Hydrogen-powered aircraft*. In: *Encyclopedia of Aircraft Engineering, Green Aviation* (ed. R. Argawal, F. Collier, et al.), 165–177. Wiley.

Seabridge, A. (2010). *Hydraulic power generation and distribution*. In: *Encyclopedia of Aerospace Engineering*, Chap. 403, vol. 8 (ed. R.H. Blockley and W. Shyy), 4939–4953. Wiley.

Spitzer, C. (1993). *Digital Avionics Systems, Principles and Practice*, 2nde. McGraw-Hill Inc.

Weller, B. (2018). *A History of the Fly-By-Wire Jaguar*. BAE Systems Heritage Department.

Xue, N., Wenbo, D., Martin, J.R.R.A., and Wei, S. (2016). *Lithium–ion batteries: thermomechanics, performance and design optimization*. In: *Encyclopedia of Aircraft Engineering, Green Aviation* (ed. R. Argawal, F. Collier, et al.), 221–237. Wiley.

Exercise

3.1 Consider the system diagram used extensively in this chapter for each system. Can you devise an alternative pictorial model that suits your own style of working. Test it by using it on representative systems until you arrive at something you would prefer to use.

4

Avionic Systems

This chapter deals with those systems of the aircraft that give the basic aircraft a purpose, a means of being useful, and performing a mission. Avionics perform the roles of world-wide navigation and provide the air crew with all the aids and information required to do this.

Many of the avionic systems are common in function to both commercial and military roles, although the means of developing and testing them may differ. In some instances, the systems are specifically applicable to military use only and this will be identified. Those systems that have a role applicable to military missions will be described in Chapter 5 (Figure 4.1).

The avionic systems are common to both civil and military aircraft. Not all aircraft types, however, will be fitted with the complete set listed below. The age and role of the aircraft will determine the exact suite of systems. The majority of the systems collect, process, transfer, and respond to data. Any energy transfer is usually performed by a command to a vehicle system. An example of this is change to aircraft attitude demanded by the flight management system (FMS), which will be performed by the auto-pilot and flight control systems (FCS).

Although both aircraft vehicle and avionics systems make extensive use of modern digital technology, processors, and data buses, these technologies are exploited in quite different ways. The fundamental differences between the tasks that each are performing for the aircraft leads to considerable differences in their design and implementation as described below.

4.1 Displays and Controls

Control signals and data are provided by data bus and some discrete wiring to a set of display management computers. The redundancy of the computers and displays and their associated power supplies must be sufficient to support a safety critical system. The system must be capable of tolerating multiple failures without

Aircraft Systems Classifications: A Handbook of Characteristics and Design Guidelines, First Edition.
Allan Seabridge and Mohammad Radaei.
© 2022 John Wiley & Sons, Inc. Published 2022 by John Wiley & Sons, Inc.

4 Avionic Systems

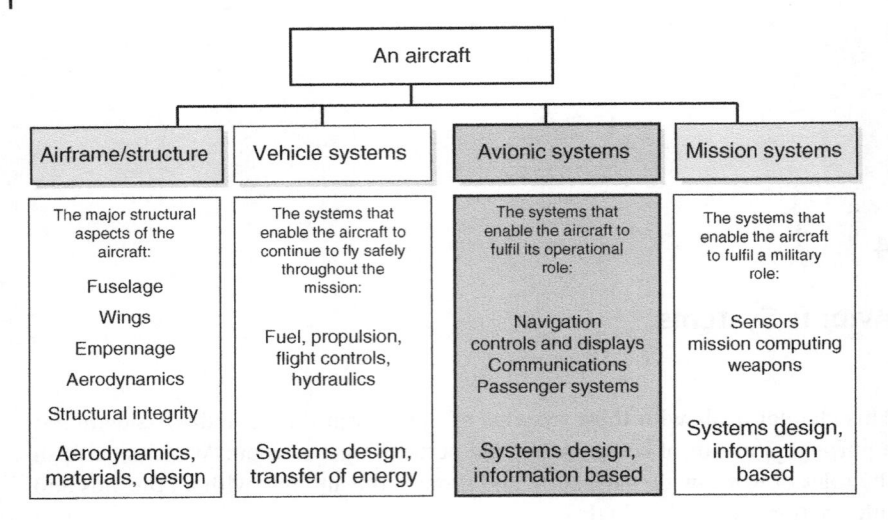

```
                            ┌─────────────┐
                            │ An aircraft │
                            └─────────────┘
```

Airframe/structure	Vehicle systems	Avionic systems	Mission systems
The major structural aspects of the aircraft:	The systems that enable the aircraft to continue to fly safely throughout the mission:	The systems that enable the aircraft to fulfil its operational role:	The systems that enable the aircraft to fulfil a military role:
Fuselage Wings Empennage Aerodynamics Structural integrity	Fuel, propulsion, flight controls, hydraulics	Navigation controls and displays Communications Passenger systems	Sensors mission computing weapons
Aerodynamics, materials, design	Systems design, transfer of energy	Systems design, information based	Systems design, information based

Figure 4.1 The systems described in this chapter.

total loss of display. A separate and independent set of displays may be provided to provide essential information such as attitude, horizontal situation, altitude, and airspeed as a get-U-home display (Figure 4.2).

To supplement the visual displays, it is common to have aural cues such as audio tones to draw the pilot's attention to warnings, and voice messages to add urgency to unusual conditions. Controls may be supplemented by direct voice input.

The controls provide an input to the systems either directly or by means of the aircraft computing architectures. Information on the performance of the systems is fed directly to the displays or from the computing architecture. This system is entirely contained in the flight deck and is designed to meet regulatory and desirable aspects of comfort, reach, and visibility in a safe environment. Human factors play an essential part of the design philosophy. Typical controls include the following:

- Control column
- Control column top (stick top)
- Throttle levers
- Communications panels
- Navigation control panel
- Auto-pilot flight director
- Discrete switches
- Soft keys or touch screens

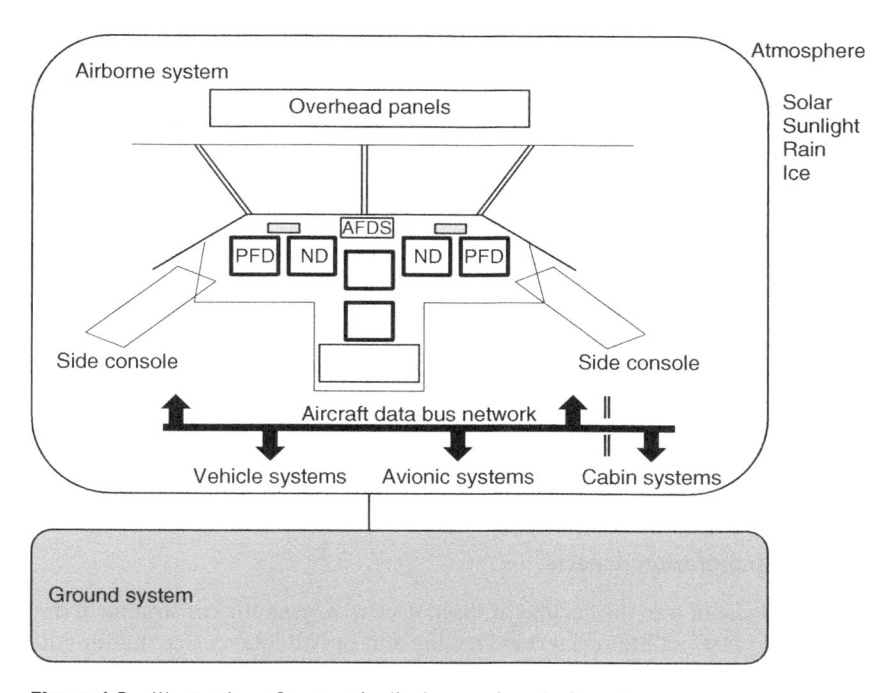

Figure 4.2 Illustration of a generic displays and controls system.

4.1.1 Purpose of System

To provide the aircrew with the means of controlling the aircraft systems by means of throttle levers, control column, pedals, switches, and rotary controls and to provide feedback on system performance by displays, warning lights, and aural means.

4.1.2 Description

The controls provide an input to the systems either directly or by means of the aircraft computing architectures. Information on the performance of the systems is fed directly to the displays or from the computing architecture. This system is entirely contained in the cockpit or flight deck and is designed to meet regulatory and desirable aspects of comfort, reach, and visibility in a safe environment. Human factors play an essential part of the design philosophy.

Control signals and data are provided by data bus and some discrete wiring to a set of display management computers. The redundancy of the computers and displays and their associated power supplies must be sufficient to support a safety critical system. The system must be capable of tolerating multiple failures without total loss of display. A separate and independent set of displays may be provided to provide essential information such as attitude, horizontal situation, altitude, and airspeed as a get-U-home display.

To supplement the visual displays, it is common to have aural cues such as audio tones to draw the pilot's attention to warnings and voice messages to add urgency to unusual conditions. Controls may be supplemented by direct voice input.

4.1.3 Safety/Integrity Aspects

The flight deck displays as a whole should be treated as safety critical to avoid situations of total display loss or misleading information. It is a good practice to provide independent get-you-home instruments to cover the event of total display failure.

4.1.4 Key Integration Aspects

The system must fit into the cockpit or flight deck to respect the constraints of the fuselage and must maintain external viewing and provide the correct lighting to allow all aspects of the cockpit to be viewable at all ambient lighting conditions. Military aircraft may need to be compatible with night vision goggles (NVG). A human factors assessment is essential.

4.1.5 Key Interfaces

- Airframe/structure: equipment installation in cockpit or flight deck.
- Other systems: exchange of information by direct wiring or data bus. The equipment associated with displays and controls will dissipate a significant amount of heat into the flight deck or cockpit and cooling will be required to ensure that the crew environment is comfortable.
- Flight deck for installation of controls and displays and human factors assessment for reach, access, lines of sight, assessment of font sizes, and colours.
- Electrical power: connection to appropriate bus bar with circuit protection devices and most suitable harness runs. Connection to bus bars to maintain power under all conditions to minimise total loss of displays.
- Installation: mounting tray for electronic line replaceable units (LRUs) and provision of suitable conditioning and cooling.
- Security of attachment of all components.
- Consideration of electro-magnetic compatibility (EMC) by radiation, induction, and susceptibility.

4.1.6 Key Design Drivers

Human factors, safety, and pilot workload.

4.1.7 Modelling

Rapid prototyping, VAPS, altitude lighting test facility, avionics integration rig, and flight simulator.

4.1.8 References

Atkin (2010), Binns (2019), Jukes (2003), Moir and Seabridge (2008) Chapter 12, Pallett (1992), Rankin and Matolak (2010), JASC Code, ATA-100 Chapter 31.

4.1.9 Sizing Considerations

Number of crew, number of display units, display computers, interfaces, redundancy, emergency displays, and head up displays.

4.1.10 Future Considerations

Synthetic vision, more automation, and integration. Gesture control.

4.2 Communications

In aviation, communications between the aircraft and the ground (air traffic/local approach/ground handling) have historically been by means of voice communication. More recently, data link communications have been introduced due to their higher data rates and in some cases, superior operating characteristics. Data links are becoming widely used in the high frequency (HF) and very high frequency (VHF) bands not only for basic communications but also to provide some of the advanced reporting features required by future air navigation systems. Data link is used on commercial aircraft to download information to a ground station to facilitate maintenance operations. On military types, various types of data links are used to transfer encrypted data pertinent to the mission (Figure 4.3).

After selecting the appropriate communications channel on the channel selector, the pilot transmits a message by pressing the transmit button which connects the pilot's microphone into the appropriate radio. The voice message is used to modulate the carrier frequency, and it is this composite signal that is transmitted.

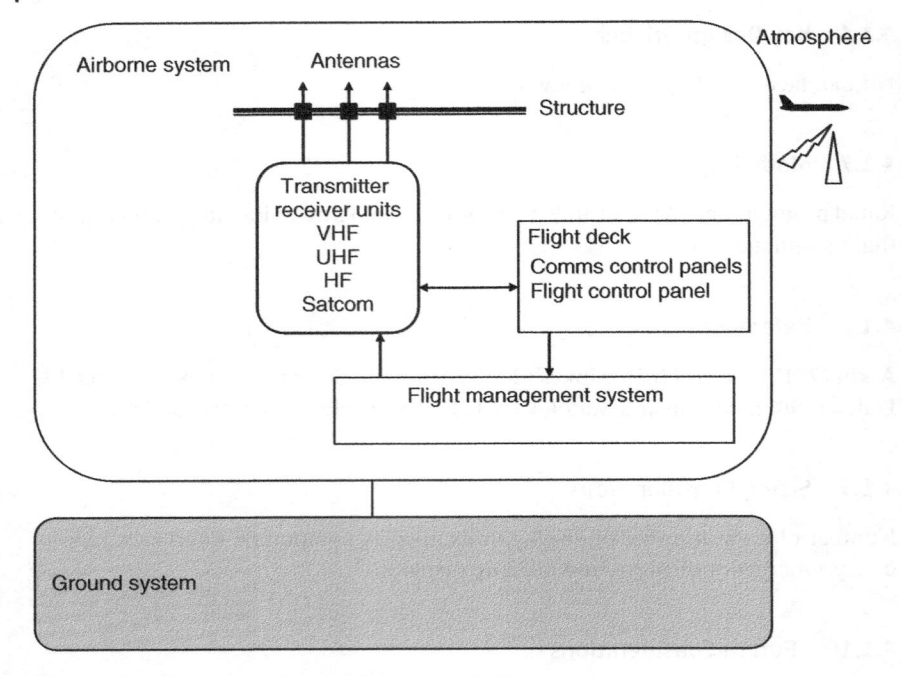

Figure 4.3 Illustration of a generic communications system.

4.2.1 Purpose of System

To allow two-way voice and data communication between the aircraft and air traffic control (ATC), ground systems, other aircraft, and co-operating forces.

4.2.2 Description

The system requires transmitting and receiving (TR) units for each waveband with appropriate antennas, personal equipment – headsets, mikes, and speakers. Each radio will be tuned by a communications control unit or radio management panel and the FMS will also perform tuning for selected systems. The following wavebands are in common use:

- VHF communications: VHF communications are used extensively by the civil aviation community. A standard fit would include two or more probably three VHF transmitter/receivers (TRs). In certain military applications, more VHF sets would be fitted. This frequency band also incorporates the instrument landing system (ILS).

- Ultra high frequency (UHF) communications: UHF communications are used exclusively by military operators. Many military communications systems will incorporate combined V/UHF sets.
- HF communications: As aircraft ranges have increased the need for HF communications have also – most aircraft have two HF systems fitted – some with high frequency data link (HFDL).
- SATCOM: Civil aircraft, particularly long-range aircraft usually have SATCOM fitted. Military aircraft will additionally have UHF SATCOM fitted operating at the top end of the UHF frequency range.

For data link applications terminals and cryptographic (crypto) devices are required to ensure that information is secure.

Maritime patrol aircraft will have additional communications sets to allow interactions with marine vessels such as short wave and marine bands.

4.2.3 Safety/Integrity Aspects

Safety critical to maintain full communications and identification of the aircraft. Mission-critical.

4.2.4 Key Integration Aspects

Antenna operability, drag, and integration with FMS for auto-tuning. Integration with the FMS for auto-selection and tuning of radio navigation aids.

4.2.5 Key Interfaces

- Airframe/structure: provision of apertures and suitable fasteners for antennas, optimum location of antennas, drag assessment, protection of the pressure shell.
- Other systems: exchange of information by direct wiring or data bus.
- Flight deck for controls and displays and human factors assessment.
- Electrical power: connection to appropriate bus bar with circuit protection devices, and most suitable harness runs.
- Installation: mounting tray for electronic LRUs and provision of suitable conditioning and cooling.
- Security of attachment of all components.
- Environment: radio frequency electromagnetic transmissions.
- Consideration of EMC by radiation, induction, and susceptibility.

4.2.6 Key Design Drivers

All weather communications, interface with emergency channels.

4.2.7 Modelling

Antenna modelling, radio frequency (RF) models, and antenna placement models.

4.2.8 References

Binns (2019), Burberry (1992), Hall and Barclay (1980), Moir and Seabridge (2013) Chapter 9, Macnamara (2010), Rankin and Matolak (2010), JASC Code, ATA-100 Chapter 23.

4.2.9 Sizing Considerations

Types of radios, redundancy, control panels, antennas, dissipation, and electrical loads.

4.2.10 Future Considerations

Fully automated tuning of all radios, more reliance on data interchange via data link.

4.3 Navigation

The crew use the navigation system to navigate safely in a fuel efficient manner to maintain their tight schedules and to avoid all other air traffic. With flight times of close to 20 hours becoming commonplace, accuracy is a major factor in modern navigation system design. The introduction of satellite navigation systems from a number of different sources has made this possible. To reduce total dependence on the satellite system, whether by corruption of data, deliberate sabotage, or equipment failures, standby navigation aids can be used (Figure 4.4).

An example of a total navigation system is shown in Figure 4.5 to illustrate where the systems described in this chapter fit into the overall picture. The key attributes of the systems are the following:

- Twin FMS computers interfacing with the navigation displays (NDs) and multi-function control and display units (MCDU).
- FMS system interfacing with the autopilot/automatic flight director system.
- Dual sources of air and inertial reference data: air data module/air data inertial reference system (ADM/ADIRS) or air data computer/inertial reference system (ADC/IRS).
- Dual fit of VHF omni-ranging/distance measuring equipment (VOR/DME); ILS; automatic direction finding (ADF), radar altimeter (Rad Alt), and global positioning system (GPS) receivers. In some installations, some of these receivers may be integrated into a single multi-mode receiver (MMR) unit.

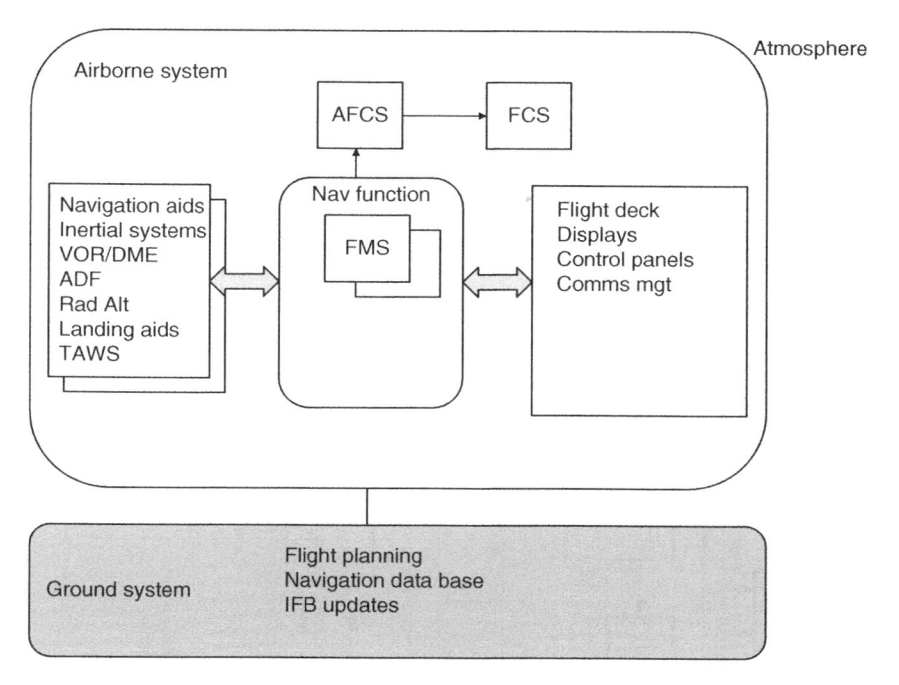

Figure 4.4 Illustration of a generic navigation system.

- Interface with traffic collision avoidance systems (TCASs) and terrain avoidance warning system (TAWS).
- An interface will also be provided to the flight data recorder and crash survivable memory.

This architecture is typical of a medium-range twin-engine transport aircraft. Longer-range variants may have a third set of ADM/ADIRS or ADC/IRS fitted for reasons of availability.

4.4 Example Navigation System Architecture

4.4.1 Purpose of System

To provide world-wide, high-accuracy navigation capability with the ability to avoid all other air traffic, to achieve fuel economy targets, and to adhere to airline schedules.

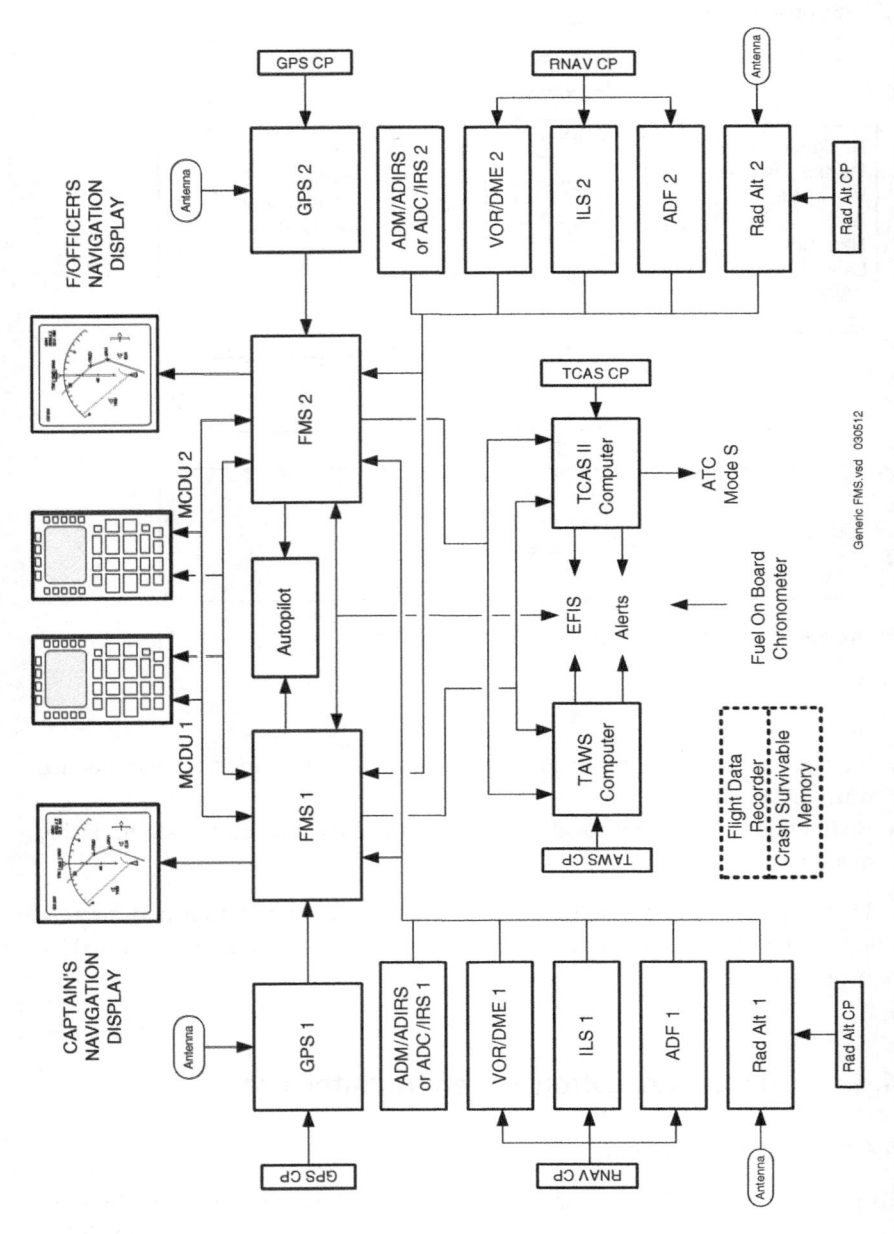

Figure 4.5 Example of a flight management system.

4.4.2 Description

Inertial, laser, or GPS-based system combined with radio navigation aids and computing to provide both manual and automated navigation.

4.4.3 Safety/Integrity Aspects

Mission-critical with safety implications.

4.4.4 Key Integration Aspects

Integration with avionics and mission system. Human factors integration in flight deck.

4.4.5 Key Interfaces

Structure, avionics:

- Airframe/structure: provision of apertures and suitable fasteners for antennas, optimum location of antennas, drag assessment, protection of the pressure shell.
- Other systems: exchange of information by direct wiring or data bus.
- Flight deck for controls and displays and human factors assessment.
- Electrical power: connection to appropriate bus bar with circuit protection devices, and most suitable harness runs.
- Installation: mounting tray for electronic LRUs and provision of suitable conditioning and cooling.
- Security of attachment of all components.
- Environment: radio frequency electromagnetic transmissions.
- Consideration of EMC by radiation, induction, and susceptibility.

4.4.6 Key Design Drivers

Accurate world-wide navigation: ATM (civil) or Global Air Traffic Management (GATM) (military).

4.4.7 Modelling

Avionics integration rig, mission system integration rig. See Chapter 7 for architecture modelling.

4.4.8 References

Binns (2019), Galotti (1998), Kayton and Fried (1997), Moir and Seabridge (2013) Chapters 9 and 11, Sabatini et al. (2016), Sen and Battach (2018), SESAR (2012), JASC Code, ATA-100 Chapter 34.

4.4.9 Sizing Considerations

Navigation sensors, Nav Aids, redundancy, antennas. Demands imposed by future global air transport navigation systems. Very long duration world wide routes.

4.4.10 Future Considerations

Use of data link for digital communications, improved methods of long-range communication. Consideration of navigation solutions for environmental (Green) sustainable operations.

4.5 Flight Management System (FMS)

The management of the overall aircraft navigation functions was becoming an increasingly complex task and modern integrated systems accuracy expectations are high, as is the need to reduce flight crew workload. The FMS functionality increased rapidly as requirements increased, and many more enhancements are in prospect as the future features required by air transport management (ATM) are added. The FMS performs the task of blending the inputs from the various navigation sensors in an optimum manner; navigating precisely to pre-programmed flight plans and procedures in space and time and providing an effective man-machine interface for the flight crew (Figure 4.6).

4.5.1 Purpose of System

To provide a means of entering and executing flight plans and allowing automatic operation of the aircraft in accordance with those plans to provide accurate navigation and to reduce crew workload.

4.5.2 Description

Flight management computers and control and display unit (CDU) to enter and modify flight plans and tune navigation aids.

The system takes inputs, usually duplicated for reasons of availability and integrity, to perform its functions such as

- Dual inertial sensors such as inertial navigation system (INS)/IRS.
- Dual navigation aids sensors: VOR/DME; DME/DME; etc.
- Dual GNSS sensors: usually GPS.
- Dual or triplex air data sensors: often combined with INS/IRS in an ADIRS.
- Dual inputs from on-board sensors relating to fuel on-board and time.

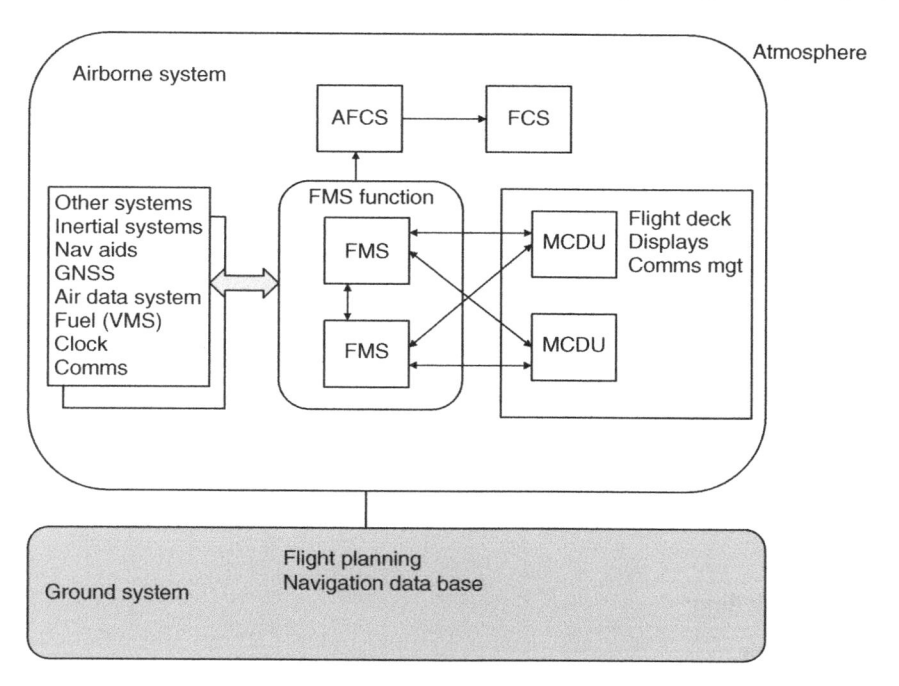

Figure 4.6 Illustration of a generic flight management system.

These inputs are used by the FMS to perform the necessary navigation calculations and provide information to the flight crew via a range of display units:

- Communications control system (CCS).
- Interface with the autopilot/flight director system to provide the flight crew with flight direction or automatic flight control in a number of pre-defined modes.

4.5.3 Safety/Integrity Aspects

Mission-critical.

4.5.4 Key Integration Aspects

Navigation system and navigation aids, cockpit lighting, and human factors.

4.5.5 Key Interfaces

Cockpit location and navigation sensors:

- Airframe/structure: provision of apertures and suitable fasteners for antennas, optimum location of antennas, drag assessment, and protection of the pressure shell.
- Other systems: exchange of information by direct wiring or data bus.
- Flight deck for controls and displays and human factors assessment.
- Electrical power: connection to appropriate bus bar with circuit protection devices, and most suitable harness runs.
- Installation: mounting tray for electronic LRUs and provision of suitable conditioning and cooling.
- Security of attachment of all components.
- Environment: radio frequency electromagnetic transmissions.
- Consideration of EMC by radiation, induction, and susceptibility.

4.5.6 Key Design Drivers

Ease of use, accessibility, pilot workload, and efficient route management.

4.5.7 Modelling

Integration rig.

4.5.8 References

Cramer (2010), Gradwell (2010), Moir and Seabridge (2013) Chapter 11, ASC Code, ATA-100 Chapter 34.

4.5.9 Sizing Considerations

Flight management CDUs on flight deck, and redundancy.

4.5.10 Future Considerations

More integrated functions to meet developments in satellite-based systems. Novel techniques to obtain independence from satellite-based systems (e.g. magneto-resonance).

4.6 Weather Radar

Aircraft must fly safely in world-wide weather conditions, the exact predominant weather depends on their geographical location and most commonly used routes.

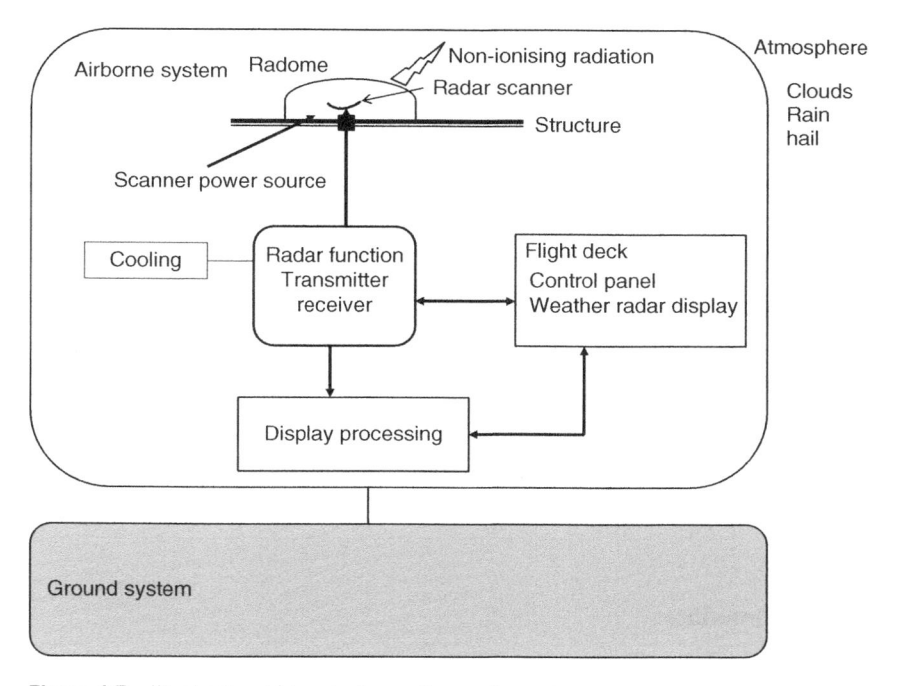

Figure 4.7 Illustration of a generic weather radar system.

Some aspects of weather are visible to radar and it is usual to identify weather conditions ahead of the aircraft so that precautions can be taken to limit damage or to select an alternative route. A simple nose mounted radar and a suitable display is used to alert the crew of adverse weather conditions (Figure 4.7).

4.6.1 Purpose of System

To alert the flight crew to the presence of adverse weather or terrain in the aircraft's flight path.

4.6.2 Description

The weather radar radiates energy in a narrow beam which may be reflected from clouds or terrain ahead of the aircraft. The radar beam is scanned either side of the aircraft centre line to give a radar picture of objects ahead of the aircraft. The antenna can also be tilted to scan areas above and below the aircraft. The antenna is contained in the nosecone radome, and the picture is presented on a display surface on the flight deck.

Typical weather conditions are displayed on a weather map with colour used to signify severity:

- Weather and precipitation with a maximum range of 320 nautical miles.
- Turbulence up to 40 nautical miles using Doppler processing.
- Wind shear up to 5 nautical miles.

4.6.3 Safety/Integrity Aspects

Safety involved and dispatch critical.

4.6.4 Key Integration Aspects

Integration of scanner with the radome and structure with clearance for all modes of scan and tilt. Integration with the cockpit displays and human machine aspects.

Military: cockpit displays, mission system consoles, weapon system, and mission computer.

4.6.5 Key Interfaces

- Airframe/structure: provision of apertures and suitable fasteners for radome – suitable materials, aerodynamic design, strength, bird strike clearance, ice and driving rain, drag assessment, and protection of the pressure shell.
- Other systems: exchange of information by direct wiring or data bus. Cockpit display system.
- Flight deck for controls and displays and human factors assessment.
- Electrical power: connection to appropriate bus bar with circuit protection devices, and most suitable harness runs.
- Installation: mounting tray for electronic LRUs and provision of suitable conditioning and cooling.
- Security of attachment of all components.
- Environment: radio frequency electromagnetic transmissions.
- Consideration of EMC by radiation, induction, and susceptibility.

4.6.6 Key Design Drivers

Areas of operations and the likelihood of severe weather conditions.

4.6.7 Modelling

Avionics integration rig.

4.6.8 References

Honeywell (2014), Moir and Seabridge (2013) Chapter 8, JASC Code, ATA-100 Chapter 31.

4.6.9 Sizing Considerations

Area of operations and world-wide deployments.

4.6.10 Future Considerations

Application of artificial intelligence (AI) to aid decisions on weather risk and avoidance strategies.

4.7 Air Traffic Control (ATC) Transponder

All aircraft flying in controlled airspace must make themselves identifiable to ATC so that the aircraft position and altitude can be established on radar plots. The ATC transponder allows ground surveillance radars to interrogate aircraft and decode data which enables correlation of a radar track with a specific aircraft. A ground-based primary surveillance radar (PSR) will transmit radar energy and will be able to detect an aircraft by means of the reflected radar energy – termed the aircraft return. This will enable the aircraft return to be displayed on an ATC console at a range and bearing commensurate with the aircraft position. Coincident with the primary radar operation, a secondary surveillance radar (SSR) will transmit a series of interrogation pulses that are received by the on-board aircraft transponder. The transponder aircraft replies with a different series of pulses which give information relating to the aircraft, normally aircraft identifier, and altitude. If the PSR and SSR are synchronised – usually by being co-boresighted, then both the presented radar returns and the aircraft transponder information may be presented together on the ATC console. Therefore, the controller will have aircraft identification (e.g. BA 123) and altitude presented alongside the aircraft radar return, thereby greatly improving the controller's situational awareness (Figure 4.8).

The system is also known as identification friend or foe (IFF/SSR), and this nomenclature is in common use in the military field.

4.7.1 Purpose of System

To provide a response to ground interrogations which identify the aircraft and provide information relating to position and height. To provide a response to aircraft equipped with TCAS Mode S transponders.

144 | *4 Avionic Systems*

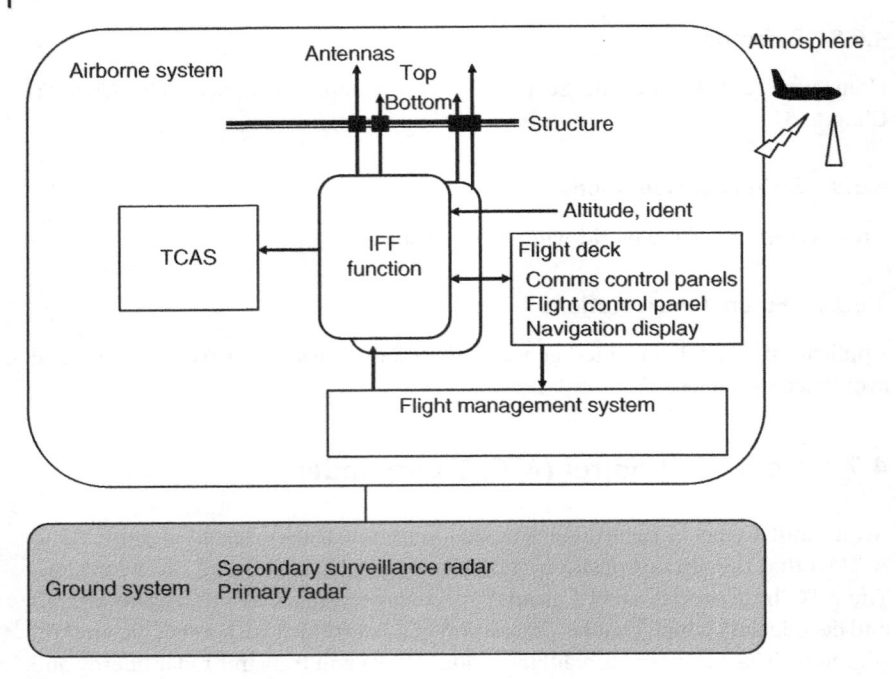

Figure 4.8 Illustration of a generic transponder system.

4.7.2 Description

The main elements of the system are the following:

- ATC transponder controller unit for setting modes and response codes.
- Dedicated ATC transponder unit.
- An ATC antenna unit with an optional second antenna. It is usual to use both upper and lower mounted antennas to prevent blanking effects as the aircraft manoeuvres.

4.7.3 Safety/Integrity Aspects

Mission critical: loss of operation will lead to air traffic violations. Military aircraft will be asked to leave the airways.

4.7.4 Key Integration Aspects

Integration with TCAS. Antenna may be shared with other RF devices using the same frequencies.

4.7.5 Key Interfaces

Communication system and ATC.

- Airframe/structure: provision of apertures and suitable fasteners for antennas, optimum location of antennas, drag assessment, protection of the pressure shell.
- Other systems: exchange of information by direct wiring or data bus, e.g. TCAS.
- Flight deck for controls and displays and human factors assessment.
- Electrical power: connection to appropriate bus bar with circuit protection devices, and most suitable harness runs.
- Installation: mounting tray for electronic LRUs and provision of suitable conditioning and cooling.
- Security of attachment of all components.
- Environment: radio frequency electromagnetic transmissions.
- Consideration of EMC by radiation, induction, and susceptibility.

4.7.6 Key Design Drivers

Confirmation of aircraft identification and height for ATC, for military aircraft – co-operative operations in combat zones.

4.7.7 Modelling

Avionics integration rig.

4.7.8 References

Moir and Seabridge (2013) Chapter 9.

4.7.9 Sizing Considerations

New functions requiring further information about aircraft position and identification which may require higher bandwidths.

4.7.10 Future Considerations

Further reductions in vertical separation minima resulting from increases in air traffic. New functions requiring further information about aircraft position and identification. Trend towards digital communications using data link. Higher density of traffic may lead to more automation.

4.8 Traffic Collision and Avoidance System (TCAS)

TCAS was developed in prototype form during the 1960s and 1970s to provide a surveillance and collision avoidance system to help aircraft avoid collisions. It was certified by the Federal Aviation Administration (FAA) in the 1980s and has been in widespread use in the United States in its initial form. TCAS is based upon beacon-interrogator and operates in a similar fashion to the ground-based SSR already described. The system comprises two elements: a surveillance system and collision avoidance system. TCAS detects the range bearing and altitude of aircraft in the near proximity for display to the pilots (Figure 4.9).

A transponder is mandated if an aircraft flies above 10 000 ft or within 30 miles of major airports; consequently all commercial aircraft and the great majority of corporate and general aviation aircraft are fitted with the equipment.

4.8.1 Purpose of System

To reduce the risk of collision with other aircraft.

4.8.2 Description

Transponder-based control unit to interrogate aircraft within a certain spherical volume of the carrier aircraft and an indication and warning system.

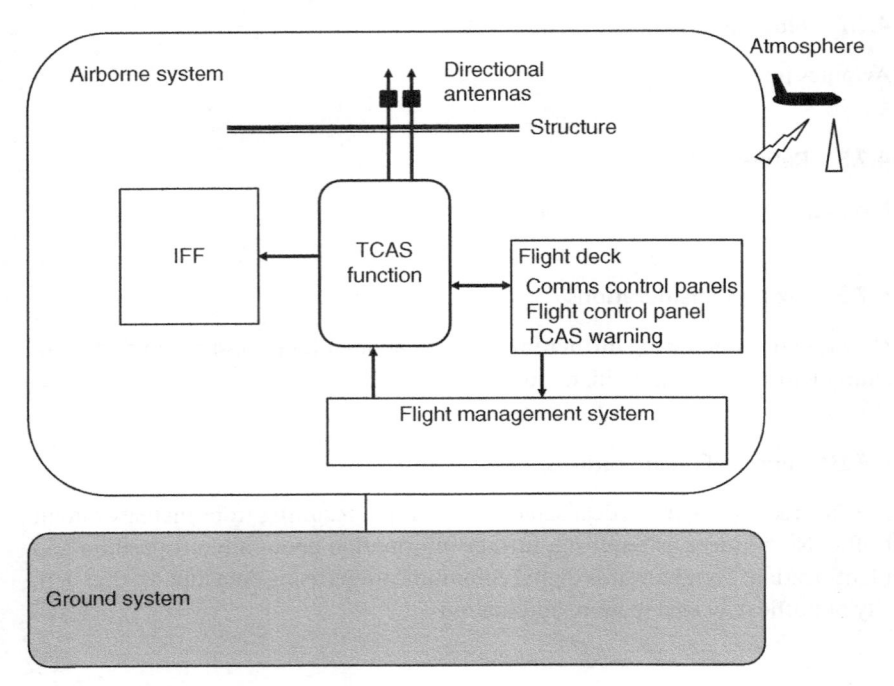

Figure 4.9 Illustration of a generic TCAS.

4.8 Traffic Collision and Avoidance System (TCAS)

TCAS transmits a Mode C interrogation search pattern for Mode A and C transponder-equipped aircraft and receives replies from all such equipped aircraft. In addition, TCAS transmits one Mode S interrogation for each Mode S transponder-equipped aircraft receiving individual responses from each one. Mode A relates to range and bearing, while Mode C relates to range, bearing, and altitude and Mode S to range, bearing, and altitude with a unique Mode S reply. The aircraft TCAS equipment comprises a radio transmitter and receiver, directional antennas, computer, and flight deck display. Whenever another aircraft receives an interrogation, it transmits a reply, and the TCAS computer is able to determine the range depending upon the time taken to receive the reply. The directional antennas enable the bearing of the responding aircraft to be measured. TCAS can track up to 30 aircraft but only display 25, the highest priority targets being the ones that are displayed.

TCAS is unable to detect aircraft which are not carrying an appropriately operating transponder or which have unserviceable equipment.

TCAS exists in two forms: TCAS I and TCAS II. TCAS I indicates the range and bearing of aircraft within a selected range; usually 15–40 nautical miles forward, 5–15 nautical miles aft, and 10–20 nautical miles on each side. The system also warns of aircraft within a defined distance (at the time of writing ±8700 ft) of the aircraft's own altitude.

The collision avoidance system element predicts the time to and separation at the intruder's closest point of approach. These calculations are undertaken using range, closure rate, altitude, and vertical speed. Should TCAS ascertain that certain safety boundaries will be violated, it will issue a Traffic Advisory (TA) to alert the crew that closing traffic is in the vicinity via the display of certain coloured symbols. Upon receiving a TA, the flight crew must visually identify the intruding aircraft and may alter their altitude by up to 300 ft. A TA will normally be advised between 20 and 48 seconds before the point of closest approach with a simple audio warning in the flight crew's headsets: 'TRAFFIC, TRAFFIC'. TCAS I does not offer any de-confliction solutions but does provide the crew with vital data in order that they may determine the best course of action. TCAS II offers a more comprehensive capability with the provision of resolution advisories (RAs). TCAS II determines the relative motion of the two aircraft and determines an appropriate course of action. The system issues a RA via Mode S advising the pilots to execute the necessary manoeuvre to avoid the other aircraft. A RA will be usually issued when the point of closest approach is within 15 and 35 seconds and the de-confliction symbology is displayed coincident with the appropriate warning.

A total of ten audio warnings may be issued. Examples are the following:

- 'CLIMB, CLIMB, CLIMB'.
- 'DESCEND, DESCEND, DESCEND'.
- 'REDUCE CLIMB, REDUCE CLIMB'

Finally when the situation is resolved:

- 'CLEAR OF CONFLICT'.

4.8.3 Safety/Integrity Aspects

Dispatch critical for certain routes.

4.8.4 Key Integration Aspects

Cockpit displays, mission computing, navigation system, navigation aids, and human factors.

4.8.5 Key Interfaces

- Airframe/structure: provision of apertures and suitable fasteners for antennas, optimum location of antennas, drag assessment, and protection of the pressure shell.
- Other systems: exchange of information by direct wiring or data bus, IFF/SSR Mode S, and cockpit displays.
- Flight deck for controls and displays and human factors assessment.
- Electrical power: connection to appropriate bus bar with circuit protection devices, and most suitable harness runs.
- Installation: mounting tray for electronic LRUs and provision of suitable conditioning and cooling.
- Security of attachment of all components.
- Environment: radio frequency electromagnetic transmissions.
- Consideration of EMC by radiation, induction, and susceptibility.

4.8.6 Key Design Drivers

Safe operation in airport terminal areas and designated air lanes.

4.8.7 Modelling

Avionics integration rig.

4.8.8 References

ICAO Doc 9863 (2006), Moir and Seabridge (2013) Chapter 9, RTCA DO-185B (2008), JASC Code, ATA-100 Chapter 34.

4.8.9 Sizing Considerations

Display type and control unit.

4.8.10 Future Considerations

Automatic response to collision warnings and robustness of algorithms.

4.9 Terrain Avoidance and Warning System (TAWS)

The ground proximity warning system (GPWS) is intended to prevent unintentional flight into the ground. Controlled flight into terrain (CFIT) is the cause of many accidents, many of them on the approach and landing phase of flight and often associated with non-precision approaches (Figure 4.10).

The installation of the original GPWS equipment for all airliners flying in US airspace was mandated by the FAA in 1974, since when the number of CFIT accidents has dramatically decreased. Within about four years, the system was adopted world-wide on the majority of passenger-carrying airliners. In time, enhanced versions have become available. Enhanced ground proximity warning system (EGPWS) offered a much greater situational awareness to the flight crew as more quantitative information is provided to the flight crew together with earlier warning of the situation arising. It used a world-wide terrain data base

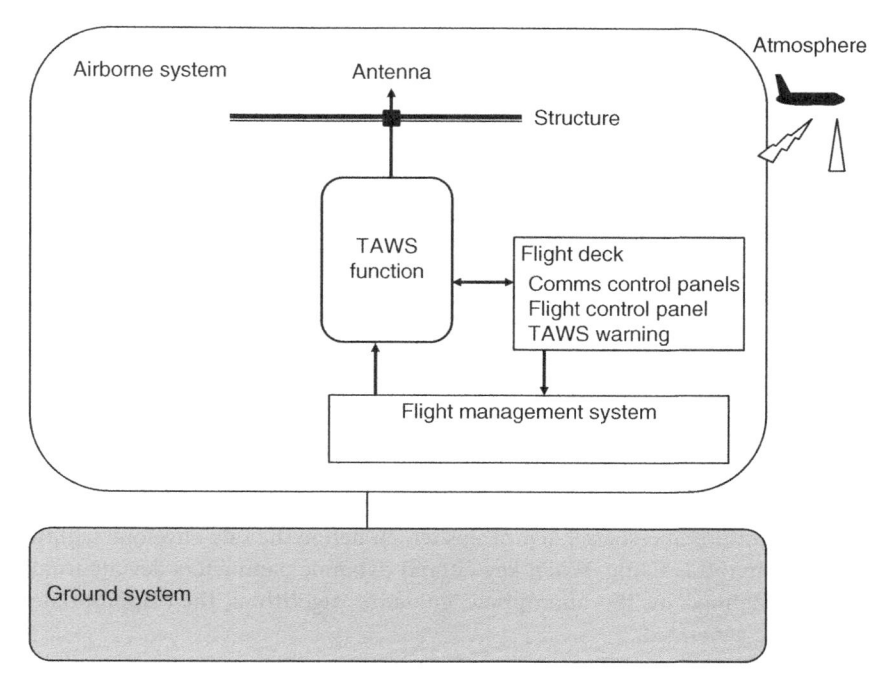

Figure 4.10 Illustration of a generic TAWS.

which is compared with the aircraft's present position and altitude. Within the terrain database, the earth's surface is divided into a grid matrix with a specific altitude assigned to each square within the grid representing the terrain at that point. EGPWS has now been superseded by TAWS.

CFIT describes conditions where the crew are in control of the aircraft; however, due to a lack of situational awareness they are unaware that they are about to crash into the terrain. GPWS originally took data from the Rad Alt and barometric vertical speed indication generated a series of audio warnings when a hazardous situation is developing. In subsequent developments has now been superseded by a more generic title: TAWS where this data is combined with that provided by other sensors.

Military aircraft with a low-level role over ground or over the sea will find this system especially useful.

4.9.1 Purpose of System

To reduce the risk of aircraft flying into the ground or into rising terrain ahead of the aircraft.

4.9.2 Description

Provides a series of advisory warnings for the flight crew when the aircraft is approaching a hazardous situation

TAWS uses Rad Alt information together with other information relating to the aircraft flight path. Warnings are generated when the following scenarios are unfolding:

- Flight below the specified descent angle during an instrument approach.
- Excessive bank angle at low altitude.
- Excessive descent rate.
- Insufficient terrain clearance.
- Inadvertent descent after take-off.
- Excessive closure rate to terrain: the aircraft is descending too quickly or approaching higher terrain.

Inputs are taken from a variety of aircraft sensors that are compared with a number of performance-based algorithms which define the safe envelope within which the aircraft is flying. When key aircraft dynamic parameters deviate from the values defined by the appropriate guidance algorithms then appropriate warnings are generated.

The aircraft's intended flight path and manoeuvre envelope for the prevailing flight conditions are compared with the terrain matrix and the result graded according to the proximity of the terrain. Terrain responses are displayed using colour imagery.

This type of portrayal using colour imagery is very similar to that for the weather radar and is usually shown on the Navigation Display. It is far more informative than the audio warnings given by earlier versions of GPWS. The TAWS also gives audio warnings but much earlier than those given by the earlier system. These earlier warnings, together with the quantitative colour display, gives the flight crew a much better overall situational awareness in respect of terrain and more time to react positively to their predicament than did previous systems.

4.9.3 Safety/Integrity Aspects

Safety involved.

4.9.4 Key Integration Aspects

Integration of display and warning into the flight deck philosophy with regard to colour, warning, and voice. Integration with sensors in other systems.

4.9.5 Key Interfaces

Rad Alts, GPS, and pilot's displays and warning systems:

- Airframe/structure: provision of apertures and suitable fasteners for antennas, optimum location of antennas, drag assessment, protection of the pressure shell.
- Other systems: exchange of information by direct wiring or data bus.
- Flight deck for controls and displays and human factors assessment.
- Electrical power: connection to appropriate bus bar with circuit protection devices, and most suitable harness runs.
- Installation: mounting tray for electronic LRUs and provision of suitable conditioning and cooling.
- Security of attachment of all components.
- Environment: radio frequency electromagnetic transmissions.
- Consideration of EMC by radiation, induction, and susceptibility.

4.9.6 Key Design Drivers

Reduction of the risk of accidents due to flight crew loss of situational awareness and subsequent CFIT by improving situational awareness during approach and landing.

Accuracy of sensors, setting of warning to allow safe avoidance action to be taken with regard to rate of descent, speed, weight, structural limitations, FCS authority, and response rate.

4.9.7 Modelling

Modelling and simulation using flight simulator, and avionics integration rig.

4.9.8 References

Moir and Seabridge (2013) Chapter 11, JASC Code, ATA-100 Chapter 34.

4.9.9 Sizing Considerations

Display type and control unit. Increasing traffic in terminal areas.

4.9.10 Future Considerations

Improved imagery and audio for better understanding of the situation.

4.10 Distance Measuring Equipment (DME)/TACAN

4.10.1 Purpose of System

DME is a method of pulse-ranging used in the 960–1215 MHz band to determine the distance of the aircraft from a designated ground station. The aircraft equipment interrogates a ground-based beacon and upon the receipt of re-transmitted pulses – unique to the on-board equipment – is able to determine the range to the DME beacon. DME beacons are able to service requests from a large number of aircraft simultaneously but are generally understood to have the capacity of handling ~200 aircraft at once. Specified DME accuracy is reportedly better than ±3% or ±0.5 nautical miles [whichever is the greater]. Precision distance measuring equipment (pDME) beacons installed at airports will usually be far more accurate (Figure 4.11).

TACAN (tactical air navigation) is a military system using very similar principles to DME but operating in a different frequency band. As a result, the beacons are smaller and are used on ships or mobile tactical units.

4.10.2 Description

The receiver is automatically tuned by the FMS to appropriate beacons along routes.

4.10.3 Safety/Integrity Aspects

May be mission-critical.

4.10 Distance Measuring Equipment (DME)/TACAN | **153**

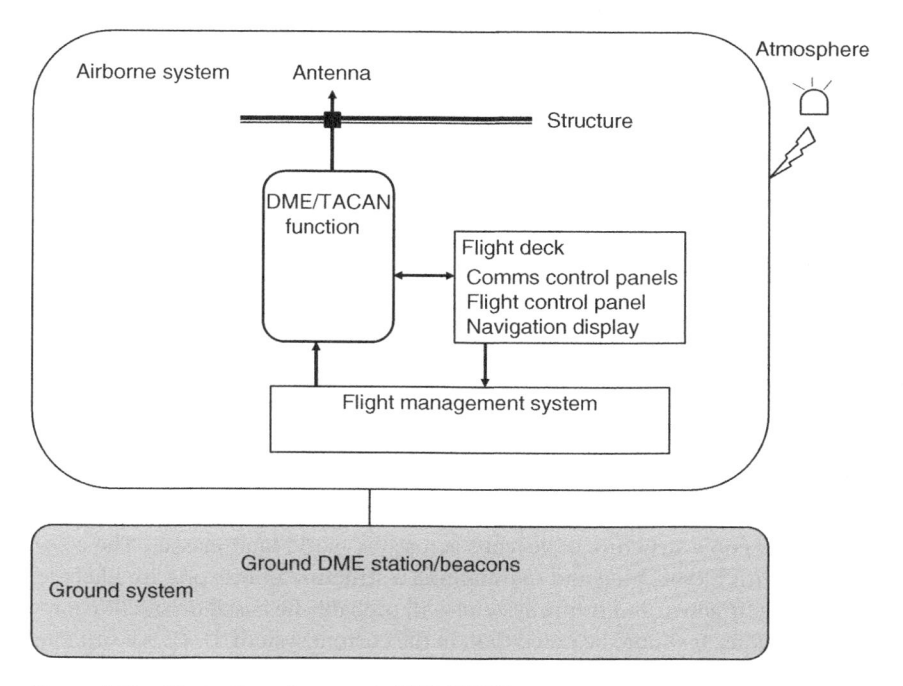

Figure 4.11 Illustration of a generic DME/TACAN system.

4.10.4 Key Integration Aspects

Cockpit displays, mission computing, navigation system, navigation aids, and human factors integration of antenna with all other antennas.

4.10.5 Key Interfaces

Tuning by FMS where an integrated system is fitted:

- Airframe/structure: provision of apertures and suitable fasteners for antennas, optimum location of antennas, drag assessment, protection of the pressure shell.
- Other systems: exchange of information by direct wiring or data bus.
- Flight deck for controls and displays and human factors assessment.
- Electrical power: connection to appropriate bus bar with circuit protection devices, and most suitable harness runs.
- Installation: mounting tray for electronic LRUs and provision of suitable conditioning and cooling.
- Security of attachment of all components.
- Environment: radio frequency electromagnetic transmissions.
- Consideration of EMC by radiation, induction, and susceptibility.

4.10.6 Key Design Drivers

Navigational accuracy and location/availability of DME beacons.

4.10.7 Modelling

Avionics integration rig.

4.10.8 References

Moir and Seabridge (2013) Chapter 9, JASC Code, ATA-100 Chapter 34.

4.10.9 Sizing Consideration

Control unit and antenna.

4.10.10 Future Considerations

DME depends on a structure of beacons across the world land masses. The costs associated with establishing and maintaining a structure of beacons are likely to make this unattractive and future systems will probably be based on satellite navigation solutions. It should be noted that, in the current system, DME is a suitable backup of adequate accuracy to cover the temporary loss of satellite navigation signals.

4.11 VHF Omni-Ranging (VOR)

4.11.1 Purpose of System

The VOR system was accepted as standard by the United States in 1946 and later adopted by the International Civil Aviation Organisation (ICAO) as an international standard. The system provides a widely used set of radio beacons operating in the VHF frequency band over the range 108–117.95 MHz with 100 kHz spacing. Each beacon emits a Morse code modulated tone which may be provided to the flight crew for the purposes of beacon identification (Figure 4.12).

4.11.2 Description

The ground station radiates a cardioid pattern that rotates at 30 rpm, generating a 30 Hz modulation at the aircraft receiver. The ground station also radiates an omni-directional signal which is frequency modulated with a 30 Hz reference tone. The phase difference between the two tones varies directly with the bearing of the aircraft. At the high frequencies at which VHF operates, there are no sky wave

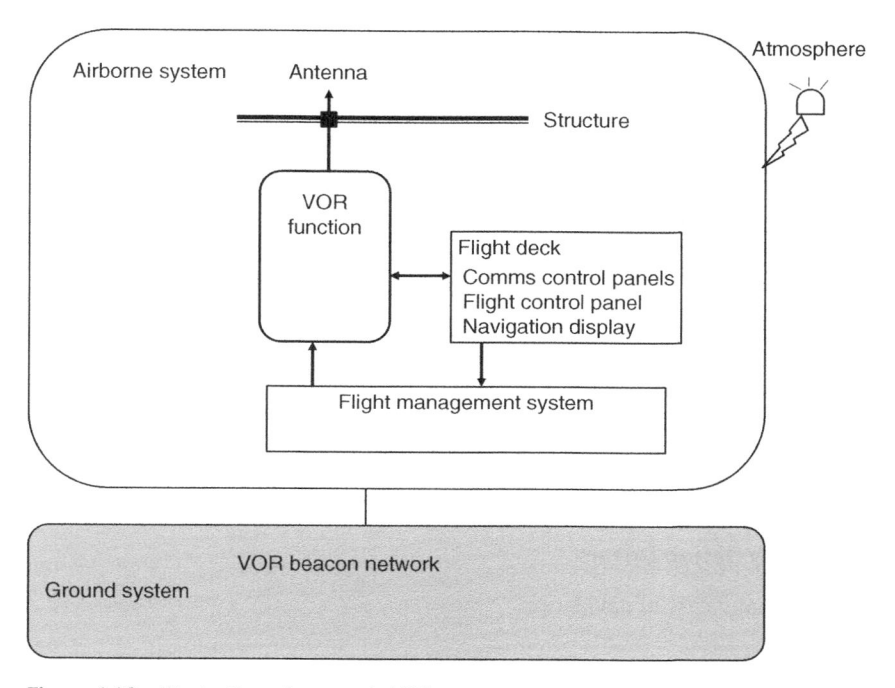

Figure 4.12 Illustration of a generic VOR.

effects and the system performance is relatively consistent. VOR has the disadvantage that it can be severely disrupted by adverse weather – particularly by electrical storms – and as such may prove unreliable on occasions.

Overland in the North American continent and Europe, VOR beacons are widely situated to provide an overall coverage of beacons. Usually, these are arranged to coincide with major airway waypoints and intersections in conjunction with DME stations – see below – such that the aircraft may navigate for the entire flight using the extensive route/beacon structure. By virtue of the transmissions within the VHF band, these beacons are subject to the line-of-sight and terrain masking limitations of VHF communications.

4.11.3 Safety/Integrity Aspects

Non-safety critical.

4.11.4 Key Integration Aspects

Cockpit displays, mission computing, navigation system, navigation aids, human factors, and communications.

4.11.5 Key Interfaces

Tuning by FMS where an integrated FMS is fitted:
- Airframe/structure: provision of apertures and suitable fasteners for antennas, optimum location of antennas, drag assessment, and protection of the pressure shell.
- Other systems: exchange of information by direct wiring or data bus.
- Flight deck for controls and displays and human factors assessment.
- Electrical power: connection to appropriate bus bar with circuit protection devices, and most suitable harness runs.
- Installation: mounting tray for electronic LRUs and provision of suitable conditioning and cooling.
- Security of attachment of all components.
- Environment: radio frequency electromagnetic transmissions.
- Consideration of EMC by radiation, induction, and susceptibility.

4.11.6 Key Design Drivers

Regulations and ease of navigation.

4.11.7 Modelling

Avionics integration rig.

4.11.8 References

Moir and Seabridge (2008) Chapter 9, Advisory Circular AC 00-31A Consult appropriate issue and date, JASC Code, ATA-100 Chapter 34 (Pratt 2000).

4.11.9 Sizing Considerations

Control unit and antenna.

4.11.10 Future Considerations

Likely to be superseded by modern navigation systems.

4.12 Automatic Flight Control System

The workload in modern flight decks, together with the precision with which the aircraft needs to be flown has led to the inclusion of systems to provide automatic flight control for the majority of the flight. The capability of these systems has been

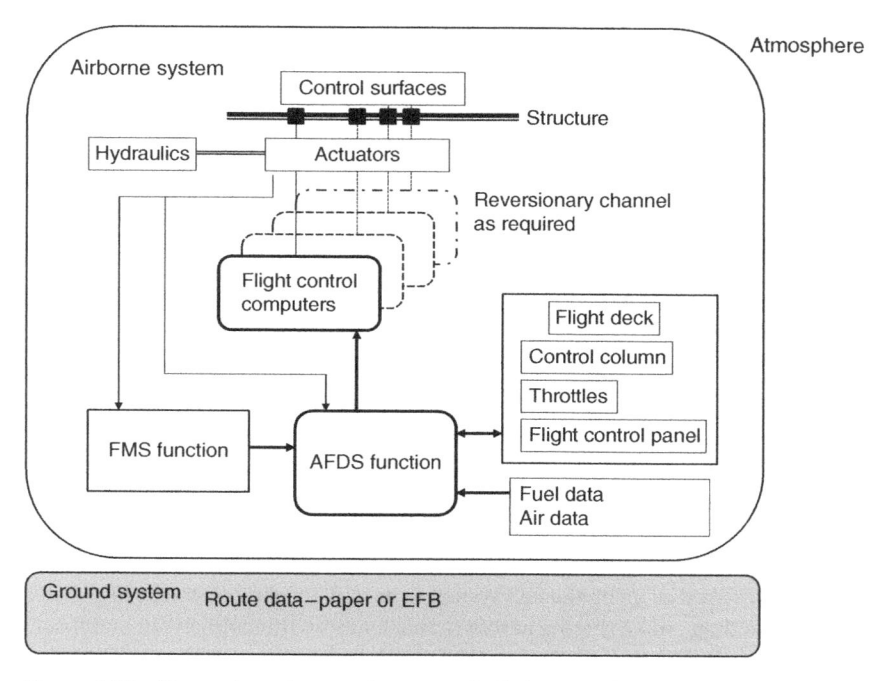

Figure 4.13 Illustration of a generic automatic flight control system.

developed over the years to allow automatic take-off landing to the extent that it is possible for a flight to be conducted entirely automatically (Figure 4.13).

The key systems in automatic flight control are the primary FCS, the FMS, and the autopilot. These systems are integrated to allow control of the attitude of the aircraft, control of the trajectory of the aircraft and control of the flight or the mission. It is important to note that each of these functions has a different integrity. Figure 4.4 above illustrates the interaction of the different systems.

The highly integrated nature of FCSs is a three-level nested control loop in a very simplified form with the following attributes:

- The FCS is an inner loop controlling aircraft attitude using a high integrity flight control (fly-by-wire) system with multiple redundancy (triplex or quadruplex) implementation.
- The AFDS is a secondary loop controlling the aircraft trajectory by means of a dual–dual autopilot system, the autopilot flight director system.
- The FMS is an outer loop using a dual FMS to control the aircraft mission from take-off to arrival at the destination airfield.

In many systems, the AFDS demands changes to engine conditions by moving the throttle levers directly so the pilot had visual feedback of operation. Similarly, demand to the FCS is made my moving the control column. In fly-by-wire systems, the demand may be directly connected to the FCS and the propulsion system, this may lead to a loss of movement feedback, especially with side stick controllers.

4.12.1 Purpose of System

To allow the aircraft to be flown safely throughout the mission with minimum pilot intervention except in emergency or failure conditions.

4.12.2 Description

The FCS comprising the inner loop is concerned with controlling of the attitude of the aircraft. Inputs from the pilot's controls: control column or side-stick, rudders, and throttles determine, via the aircraft dynamics, how the aircraft will respond at various speeds and altitudes throughout the flight envelope. Inertial and air data sensors determine the aircraft response and close the pitch, roll and yaw control loops to ensure that the aircraft possess well harmonised control characteristics throughout the flight regime. In some aircrafts, relaxed stability modes of operation may be invoked by using the fuel system to modify the aircraft centre of gravity, reducing trim drag, and reducing aerodynamic loads on the tailplane or stabiliser. The aircraft pitch, roll and yaw (azimuth) attitude or heading are presented on the primary flight display and navigation display.

The autopilot flight director system (AFDS) performs additional control loop closure to control the aircraft trajectory. The AFDS controls the speed, height and heading at which the aircraft flies. Navigation functions associated with specific operations such as heading hold and heading acquire are also included. Approach and landing guidance is provided by coupling the autopilot to the ILS or MLS approach systems. The control and indication associated with these multiple autopilot modes is provided by a flight mode selector panel (FMSP) or flight control panel (FCP) which enables the selection of the principal modes and also provides information confirming that the modes are correctly engaged and functioning properly.

The final outer loop closure is that undertaken by the FMS that performs the navigation or mission function, ensuring that the fly by wire (FBW) and AFDS systems position aircraft at the correct point in the sky to coincide with the multiple way points which characterise the aircraft route from departure to destination airfield. The pilot interface with the FMS to initiate and monitor the aircraft progress is via a MCDU, also known more loosely as the CDU.

4.12.3 Safety/Integrity Aspects

Not safety critical.

The absence of movement feedback of control column and throttle demands may lead to loss of feedback to the crew.

An instinctive cut-out (ICO) button will be provided amongst the pilot's controls to allow the pilot to rapidly disengage the autopilot in an emergency such as an un-commanded pitch movement or trim runaway.

4.12.4 Key Integration Aspects

Cockpit displays, mission computing, navigation system, navigation aids, human factors, and communications.

For the FCS and propulsion control system, the integrity of the systems must not be compromised in any way.

4.12.5 Key Interfaces

Tuning by FMS where an integrated FMS is fitted:

- Airframe/structure: provision of apertures and suitable fasteners for antennas, optimum location of antennas, drag assessment, and protection of the pressure shell.
- Other systems: exchange of information by direct wiring or data bus. Interface with FCS with suitable integrity.
- Flight deck for controls and displays and human factors assessment.
- Electrical power: connection to appropriate bus bar with circuit protection devices, and most suitable harness runs.
- Installation: mounting tray for electronic LRUs and provision of suitable conditioning and cooling.
- Security of attachment of all components.
- Environment: radio frequency electromagnetic transmissions.
- Consideration of EMC by radiation, induction, and susceptibility.

4.12.6 Key Design Drivers

Regulations, ease of navigation, and confidence in safe automatic operation.

4.12.7 Modelling

Avionics integration rig.

4.12.8 References

McLean (1990), Moir and Seabridge (2013) Chapter 10, Roskam (2007), JASC Code, ATA-100 Chapter 22.

4.12.9 Sizing Considerations

Control unit and antenna.

4.12.10 Future Considerations

GPS augmentation systems will be capable of providing auto-land quality flight guidance once the systems are fully operational. It is intended that the local area augmentation system (LAAS) and wide area augmentation system (WAAS) GPS augmentation systems will be capable of providing Cat II/III and Cat I guidance, respectively.

Consideration must be given to ensuring that the number of auto-pilot modes is not allowed to rise to the extent that the crew become confused. There have been instances where a large number of modes have led to confusion (see Asiana B-777 incident 2013, Seabridge and Moir 2020, Chapter 11).

4.13 Radar Altimeter (Rad Alt)

The Rad Alt uses radar transmissions to reflect off the surface of the sea or the ground immediately below the aircraft. It therefore provides an absolute reading of altitude with regard to the terrain directly beneath the aircraft – absolute distance above terrain. This contrasts with the barometric or air data altimeter where the altitude may be referenced to sea level or some other datum such as the local terrain where it is referred to as height. Rad Alt is therefore of particular value in warning the pilot that he is close to the terrain and to warn him if necessary to take corrective action. Alternatively, the Rad Alt may provide the flight crew with accurate altitude with respect to terrain during the final stages of a precision approach (Figure 4.14).

Rad Alts may operate using continuous wave/frequency modulated (CW/FM) or pulsed signal techniques. CW/FM techniques lose accuracy above a range of 5000 ft so many radio altimeters are limited to an operational range of 5000 ft. Above this pulsed techniques have to be used.

4.13.1 Purpose of System

To provide an absolute reading of height above the ground or sea for display in the cockpit or for use by other systems such as FMS or TAWS. For maritime patrol aircraft operating at very low level over the sea, the Rad Alt is provided with antennas suitably placed to provide height information in high g turn manoeuvres. This is especially valuable in low flying operations over the sea where visual observation is reduced, for example night, fog, mist, and glare.

4.13 Radar Altimeter (Rad Alt) | 161

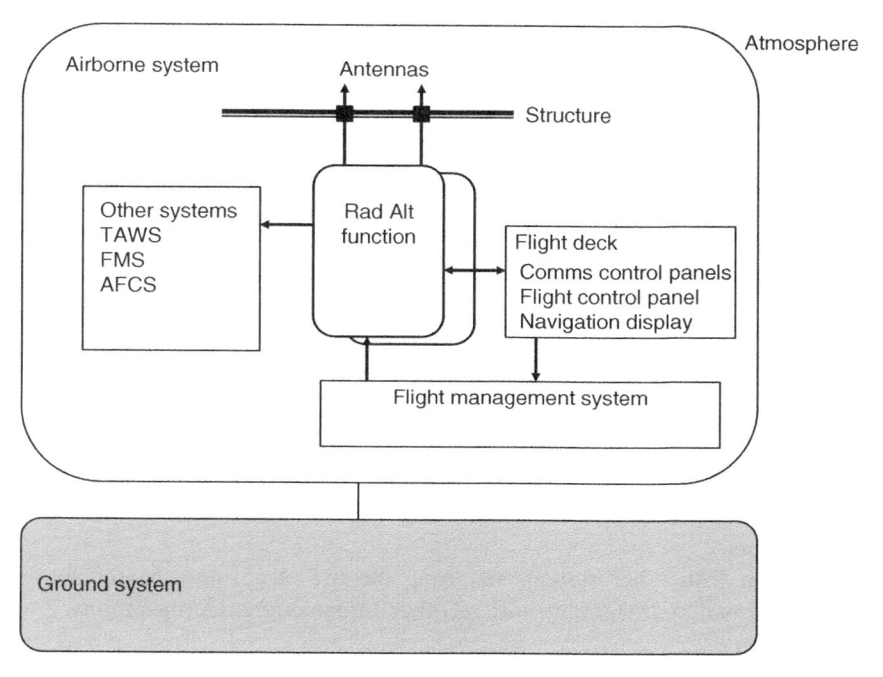

Figure 4.14 Illustration of a generic radar altimeter system.

4.13.2 Description

One or more antennas send a signal to the surface and the system reads the return signal to calculate height above the surface; this is used for display or by other systems.

An oscillator and modulator provide the necessary signals to the transmitter and transmit antennae which direct radar energy towards the terrain beneath aircraft. Reflected energy is received by the receive antenna and passed to a frequency counter. The frequency counter demodulates the received signal and provides a Rad Alt reading to the cockpit displays. In modern systems, radar altitude will be provided to other systems such as FMS, TAWS, and autopilot as well as being displayed directly to the flight crew. Rad Alts usually operate over a maximum range of 0–5000 ft.

Most Rad Alts use a triangular modulated frequency technique on the transmitted energy though this typically limits the operational range to 5000 ft. The transmitter/receiver generates a CW signal varying from 4250 to 4350 MHz modulated at 100 Hz. Comparison of the frequency of the reflected energy with the transmitted energy yields a frequency difference that is proportional to the time taken for the radiated energy to return and hence radar altitude is calculated.

Rad Alt installations are calibrated to allow for the aircraft installation delay which varies from aircraft to aircraft. This allows compensation for the height of the antenna above the landing gear and any lengthy runs of co-axial cable in the aircraft electrical installation. The zero reading of the Rad Alt is set such it coincides with the point at which the aircraft landing gear is just making contact with the runway.

4.13.3 Safety/Integrity Aspects

Safety involved. Low flying, manoeuvrable aircraft will need antennas to be sited on the lower fuselage surface so that one antenna is always directed at the ground or sea surface in high-g turns.

4.13.4 Key Integration Aspects

Cockpit displays, mission computing, navigation system, navigation aids, and human factors.

Integration with other systems requiring the use of absolute height above ground information. Integration with all other antennas and interoperability.

4.13.5 Key Interfaces

- Airframe/structure: provision of apertures and suitable fasteners for antennas, optimum location of antennas, drag assessment, and protection of the pressure shell.
- Other systems: exchange of information by direct wiring or data bus.
- Flight deck for controls and displays and human factors assessment.
- Electrical power: connection to appropriate bus bar with circuit protection devices, and most suitable harness runs.
- Installation: mounting tray for electronic LRUs and provision of suitable conditioning and cooling.
- Security of attachment of all components.
- Environment: radio frequency electromagnetic transmissions.
- Consideration of EMC by radiation, induction, and susceptibility.

4.13.6 Key Design Drivers

Accuracy of height measurement and independence from barometric conditions.

4.13.7 Modelling

Avionics integration rig and antenna placement modelling.

4.13.8 References

Moir and Seabridge (2013) Chapter 8, JASC Code, ATA-100 Chapter 34.

4.13.9 Sizing Considerations

Antennas, display.

4.13.10 Future Considerations

Use of multiple transmitter/receivers for redundancy for safer operations at very low levels. Alternative means of detecting height above ground may be developed.

4.14 Automated Landing Aids

As the importance of maintaining airline schedules in adverse weather conditions has increased, so too has the importance of automatic landing. At its most refined, automatic landing allows the aircraft to land in virtually zero visibility conditions (Figure 4.15). The following categories apply:

- Category I (Cat I): This relates to a decision height (DH) not less than 200 ft and visibility not less than 2600 ft (or not less than 1800 ft (545 m)) runway visual

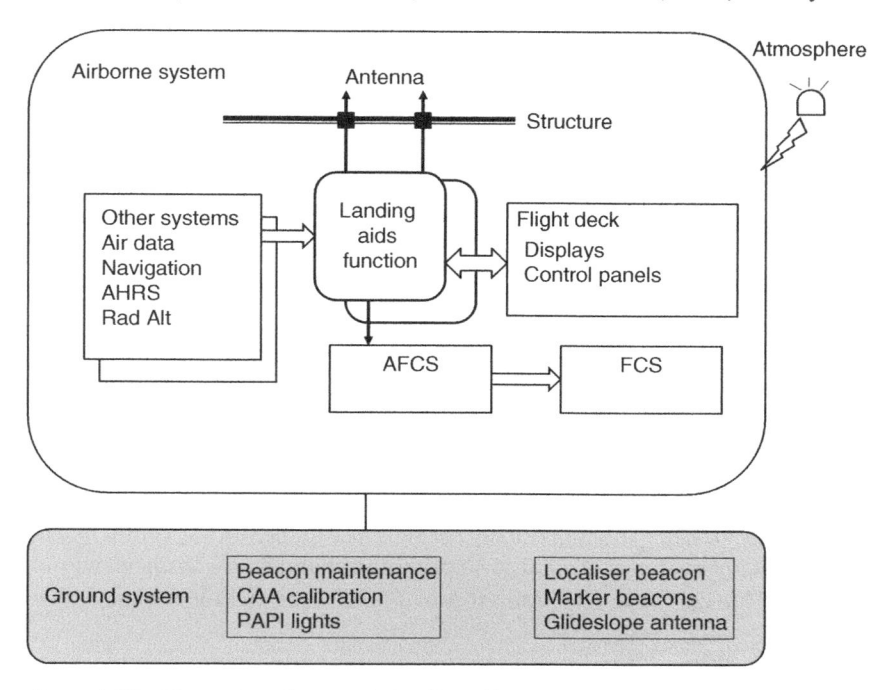

Figure 4.15 Illustration of a generic landing aids system.

range (RVR)) where the airfield/runway is equipped with dedicated measuring devices).
- Category II (Cat II): DH not less than 100 ft and RVR not less than 1200 ft (364 m).
- Category IIIA (Cat IIIA): DH less than 100 ft and RVR not less than 700 ft (212 m). Also described as 'see to land'.
- Category IIIB (Cat IIIB): DH less than 50 ft, RVR not less than 150 ft (45.4 m). Also described as 'see to taxi'.

NOTE. The introduction of a Cat III auto-land capability poses further constraints for the aircraft systems:
- Three independent AC and DC electrical power channels.
- Three independent flight control channels.
- Precision approach guidance: ILS or MLS. The categorisation of the ground-based equipment is equally as important as the aircraft mounted equipment.
- Auto-trim: it is vital in all autopilot applications that the autopilot has to be 'in trim'.

Automated landing systems are available to assist aircraft to perform precision approaches in bad weather or low visibility – typical systems are the following:
- ILS.
- Microwave landing system (MLS).
- Global navigation satellite system (GNSS) aided systems or GPS overlays.

4.14.1 Purpose of System

To provide the capability to make an automated approach and landing under poor visibility conditions using an ILS provided at the airport.

4.14.2 Description

Ground-based antennas providing standard radio frequency beam at an angle and direction that facilitates a safe approach and landing pattern, associated beacons and markers. The airborne system detects the beam and warns of deviations from the beam.

ILS is an approach and landing aid that has been in widespread use since the 1950s. The main elements of ILS include the following:
- Localiser: A localiser antenna usually located close to the runway centre line to provide lateral guidance. A total of 40 operating channels are available within the band 108–112 MHz. The localiser provides left- and right-lobe signals that

are modulated by different frequencies – 90 and 150 Hz such that one signal or the other will dominate when the aircraft is off the runway centre line. The beams are arranged such that the 90 Hz modulated signal will pre-dominate when the aircraft is to the left while the 150 Hz signal will be strongest to the right. The difference in the depth of signal modulation is used to drive a cross-pointer deviation needle such that the pilot is instructed to 'fly right' when the 90 Hz signal is strongest and 'fly left' when the 150 Hz signal dominates. When the aircraft is on the centre line, the cross-pointer deviation needle is positioned in the central position. This deviation signal is proportional to azimuth out to $\pm 10°$ of the centre line.

- Glideslope: A glide slope antenna located beside the runway threshold to provide lateral guidance. Forty operating channels are available within the frequency band 329–335 MHz. As for the localiser, two beams are located such that the null position is aligned with the desired glide slope, usually set at a nominal 3° (usually ~3°). In the case of the glide slope, the 150 Hz modulated signal pre-dominates below the glide slope and the 90 Hz signal is stronger above. When the signals are balanced, the aircraft is correctly positioned on the glide slope and the glide slope deviation needle is positioned in a central position. As for the localiser needle, the pilot is provided with 'fly up' or fly down' guidance to help him to acquire and maintain the glide slope. On older aircraft, this would be shown on a dedicated deflection display, on modern aircraft with digital cockpits this information is displayed on the primary flight display (PFD). More aggressive ILS approaches have recently been developed with steeper glide slopes such as at London City airport.

The ILS localiser, glide slope, and DME channels are connected such that only the localiser channel needs to be tuned for all three channels to be correctly aligned.

- Marker beacons are located at various points down the approach path to give the pilot information as to what stage on the approach has been reached. These are the outer, middle, and inner markers. Location of the marker beacons are the following:
  - Outer marker approximately 4–7 nautical miles from the runway thresh-hold
  - Middle marker ~3000 ft from touchdown.
  - Inner marker ~1000 ft from touchdown.
  - The high approach speeds of most modern aircrafts render the inner marker almost superfluous, and it is seldom installed.
- The marker beacons are all fan beams radiating on 75 MHz and provide different Morse code modulation tones which can be heard through the pilot's headset and may also cause visual cues on the aircraft direction indicator (ADI).

The beam pattern is ±40° along track and ±85° across track. The overall audio effect of the marker beacons is to convey an increasing sense of urgency to the pilot as the aircraft nears the runway threshold.

A significant disadvantage of the ILS system is its susceptibility to beam distortion and multi-path effects. This distortion can be caused by local terrain effects, large man-made structures, or even taxiing aircraft can cause unacceptable beam distortion, with the glide slope being the most sensitive. At times on busy airfields and during periods of limited visibility, this may preclude the movement of aircraft in sensitive areas that in turn can lead to a reduction in airfield capacity. More recently, interference by high-power local FM radio stations has presented an additional problem, although this has been overcome by including improved discrimination circuits in the aircraft ILS receiver.

Standard ILS glide slope approaches have been based upon a nominal 3.0° angle of approach. Recent developments have led to the approval of steeper and more integrated approaches.

MLS is an approach aid that was conceived to redress some of the shortcomings of ILS. The specification of a time-reference scanning beam MLS was developed through the late 1970s/early 1980s, and a transition to MLS was envisaged to begin in 1998. However, with the emergence of satellite systems such as GPS there was also a realisation that both ILS and MLS could be rendered obsolete when such systems reach maturity. In the event, the US civil community is embarking upon higher-accuracy developments of the basic GPS system: WAAS and LAAS. In Europe, the United Kingdom, the Netherlands, and Denmark have embarked upon a modest programme of MLS installations at major airports.

MLS operates in the frequency band of 5031.0–5190.7 MHz and offers some 200 channels of operation. It has wider field of view than ILS, covering ±40° in azimuth and up to 20° in elevation – with 15° useful range coverage. Coverage is out to 20 nautical miles for a normal approach and up to 7 nautical miles for back azimuth/go-around. The co-location of a DME beacon permits 3D positioning with regard to the runway and the combination of higher data rates mean that curved arc approaches may be made, as opposed to the straightforward linear approach offered by ILS. This offers advantages when operating into airfields with confined approach geometry and tactical approaches favoured by the military. For safe operation during go-around, pDME is usually used for a more precise back azimuth signal.

A ground-based MLS installation comprises azimuth and elevation ground stations each of which transmit angle and data functions that are frequency shift key (FSK) modulated and which are scanned within the volume of coverage already described. The MLS scanning function is characterised by narrow beam widths of around 1–2° scanning at high slew rates. Scanning rates are extremely high at

20 000°/s which provides data rates which are around ten times greater than is necessary to control the aircraft. These high data rates are very useful in being able to reject spurious and unwanted effects due to multiple reflections, etc.

GNSSs offer the prospect of precision approaches; however, GPS along with other GNSS systems are not permitted to be used as sole source sensors: they have to be augmented by more conventional ground based techniques. In such a fashion, GPS overlay procedures may be flown which overlay the precision that GPS offers with integrity being provided by other sensors. For example, a GPS overlay procedure may be executed together with a dual DME procedure. These procedures may incorporate differential GPS techniques using either a space based augmentation system (SBAS) using space-based assets or a ground-based augmentation system (GBAS) using local assets.

4.14.3 Safety/Integrity Aspects

Not safety critical: The automated system is supplemented by a ground-based system – precision approach path indicator (PAPI) lights can be used as a reversionary aid.

4.14.4 Key Integration Aspects

Integration with FMS, auto-pilot or flight director, and ground-based landing system.

4.14.5 Key Interfaces

- Airframe/structure: provision of apertures and suitable fasteners for antennas, optimum location of antennas, drag assessment, protection of the pressure shell.
- Other systems: exchange of information by direct wiring or data bus. FMS, FCS, and location of aircraft antennas. Civil Aviation Authority (CAA) calibration flights are required and will be managed by the airport authority. Ground-based beacons to be maintained on a regular basis.
- Flight deck for controls and displays and human factors assessment.
- Electrical power: connection to appropriate bus bar with circuit protection devices, and most suitable harness runs.
- Installation: mounting tray for electronic LRUs and provision of suitable conditioning and cooling.
- Security of attachment of all components.
- Environment: radio frequency electromagnetic transmissions.
- Consideration of EMC by radiation, induction, and susceptibility.

4.14.6 Key Design Drivers

Safety and category of approach involving decision height and visibility.

4.14.7 Modelling

Avionics integration rig.

4.14.8 References

Moir and Seabridge (2013) Chapter 9, JASC Code, ATA-100 Chapter 34.

4.14.9 Sizing Considerations

Type of landing aid and antennas.

4.14.10 Future Considerations

GPS augmentation systems will be capable of providing auto-land quality flight guidance once the systems are fully operational. It is intended that the LAAS and WAAS GPS augmentation systems will be capable of providing Cat II/III and Cat I guidance, respectively.

4.15 Air Data System (ADS)

The aircraft, whether on the ground or in flight, is in an atmosphere that is in a state of constant change. There are variations in temperature, pressure, wind speed, and precipitation that affect the aircraft on the ground; in flight these parameters are joined by attitude in six degrees of freedom, rates of change of attitude, and changes in direction. The atmosphere and the effects of temperature and pressure are well understood, the changes and rates of change of altitude and attitude vary from flight to flight (Figure 4.16).

The air data system employs various sensors to measure these known parameters throughout the flight envelope. A computing function processes the data to provide information to the crew to display what the aircraft is doing, and to other systems so that they can modify what the aircraft is doing. The whole process is to ensure that the aircraft is always in a controlled state and is safe.

4.15.1 Purpose of System

To provide information to aircraft systems on air pressures – total pressure and static pressure, and to convert these pressures into signals representing airspeed, altitude, and Mach number.

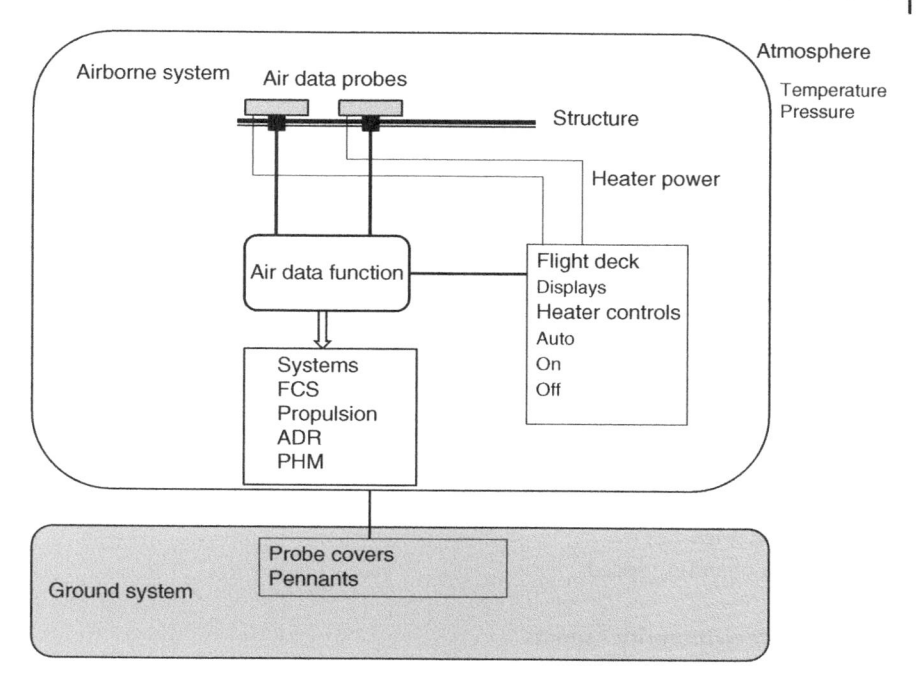

Figure 4.16 Illustration of a generic air data system.

4.15.2 Description

The air data system involves the sensing of the medium through which the aircraft is flying. Typical sensed parameters are dynamic pressure, static pressure, rate of change of pressure, and temperature. Derived data includes the following:

- Barometric altitude (ALT)
- Indicated airspeed (IAS)
- Vertical speed (VS)
- Mach (M)
- Static air temperature (SAT)
- Total air temperature (TAT)
- True air speed (TAS).

The simplest system provides ALT and IAS as a minimum, modern jet aircraft require Mach, VS, maximum operating speed (V_{mo}), maximum operating Mach (M_{mo}), SAT, TAT, and TAS to satisfy the aircraft requirements.

The system has one or more ADCs, which centrally measure air data and provide corrected data to the recipient subsystems. This has the advantage that while the pilot still has the necessary air data presented to him whilst more accurate and more relevant forms of the data are available to other aircraft systems. Typical parameters provided by an ADC are the following:

- Barometric correction.
- Barometric corrected altitude.
- Altitude rate.
- Pressure altitude.
- Computed airspeed.
- Mach number.
- TAS.
- SAT.
- TAT.
- Impact pressure.
- Total pressure.
- Static pressure.
- Indicated angle of attack.
- Over-speed warnings.
- Maximum operating speeds.

4.15.3 Safety/Integrity Aspects

Safety critical: used by FCS, propulsion system, navigation, and cockpit displays.

Historical note: Ice has been a potentially serious hazard in the air data system since the early days of flying. In some heritage aircraft, the probes were connected by small bore pneumatic pipes to an ADC which contained the capsules necessary to measure the raw air data parameters and the computing means to calculate the necessary corrections. The advent of digital computing and digital data buses such as ARINC 429 meant that computation of the various air data parameters could be accomplished in air data modules closer to the pitot-static sensing points. Widespread use of the ARINC 429 data buses enabled this data to be rapidly disseminated throughout all the necessary aircraft systems. The emergence of 'Smart Probes' means that some of the ADM features are contained in the probe housing, thereby eliminating the need for small bore tubing in the aircraft.

The small bore pneumatic-sensing tubes which route the sensed pitot or static pressure throughout the aircraft poses significant engineering and maintenance penalties. As the air data system is critical to the safe operation of the aircraft, it is typical for three or four or more alternate systems to be provided. The narrow bore of the sensing tubes necessitates the positioning of water drain traps at low points in the system where condensation may be drained off – avoiding the blockage of the lines due to moisture accumulation. Finally, following the replacement of an instrument or disturbance of any section of the tubing; pitot-static leak and independent checks are mandated to ensure that the sensing lines were intact and leak-free and that no corresponding instrumentation sensing errors are likely to occur.

The air data probes contain heaters to prevent icing. The heaters will be set to a low-power level for ground operations and automatically set to a higher-power level when the aircraft takes off using a weight on wheels signal.

Covers are provided to protect the probes on the ground. The covers must be fitted with banners to enable the ground crew to identify them and to remove them before flight, or before engine running.

4.15.4 Key Integration Aspects

Integrated with navigation system, guidance and control, and sole source of critical air data.

4.15.5 Key Interfaces

- Other systems: exchange of information by direct wiring or data bus, probe heating.
- Flight deck for controls and displays and human factors assessment.
- Electrical power: connection to appropriate bus bar with circuit protection devices, and most suitable harness runs.
- Mounting tray for electronic LRUs and provision of suitable conditioning and cooling.
- Security of attachment of all components.
- Consideration of EMC by radiation, induction, and susceptibility.
- Ground systems for provision of covers and banners and removal and stowage before flight.

4.15.6 Key Design Drivers

Air data accuracy: Robustness for FCS interface.

4.15.7 Modelling

Location of probes on fuselage using CAD and digital mock-up, confirmation in wind tunnel.

4.15.8 References

Moir and Seabridge (2008), (2013) Chapter 8, JASC Code, ATA-100 Chapter 22.

4.15.9 Sizing Considerations

Probes and electrical loads for heaters.

4.15.10 Future Considerations

Subject to changes to statutory requirements – changes to separation minima and to requirements for parameters for in-flight monitoring for TCAS and TAWS must be constantly monitored. This will demand higher accuracy of height sensing and methods of maintaining height within strict limits to meet the statutory demands for reduced vertical separation minima (RVSM).

Developments in technology may lead to alternative methods of obtaining and calculating air data parameters. 'Smart' probes will continue to develop.

4.16 Accident Data Recording System (ADRS)

Early accident data recorders (ADR) were connected to certain components in the aircraft systems that were believed to be most relevant to an investigation following an accident. This was supplemented by recording of the microphones and headset on the flight deck intercom to record internal and external communications. The recording was on a magnetic wire which was overwritten at intervals. Current designs will make use of the data bus network to obtain relevant information and storage will be on hardened solid-state memory (Figure 4.17).

4.16.1 Purpose of System

To continuously record specified aircraft parameters together with recording of speech and video for use in analysis of serious incidents and to support a board of enquiry.

4.16.2 Description

Data is acquired by suitable interfaces to relevant systems, and this data is continuous recorded on magnetic wire or solid-state bulk memory store. The recorder will be fire and water-resistant and will be provided with locator beacon to aid recovery. There may be a capability to eject the beacon from the aircraft to assist recovery. The data is analysed after recovery of the recorder and will be used to support any subsequent enquiry.

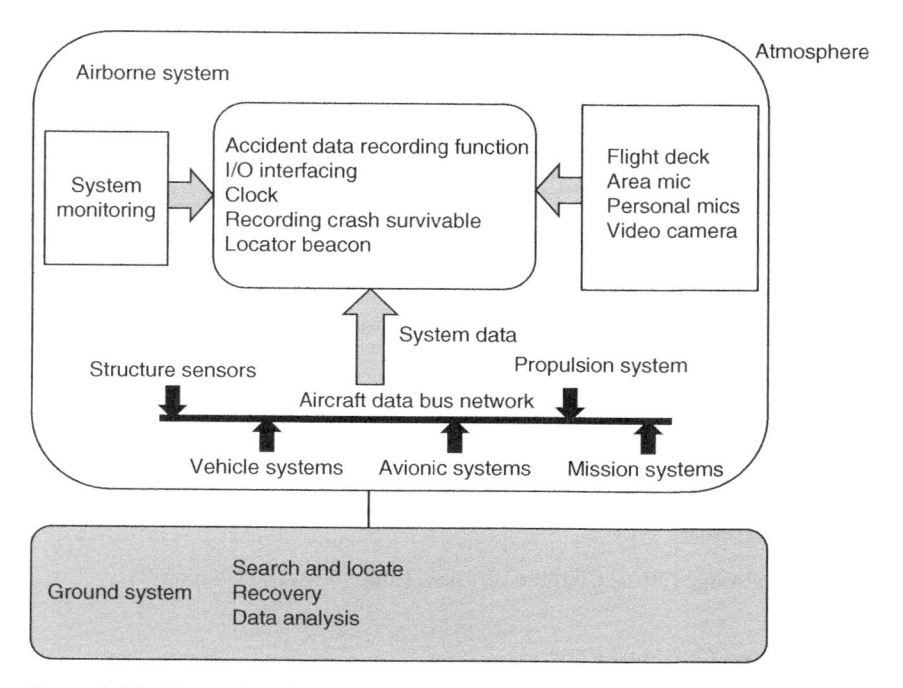

Figure 4.17　Illustration of a generic accident data recording system.

4.16.3　Safety/Integrity Aspects

Dispatch critical: It is essential that any connection to the aircraft systems' components and to the data bus network cannot cause any degradation to the performance of the systems.

4.16.4　Key Integration Aspects

Integration with relevant data bus types as a listen only device.

4.16.5　Key Interfaces

- Adherence to project and international standards (ICAO) for relevant mandated parameters, information identification, and format.
- Airframe/structure: provision of apertures and suitable fasteners for recorder and mechanism for automatic ejection in the event of an accident.
- Other systems: exchange of information by direct wiring or data bus.
- Electrical power: connection to appropriate bus bar with circuit protection devices, and most suitable harness runs.

- Installation: mounting tray for electronic LRUs and provision of suitable conditioning and cooling. Mechanism for ejecting the recorder on impact.
- Security of attachment of all components.
- Consideration of EMC by radiation, induction, and susceptibility.
- Ground systems for removal for analysis or for recovery after serious accident.

4.16.6 Key Design Drivers

Regulations, crash survivable memory – impact, immersion, and fire. Incorporation of suitable devices to assist in location, e.g. beacon.

4.16.7 Modelling

Avionics integration rig.

4.16.8 References

Moir and Seabridge (2013) Chapter 10, JASC Code, ATA-100 Chapter 31.

4.16.9 Sizing Considerations

Recording unit and special sensors.

4.16.10 Future Considerations

The current implementation is effective, although there have been incidents where the recorder (and the host aircraft) has been totally lost or where a long time has elapsed until the recorder has been found. Continuous transmission by data link (similar to ACARS) may be useful, but it will result in an enormous recording task.

4.17 Electronic Flight Bag (EFB)

The electronic flight bag (EFB) is a virtual aircraft system in the sense that it can be used to replace the crew bag with notes, manuals, and calculations. The EFB saves weight, provides speedy access to information, and reduces the cost of information distribution. It is carried on board under certain rules to minimise any incorrect actions. The ready availability of portable data solutions offers an opportunity for the flight crew to dispense with their heavy flight bags and also to gain ready access to databases of information previously only available in manuals, paper maps, files of leaflets, and NOTAMs and aircraft status reports. To keep up with this

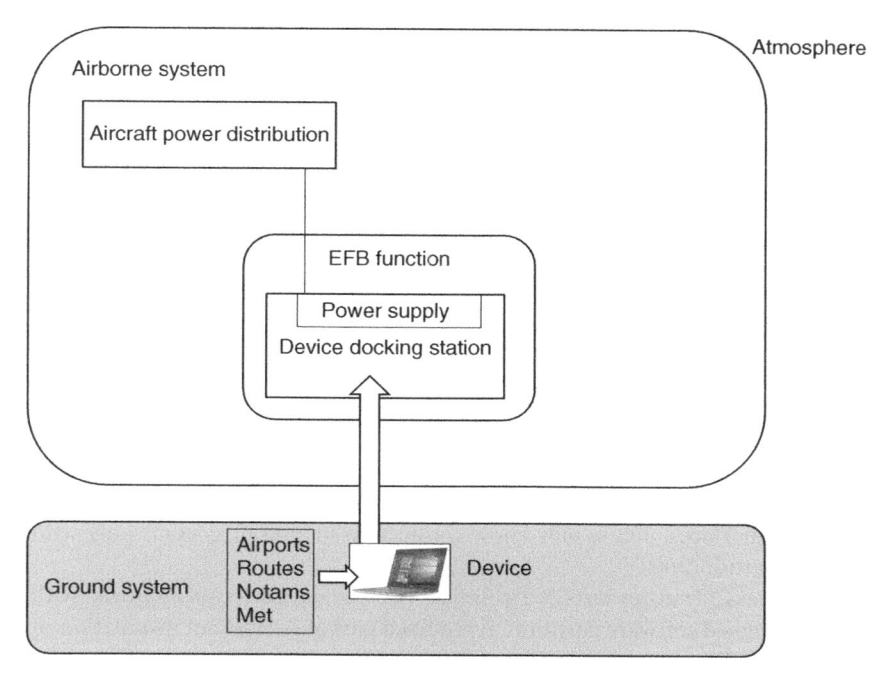

Figure 4.18 Illustration of a generic EFB.

rapidly changing technology, the FAA released Advisory Circular 120-76B 'Guidelines for the certification, airworthiness, and operational use of EFBs in June 2012' (Figure 4.18).

EFBs are intended primarily for flight deck use and include the hardware and software necessary to support their intended function. EFB devices can display a variety of aviation data and perform basic calculations such as performance, weight and balance, fuel calculations. A list of EFB functions, providing display of paper replacement, planning, advisory-use, and various other functions, can be found in the Type A and Type B software applications appendices of AC 120-76B.

Type A applications are those paper replacement applications primarily intended for use during flight planning, on the ground, or during noncritical phases of flight. Type B applications are those paper replacement applications that provide the aeronautical information required for each flight at the pilot station and are primarily intended for use during flight planning and all phases of flight. In the past, many of these functions were accomplished using paper references or were based on data provided to the flight crew by an airline's flight dispatch function. Most importantly, by definition, an EFB must be able to host Type A and/or Type B software applications.

4.17.1 Purpose of System

To give the aircrew access to flight-planning information by means of portable data solutions carried onto the aircraft and powered by the aircraft power supply.

4.17.2 Description

The definitions of an EFB in terms of hardware implementation encompass the following general attributes:

- Class 1: A portable Class 1 EFB, such as a laptop, which generally only has an interface to airplane power and must be stowed below 10 000 ft. This represents the lowest level of capability.
- Class 2: The hardware does not allow direct connection to airplane systems. It must have the ability to be removed by the crew without tools and without leaving the flight deck – typically modern solutions offer a docking solution. As before: this is electronic portable device and not a fixed installation. A separate airplane interface module may allow connection to aircraft systems only with suitable partitioning.
- Class 3: Class 3 features installed avionics covered by type design approval, with a type-designed software partition. It is a fixed part of the aircraft installation. It may have a Class 2 partition of functions but depending upon the implementation may not necessarily require high-integrity software.

Typical functions provided by an EFB may include the following:

- Airport moving map: An airport moving map may provide high-resolution airport diagrams with the aircraft position super-imposed. Airport databases with GPS system-level tracking accuracy depict the airport ground environment and layout with a high degree of fidelity and visual detail.
- Electronic charts: Electronic charts offer clear concise airport charts and procedures such as standard instrument departure routes (SIDs), en-route procedures, standard terminal arrival routes (STARs), as have already been outlined in the FMS description. World-wide coverage may be customised to form a dedicated subset for individual airlines/operators or aircraft type.
- Electronic documents: Electronic documents offer the potential to call up a wide range of aircraft documentation whether it be operational, procedural, or technical in nature.
- On-board performance: Databases with look-up tables and performance variations or limitations may be provided for a given aircraft status or configuration that enable any vital performance issues to be noted and progressed.
- Electronic logbook: The electronic logbook works in association with on-board technical fault logs – often down to sub-module level, system performance, and operational logs recording aircraft performance.

- Electronic flight folder: The concept of an electronic flight folder enables the concept of digitising all of the specific briefing information relative to a specific flight: NOTAMs, weather, or other key information pertaining to that particular flight.
- Video surveillance: The ability of the flight crew to call up and display video sensor data from cameras located around the aircraft either to show external or internal video data.

4.17.3 Safety/Integrity Aspects

Dispatch critical.

Rigid control is to be provided over all carry-on devices to ensure that no malware or viruses can enter the aircraft systems. Any calculations performed that influence decisions affecting flight safety must be validated to ensure that algorithms are correct and that no errors are tolerated. A robust design and validation process is required for any EFB solution.

4.17.4 Key Integration Aspects

Integration with ground systems procedures for downloading of correct relevant information and subsequent analysis. Adherence to international standards to identification of recorded data.

4.17.5 Key Interfaces

- Other systems: no connection to EFB is permitted.
- Flight deck for controls and displays and human factors assessment.
- Electrical power: connection to appropriate bus bar with circuit protection devices.
- Mounting tray/table for laptop with secure attachment.
- Security of attachment of all components.
- Ground systems for provision flight data directly to aircrew.

4.17.6 Key Design Drivers

Robustness of the solution and restriction of access to aircraft systems.

4.17.7 Modelling

Laboratory modelling of data loading and display.

4.17.8 References

Fitzsimmons (2002), FAA (2017), Moir and Seabridge (2013), Advisory Circular AC 129-76B Chapter 11.

4.17.9 Sizing Considerations

Size of system components and safe stowage on the fight deck. Memory capacity to meet the requirements of data required.

4.17.10 Future Considerations

Integration of EFB functions safely into flight deck displays.

4.18 Prognostics and Health Management System (PHM)

Early implementations of maintenance data collection provided a log of components and systems which had failed in flight. This was often in the form of a maintenance data panel with a fixed set of lights that illuminated on detection of a failure; post-flight examination and maintenance manuals guided maintenance crews to the potential source of the failure. Technology enabled this function to be elaborated so that the indication was in the form of key words and equipment numbers which reduced the time required to locate a faulty component. The emergence of on-board digital computing and data bus provided access to more data and algorithms in the systems allowed a more detailed analysis of system performance degradation. With this data maintenance, personnel were able to make decisions on which components to replace immediately and which could be delayed until a more convenient location was reached. This led to savings in time and cost of maintenance. Information could be recorded on-board or could be made available to ground stations by data link (Figure 4.19).

4.18.1 Purpose of System

To provide a continuous record of systems performance and failures during flight. To analyse this information either in flight or on the ground and to use the results to determine trends and declining system health and aid decision-making about immediate repair or deferred repair.

4.18 Prognostics and Health Management System (PHM)

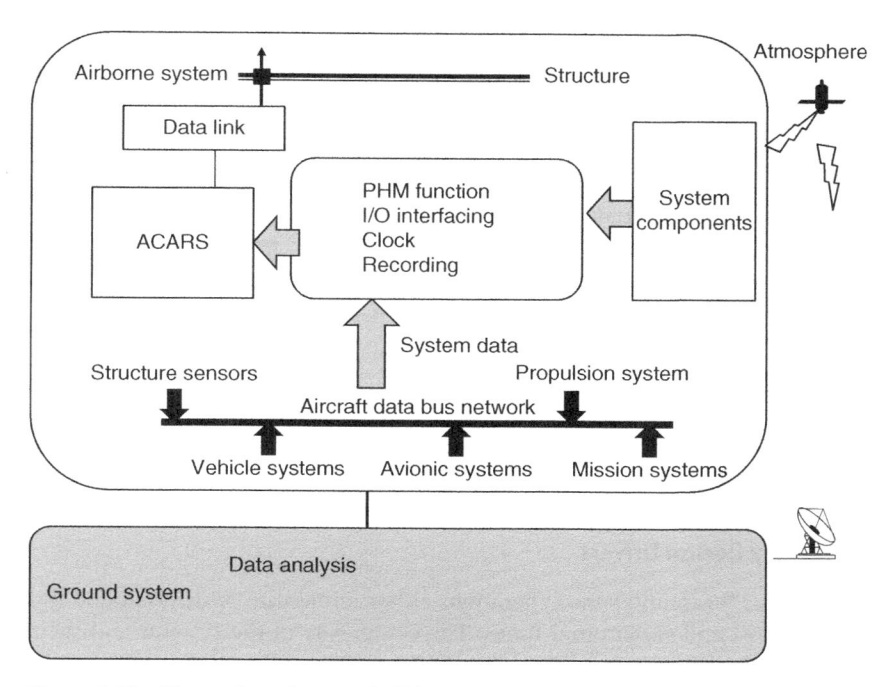

Figure 4.19 Illustration of a generic PHM system.

4.18.2 Description

The prognostics and health management (PHM) function is connected to data buses and system line replaceable items (LRIs) to extract information on the status of systems and components and to perform appropriate algorithms and output results to data storage or for transmission to the ground.

4.18.3 Safety/Integrity Aspects

Non-safety critical may provide information which is dispatch critical.

4.18.4 Key Integration Aspects

All systems and ground aspect of maintenance management. Integration with data link for transmission of data to ground.

4.18.5 Key Interfaces

- Adherence to project and international standards (ICAO) for relevant mandated parameters, information identification, and format.
- Other systems: exchange of information by direct wiring or data bus. Compatibility with all types of data bus in use on the aircraft. Connection to data link for data down load to ground stations (ACARS).
- Electrical power: connection to appropriate bus bar with circuit protection devices, and most suitable harness runs.
- Installation: mounting tray for electronic LRUs and provision of suitable conditioning and cooling.
- Security of attachment of all components.
- Consideration of EMC by radiation, induction, and susceptibility.
- Ground systems for removal of data storage or receipt of data link information for analysis or for historical records or for maintenance action.

4.18.6 Key Design Drivers

Requirements for maintenance operations either immediate or deferred and the impact on cost and turnaround times. The complexity of the systems and their components will determine the amount of data be acquired and analysed in real time.

4.18.7 Modelling

Logistics modelling, avionics integration rig, modelling of support scenario, and cost–benefit analysis.

4.18.8 References

Johnson et al. (2011), Moir and Seabridge (2013) Chapter 7, Srivastava (2010), JASC Code, TA-100 Chapter 45.

4.18.9 Sizing Considerations

Recording unit and data download requirements.

4.18.10 Future Considerations

The use of knowledge-based systems and AI will lead to more 'intelligent' analysis of recorded data to locate faults.

4.19 Internal Lighting

The flight deck or cockpit is provided with lighting for the crew to read documents and to allow all instruments to be visible under all conditions of external light conditions without causing strain to the eyes. The lighting must be 'balanced' so that all instruments, panels, and screens appear to be evenly illuminated with no accentuated areas. In some cases, 'wander' lights may be provided to allow detailed local lighting. All lighting solutions must preserve night vision and must be useable to military operators using NVG (Figure 4.20).

4.19.1 Purpose of System

To provide a balanced illumination of cockpit or flight deck panels to aid flight in poor or bright ambient lighting conditions and at night.

4.19.2 Description

Modern cockpits with liquid crystal displays and full colour can be made to be adjustable to ambient lighting conditions to suit the individual pilot. The design

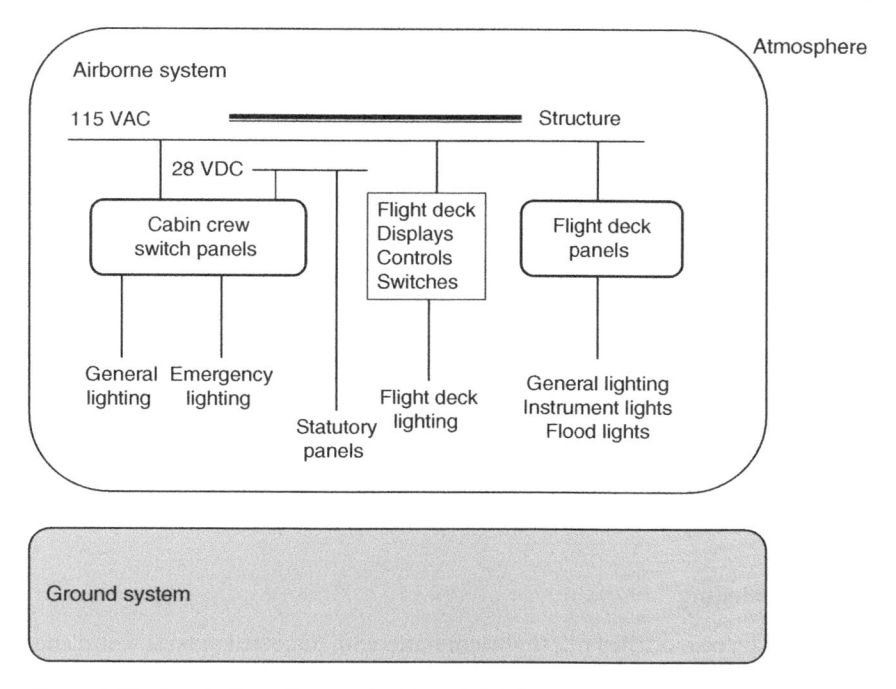

Figure 4.20 Illustration of a generic internal lighting system.

of the glass must be such that it is non-reflective. Individual switch panels and control panels on the side consoles and overhead must be designed to be integrally lit with the correct contrast and dimmable. It may be necessary to include flood lighting, especially red floods for night use, and wander lights for situations where it may be necessary to search for items in the cockpit. General overhead lighting will also be provided.

4.19.3 Safety/Integrity Aspects

Emergency lighting is required to power floodlights and wander lights in the event of total loss of panel lighting.

4.19.4 Key Integration Aspects

The lighting solution must be integrated into cockpit design and lighting control system in accordance with the cockpit lighting philosophy and human factors design. There will be a need for the overall solution to be to be compatible with NVG for military use.

4.19.5 Key Interfaces

- Airframe/structure: incorporation of lighting into cabin furnishing.
- Other systems: exchange of information by direct wiring or data bus.
- Flight deck and passenger cabin for controls and displays and human factors assessment.
- Electrical power: connection to appropriate bus bar with circuit protection devices, and most suitable harness runs.
- Security of attachment of all components.
- Ground systems for inspection and replacement of failed lights.

4.19.6 Key Design Drivers

Human factors, cockpit/flight deck lighting philosophy.

4.19.7 Modelling

The system can be modelled in a flight simulator with projected outside world and realistic lighting or a mock-up in lighting test facility with a facility for altitude lighting simulation.

4.19.8 References

JASC Code, TA-100 Chapter 33.

4.19.9 Sizing Considerations

Electrical loads.

4.19.10 Future Considerations

Long-life lighting built into the cabin furnishings and structure.

4.20 Integrated Modular Architecture (IMA)

Figure 4.21 provides a generic illustration of the integrated modular architecture (IMA) used to provide a computing and management framework of many modern airliners. IMA principles promote a general-purpose centralised computing resource comprising a set of common hardware computing modules. The avionics

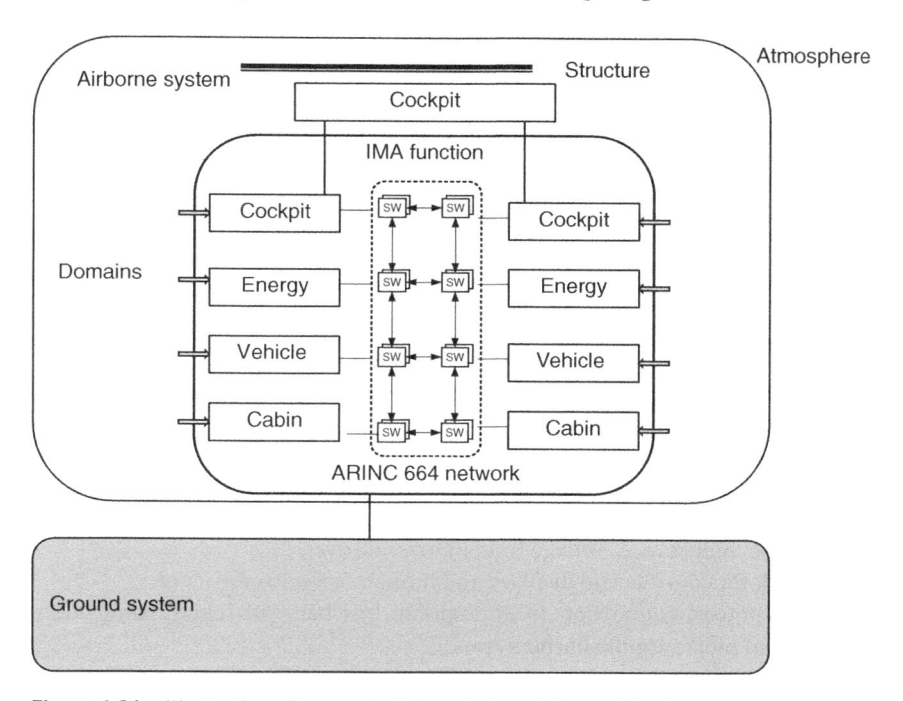

Figure 4.21 Illustration of a generic integrated modular architecture.

application software previously embedded in proprietary task-oriented computers in a federated architecture is now hosted on the general-purpose processors in the common core computing resource. Application software is provided and certificated by the subsystem developer, independently of the IMA hardware platform. Data is exchanged between applications via a dual redundant network that is based on commercial off the shelf (COTS) Ethernet technology with enhancements to provide real-time features needed for safety-critical avionics applications. Originally developed as aviation full duplex (AFDX), this network is known as ARINC 664 part 7. It provides full duplex, bidirectional communication between network resources at 100 Mbps.

4.20.1 Purpose of System

To provide a computing framework for the avionic systems, often divided into domains for cabin systems, energy management, and vehicle systems.

4.20.2 Description

Computing system with remote data concentrators to provide interfacing and control of systems based on a data bus network, commonly ARINC 664.

4.20.3 Safety/Integrity Aspects

Segregation and redundancy as required to maintain integrity of redundant systems.

4.20.4 Key Integration Aspects

Performs as integrating medium and control for all avionics and utility systems.

4.20.5 Key Interfaces

- Other systems: exchange of information by direct wiring or data bus. Interfaces to vehicle systems components such as sensors and effectors of very different interface characteristics such as impedance, rate of change, slewing rate, capacitance, inductance, voltage levels, current drives.
- Flight deck for controls and displays and human factors assessment.
- Electrical power: connection to appropriate bus bar with circuit protection devices, and most suitable harness runs.
- Installation: mounting trays or equipment cabinet for IMA modules and provision of suitable conditioning and cooling.

- Security of attachment of all components.
- Consideration of EMC by radiation, induction, and susceptibility.
- Ground systems for download of data, provision of services on turn around.

4.20.6 Key Design Drivers

Safety and control.

4.20.7 Modelling

Modelling of individual systems and integrated model on test bench.

4.20.8 References

Moir and Seabridge (2013) Chapter 5, Srivastava et al. (2010), JASC Code, ATA-100 Chapter 42.

4.20.9 Sizing Considerations

Number of interfaces, processing requirement, and throughput.

4.20.10 Future Considerations

IMA has only fairly recently been adopted and the concept will continue to develop using high-speed processing.

4.21 Antennas

This paragraph will take the slightly unusual step of looking at the total set of antennas as a system. The reason for this is that although each antenna is carefully matched to its own system's needs in terms of impedance, frequency, power, and range, when it is installed on the fuselage exterior it is operating in a dynamic and complex environment. It will be placed with other antennas each transmitting or receiving independently with little or no relationship to other antennas. An indication of the frequency range used by the avionic systems is shown in the simplified spectrum in Figure 4.22 (Moir and Seabridge 2013).

Figure 4.23 shows an example of the antennas required by the systems described in this chapter for a long-haul commercial airliner operating today, particularly when operating on trans-oceanic routes. The location shown are indicative only,

186 | *4 Avionic Systems*

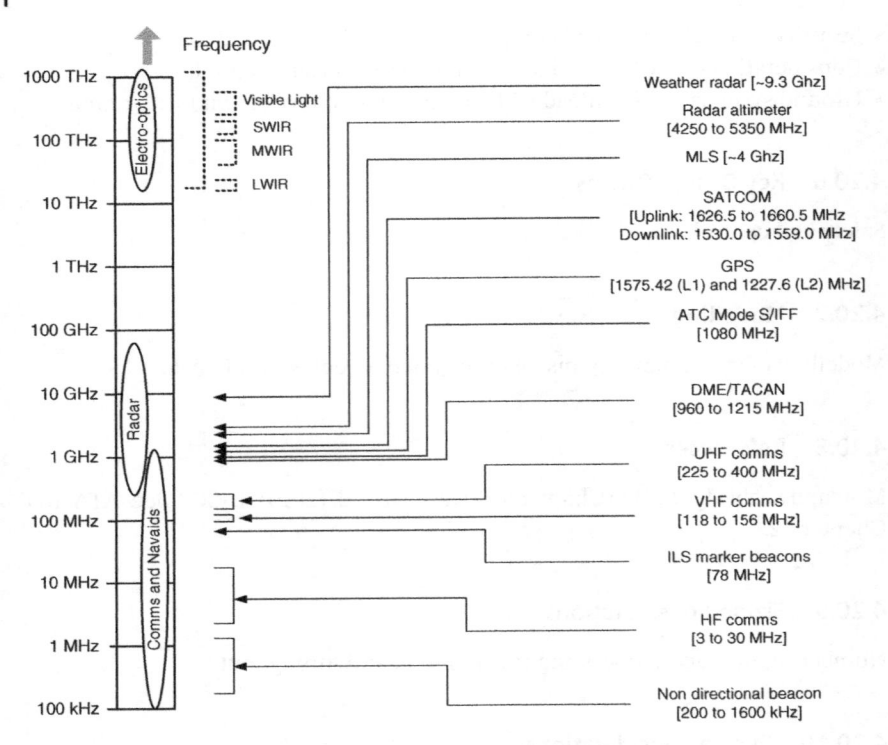

Figure 4.22 Simplified radio frequency spectrum – civil use.

the exact location will depend on the size of the airframe and the nature of the transmissions from each antenna with relationship to its neighbours.

The number of antennas required on-board an aircraft to handle all the sensors, communications and navigation aids is considerable. This is compounded by the fact that the many of the key equipment may be replicated in duplicated or triplicated form. This is especially true of VHF, HF, VOR, and DME equipment. Due to their operating characteristics and transmission properties, many of these antennae have their own installation criteria. SATCOM antennae that communicate with satellites will have the antennae mounted on the top of the aircraft so as to have the best coverage of the sky. ILS antennae associated with the approach and landing phase will be located on the forward, lower side of the fuselage. Others may require continuous coverage while the aircraft is manoeuvring and may have antennae located on both upper and lower parts of the aircraft; multiple installations are commonplace.

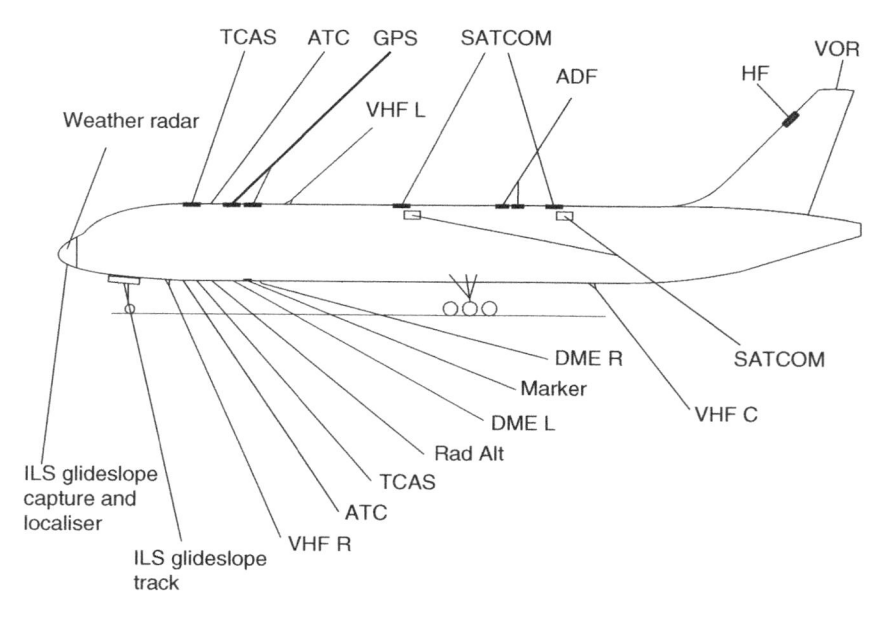

Figure 4.23 Example antenna configuration.

Early aircraft often suffered from mutual interference, especially with systems that operate on adjacent parts of the spectrum. Some aircraft types included a system to ensure effective communications by suspending transmission or reception, known as blanking and suppression.

4.21.1 Purpose of System

To provide a means of transmission and reception of radio frequency signals for all relevant systems on the aircraft and to ensure interoperability of the systems throughout the flight with no mutual interference, despite weather, or meteorological conditions. The system also provides ready access to internationally accepted emergency channels for transmission or for continuous monitoring.

4.21.2 Description

Many of the systems described above have a need to transmit information to other aircraft or ground systems, and also to receive information from similar sources. The selection of wave band and radio type will be performed by the pilots either directly or by means of the FMS matching waveband and frequency to the navigation aid selected.

4.21.3 Safety/Integrity Aspects

Some antennas will be considered safety-critical, the total loss of transmission, or reception can hazard the aircraft and due consideration must be given to suitable redundancy.

4.21.4 Key Integration Aspects

Integration with individual systems and with all other antennas to avoid mutual interference or blanking/disturbance of transmission.

4.21.5 Key Interfaces

- Airframe/structure: provision of apertures and suitable fasteners for antennas, optimum location of antennas, drag assessment, and protection of the pressure shell.
- Other systems: exchange of information by direct wiring or data bus.
- Flight deck for controls and displays and human factors assessment.
- Electrical power: connection to appropriate bus bar with circuit protection devices, and most suitable harness runs.
- Environment: radio frequency electromagnetic transmissions.
- Consideration of EMC by radiation, induction, and susceptibility.

4.21.6 Key Design Drivers

Compatibility of individual system and antenna, compatibility with complete antenna complement, and interoperability. Low drag antenna designs.

4.21.7 Modelling

Avionics integration rig, antenna placement modelling, realistic modelling of sub-scale mock-ups, and sub-scale wind tunnel models for drag assessment.

4.21.8 References

Macnamara (2010) Chapter 7, Moir and Seabridge (2013) Chapter 9.

4.21.9 Sizing Considerations

Type of aircraft, complement of antennas, and wavebands.

4.21.10 Future Considerations

Conformal antennas and multiple user antennas to reduce drag.

References

Advisory Circular AC 00-31A. National Aviation Standard for the Very High Frequency Omni-directional Radio Range (VOR)/Distance Measuring Equipment (DME)/Tactical Air Navigation (TACAN) Systems. Consult appropriate issue and date.

Advisory Circular 120-76B. 'Guidelines for the certification, airworthiness and operational use of electronic flight bags'. Consult latest issue.

Air Transport Association of America (ATA). Specification 100 code.

Atkin, E.M. (2010). Chapter 391: Aerospace avionics systems. In: *Encyclopedia of Aerospace Engineering*, vol. Vol. 8 (ed. R.H. Blockley and W. Shyy), 4787–4797. Wiley.

Binns, C. (2019). *Aircraft Systems: Instruments, Communications, Navigation and Control*. Wiley-IEEE.

Burberry, R.A. (1992). *VHF and UHF Antennas*. Peter Pergrinus.

Cramer, M.R., Herndon, A., Steinbach, D., and Mayer, R.H. (2010). Chapter 397: *Modern Aircraft Flight Management Systems*. In: *Encyclopedia of Aerospace Engineering*, vol. Vol. 8 (ed. R.H. Blockley and W. Shyy), 4861–4872. Wiley.

Fitzsimmons, F. (2002). *The Electronic Flight Bag: A Multi-Function Tool for the Modern Cockpit*. IITA Research Publication 2 Information Series.

Federal Aviation Administration (FAA) (2017). *AC 120-76D. Authorisation for the use of Electronic Flight Bags*. US Department of Transportation.

Galotti, V.P. Jr., (1998). *The Future Air Navigation System (FANS)*. Ashgate Publishing Company Limited.

Gradwell, D.P. (2010). Physiology of the flight environment. In: *Encyclopedia of Aerospace Engineering*, vol. 8: Chapter 382 (ed. R.H. Blockley and W. Shyy), 4693–4702. Wiley.

Hall, M.R.M. and Barclay, L.W. (1980). *Radiowave Propagation*. Peter Pergrinus.

Honeywell (2014). *IntuVue 3-D Automatic Weather Radar System with Forward Looking Windshear Detection*. Honeywell International Inc.

ICAO Doc 9863. (2006). *Airborne Collision Avoidance Systems Management*.

JASC Joint Aircraft System/Component Code (n.d.). *FAA and Joint Aviation Authority (European Civil Aviation Conference) Code Table for Printed and Electronic Manuals*.

Johnson, S.B., Gormley, T., Kessler, S. et al. (2011). *System Health Management with Aerospace Applications*. Wiley.

Jukes, M. (2003). *Aircraft Display Systems*. Professional Engineering Publishing.
Kayton, M. and Fried, W.R. (1997). *Avionics Navigation Systems*. Wiley.
McLean, D. (1990). *Automatic Flight Control Systems*. Prentice Hall.
Macnamara, T.M. (2010). *An Introduction to Antenna Placement and Installation*. Wiley.
Moir, I. and Seabridge, A. (2008). *Aircraft Systems*, 3e. Wiley.
Moir, I. and Seabridge, A. (2013). *Civil Avionics*, 2e. Wiley.
Pallett, E.H.J. (1992). *Aircraft Instruments & Integrated Systems*. Longmans Group Limited.
Pratt, R. (2000). *Flight Control Systems: Practical Issues in Design & Implementation*. Institution of Engineering and Technology (IET).
Rankin, J.M. and Matolak, D. (2010). Chapter 394: *Aircraft communications and networking*. In: *Encyclopedia of Aerospace Engineering*, vol. Vol. 8 (ed. R.H. Blockley and W. Shyy), 4829–4852. Wiley.
Roskam, J. (2007). *Airplane Flight Dynamics and Automatic Flight Controls*. DARCO.
RTCA DO-185B. (2008). *Minimum operational standards for TCAS II*.
Sabatini, S., Gardi, A., Ramasamy, S. et al. (2016). Modern avionics and ATM systems for green operations. In: *Encyclopedia of Aircraft Engineering, Green Aviation* (ed. R. Agarwal, F. Collier, A. Schäfer and A. Seabridge), 323–342. Wiley.
Seabridge, A. and Moir, I. (2020). *Design and Development of Aircraft Systems*, 3e. Wiley.
Sen, A.K. and Battach, A.B. (2018). *Radar Systems and Radio Aids to Navigation*. Mercury Learning and Information.
SESAR (2012). *European Air Transport Management Master Plan*, 2e. Eurocontrol SESAR.
Srivastava, A.N., Meyer, C., and Mah, R.W. (2010). Chapter 436: *In-flight vehicle health management*. In: *Encyclopedia of Aerospace Engineering*, vol. Vol. 8 (ed. R.H. Blockley and W. Shyy), 5327–5338. Wiley.

5

Mission Systems

The mission system encompasses those systems that provide navigation, communications, sensors, and weapons required to perform a military mission. In some cases, the aircraft type will have been specified and designed for particular roles, in other cases, the aircraft is a reused commercial aircraft type that has been adapted and equipped with a set of systems for a particular role. In this book, the systems will be referred to as mission systems and their position in the overall aircraft system structure is illustrated in Figure 5.1.

The military aircraft requires a range of sensors and computing to enable the crew to prosecute designated missions. The mission systems gain information about the outside world from active and passive sensors and process this information to form intelligence. This is used by the crew, sometimes in conjunction with remote analysts on the ground, to make decisions that may involve aggressive action. These decisions may, therefore, result in the release of weapons or defensive aids, an action which requires a particular set of safety and integrity design considerations.

As an extensive use of digital data technology as already outlined, mission systems utilise a wide range of electronic sensors covering up to 10 decades of the electromagnetic spectrum ranging from 100 kHz (1×10^5 Hz) up to 1000 THz (1×10^{15} Hz). This covers those areas of the electromagnetic spectrum in which communications, radar, and electro-optic (EO) equipment operate. This is a highly complex topic and readers are referred to the Military Avionics Systems book of Moir and Seabridge (2006) for further information.

A system diagram of the mission system is shown in Figure 5.2. This is intended to show a total hypothetical mission system to give an impression of the complexity of sensors, processing and crew interfaces for fast jet combat aircraft, and surveillance type aircraft. A more realistic impression of the sensor fit and crew interfaces is illustrated in Section 5.15.

Aircraft Systems Classifications: A Handbook of Characteristics and Design Guidelines, First Edition.
Allan Seabridge and Mohammad Radaei.
© 2022 John Wiley & Sons, Inc. Published 2022 by John Wiley & Sons, Inc.

5 Mission Systems

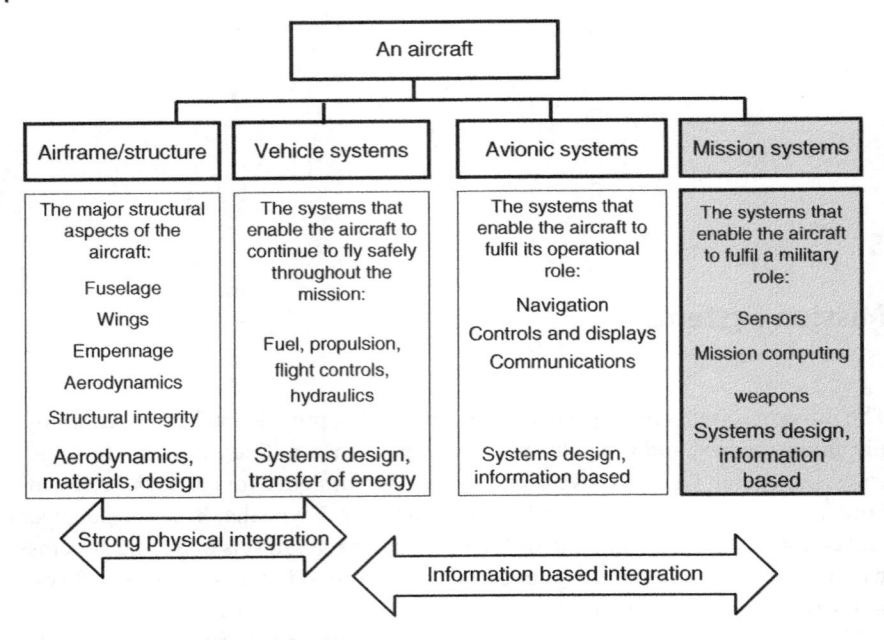

Figure 5.1 The systems described in this chapter.

5.1 Radar System

Most military aircraft will be equipped with radar as its main sensor specified and developed to suit the role and with the appropriate performance characteristics. The information obtained from the radar will be used in conjunction with other sensor information to provide a composite picture of the operational scenario and to enable the crew to determine the most appropriate action (Figure 5.3).

A number of radar modes are performed by different types of radars, either by a radar system designed specifically for a single mode or as part of a multi-mode design. The modes include the following:

- **Airborne intercept (AI):** used by air superiority (fighter) aircraft to search large volumes of airspace to detect targets. Using a variety of scan patterns the radar system is capable of tracking single or multiple targets whilst continuing to scan, known as track-while-scan mode. Usually, installed in the nose of the aircraft.
- **Ground attack:** used by ground attack or close air support aircraft to identify targets and to discriminate friendly forces and enemy forces.
- **Ground mapping:** a mode to map the terrain ahead of the aircraft using the different reflective characteristics of land, water, buildings, and centres of

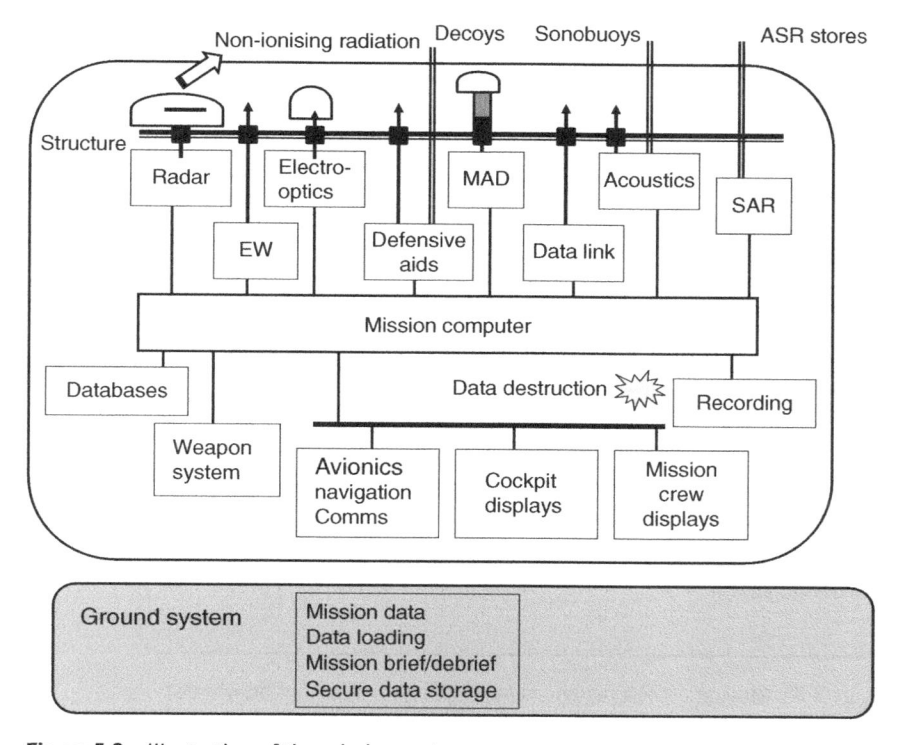

Figure 5.2 Illustration of the mission system.

occupation to paint a representative map identifying major features. Technology and advanced signal processing techniques provide the ability to resolve small features to provide a high-resolution map.

- **Target tracking:** locking on to a target and to continue tracking in range and angle in azimuth and elevation.
- **Ranging:** to detect the range of targets to improve weapon aiming.
- **Terrain following:** to allow a low-level attack by combat aircraft by steering the aircraft at high speed and avoiding all obstacles to enable aircraft to fly below radar coverage.
- **Sideways looking radar (SLR):** to provide an oblique view of terrain in enemy territory allowing the surveillance aircraft to patrol beyond enemy missile engagement range.
- Doppler radar.
- **Maritime:** a mode used by a specific maritime radar optimised for detecting small fleeting targets (such as periscope and masts) and discriminating the target in the noisy environment of open ocean wave reflections. This mode also includes weather and periodic look up mode to detect airborne threats.

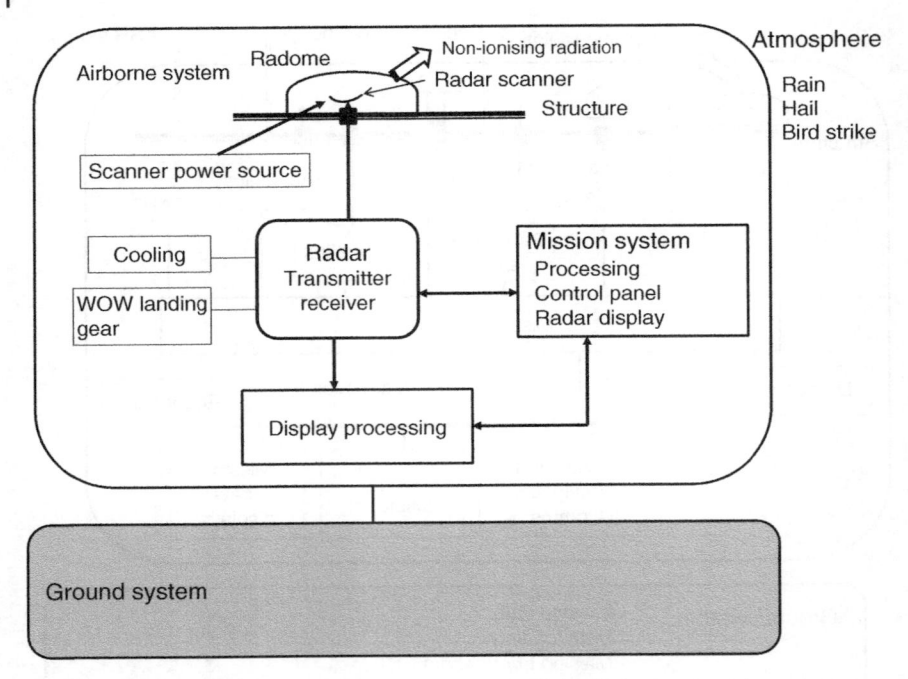

Figure 5.3 Illustration of a generic radar system (WOW, weight on wheels).

- **Airborne early warning (AEW):** a mode used to provide early warning of airborne threats with the ability to discriminate very small radar cross-section targets.
- **Weather radar:** used to detect meteorological conditions that can pose a risk to the aircraft (see also Section 5.4.5).
- **Synthetic aperture (SAR):** is used to create two-dimensional images or three-dimensional reconstructions of objects, such as landscapes using the motion of the radar antenna over a target region to provide finer spatial resolution than conventional beam-scanning radars. SAR is typically mounted on a moving platform on the side of the aircraft or in a pod on the wings or beneath the fuselage.

5.1.1 Purpose of System

To provide information on hostile and friendly targets for attack, to provide AEW, or to perform detailed surface surveillance and mapping.

5.1.2 Description

A radar antenna and transmitter/receiver with appropriate displays. Attack aircraft house the antenna in the nose, whilst surveillance aircraft may have the antenna mounted in the nose, nose and tail, or in radomes mounted on the upper or lower surface of the aircraft.

5.1.3 Safety/Integrity Aspects

Mission critical – radar is an active sensor and in covert operation, its use must be limited to avoid detection by enemy electronic support measures (ESM) or electronic warfare (EW) systems.

Aircrew must be protected from radiation and warning notices posted to protect ground crew. For aircraft operating in restricted environments such as a carrier deck, it may be necessary to limit high-power transmission on the ground using a weight on wheels (WOW) signal from the landing gear system.

5.1.4 Key Integration Aspects

Integration with mission computing, display systems, and weapon aiming systems.

5.1.5 Key Interfaces

The location of the radome on the airframe needs careful consideration to ensure optimum performance of field of regard, drag, security of attachment, and maintenance access.

The radome must be designed to be the correct aerodynamic shape, the material must provide the optimum transmission characteristics, be strong enough to withstand a bird strike, driving rain, and hail, yet be light. Any radiation into the aircraft must be minimised to prevent exposure of the crew to non-ionising radiation, and this may require the use of radar absorbent material (RAM) to be used as a barrier.

- **Airframe/structure:** provision of apertures and suitable fasteners for antennas, optimum location of antennas – nose, side fuselage, top of fuselage, under-fuselage pod, etc., drag assessment, protection of the pressure shell.
- **Other systems:** exchange of information by direct wiring or data bus. The most appropriate power source for the scanner will require an interface with hydraulic or electrical supplies. An alternative is the use of a non-moving

5 Mission Systems

electrically scanned radar (e-scan) which may require a high electrical load and cooling.
- Flight deck and mission crew stations for controls and displays and human factors assessment.
- **Electrical power:** connection to appropriate bus bar with circuit protection devices, and most suitable harness runs.
- **Installation:** mounting tray for electronic line replaceable units (LRUs) and provision of suitable conditioning and cooling. Design of a radome of suitable materials, weight, aerodynamic shape, strength, to contain the swept volume of the antenna with suitable hinges for access for ground crew. Installation of RAM to protect occupants as required.
- Security of attachment of all components especially fuselage-mounted radomes – stress check required.
- **Environment:** radio frequency (RF) electromagnetic transmissions from high-power radar will require an interlock to reduce or prevent transmission on the ground to safeguard ground crew and neighbouring aircraft.
- Consideration of electro-magnetic compatibility (EMC) by radiation, induction, and susceptibility.

5.1.6 Key Design Drivers

Mission success, cost, and performance.

5.1.7 Modelling

Laboratory models.

5.1.8 References

Avery and Berlin (1985), Hannen (2013), Lacomme et al. (2001), Moir and Seabridge (2006), Oxlee (1997), Richards (2014), Richards et al. (2014), Rigby (2010), Skolnik (1980), Schleher (1978), Stendby (2012), Stimson et al. (2014), Walton (1970).

5.1.9 Sizing Considerations

Number of targets to be tracked, requirements for longer range and higher resolution, improvements in detection and jamming techniques, need for greater resolution of maps, and ground surveillance.

Antenna, antenna-driven mechanism, radome, transmitter/receiver, cooling system, and display.

5.1.10 Future Considerations

Electronic scan (e-scan) has many advantages by not using a scanner moving through a wide angle in azimuth and elevation. The development of e-scan radar low-power elements will lead to simpler radome design and control requirements and may reduce the need for cooling.

Development of low probability intercept (LPI) techniques to evade enemy ESM and improvements in jamming techniques.

Image processing improvements to obtain very high-resolution images.

5.2 Electro-optical System

An optical sensor provides a recognisable picture of warm-blooded creatures or variable temperature physical structures that can be used by observers to identify and track. Optical sensors are useful in a range of atmospheric conditions where unaided vision would be degraded, and they also provide an image that can be recorded. Typical conditions include dawn, daylight, overcast skies, dusk, and night, although some degradation is apparent in mist, fog, drizzle, and rain. These are typically the conditions in which searches are carried for lost people, for search and rescue operations on land or sea, and to detect criminal or covert military activities (Figure 5.4).

The electro-optical system can be assembled from a number of sensor systems depending on the role of the aircraft. These may include one or more of the following:

- Television (TV)
- Low light television (LLTV)
- Thermal imaging
- Infrared (IR) imaging
- Night vision systems
- Laser systems

The electro-optical system provides a passive method of searching for contacts in day and night conditions based on the thermal emission of the contact. The system performance is affected by atmospheric conditions such as mist or fog. The system is widely used by the military for identifying targets as well as emergency and law enforcement forces to identify victims or contacts of interest. It is an important sensor in search-and-rescue operations.

5.2.1 Purpose of System

To provide passive surveillance of targets or contacts based on their thermal characteristics.

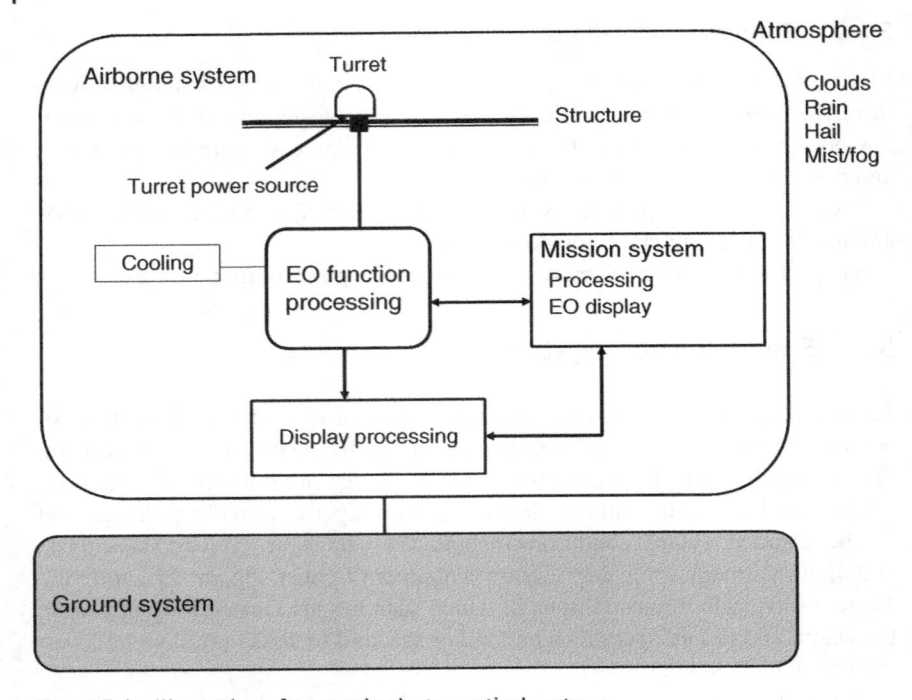

Figure 5.4 Illustration of a generic electro-optical system.

5.2.2 Description

Electro-optical sensors installed in a fuselage-mounted, steerable turret or in an under-wing pod. IR, ultraviolet (UV), and TV sensors are able to provide images in poor visibility.

- The EO turret provides a means of housing a number of sensors in one single housing that can be steered in azimuth and tilted by the pilot. The turret is usually located on the airframe of a fixed wing aircraft or helicopter to achieve a 360° field of view.
- Forward-looking infrared (FLIR) mounted in the aircraft or in an externally carried pod.
- Night vision goggles (NVG) can be used like a binocular, worn around the head, in a helmet, or integrated into the helmet/visor.
- Thermal imaging and laser designator (TIALD) enables one aircraft to identify and confirm a target using IR and to designate the target to another aircraft using laser.

5.2.3 Safety/Integrity Aspects

Mission-critical.

5.2.4 Key Integration Aspects

Integration with mission computing and displays. Integration with airframe to achieve best field of regard with low drag.

5.2.5 Key Interfaces

Turret to fuselage or pod to pylon station:

- **Airframe/structure** provision of apertures and suitable fasteners for turrets, external pods for optimum field of view. Drag assessment, protection of the pressure shell.
- **Other systems:** exchange of information by direct wiring or data bus. Some optical systems may require cooling using their own refrigeration cycle machine.
- Flight deck and mission-crew stations for controls and displays and human factors assessment.
- **Electrical power:** connection to appropriate bus bar with circuit protection devices, and most suitable harness runs.
- **Installation:** mounting tray for electronic LRUs and provision of suitable conditioning and cooling. Connections to external turrets and pods.
- Security of attachment of all components especially fuselage mounted turrets – stress check required.
- Consideration of EMC by radiation, induction, and susceptibility.

5.2.6 Key Design Drivers

Optimum field of view, mission success, cost, and performance.

5.2.7 Modelling

Mission system test rig.

5.2.8 References

Gething (2004), Moir and Seabridge (2006).

5.2.9 Sizing Considerations

Sensor turret or pod (drag), cooling system, deployment, and steering mechanism.

5.2.10 Future Considerations

Developments in sensor technology and display technology and image enhancement techniques will lead to long-range, high-resolution systems. The adoption of sensors in unmanned air vehicles (UAVs) in military and law enforcement sectors will lead to development of miniature lightweight sensors.

5.3 Electronic Support Measures (ESM)

ESM is an electronic warfare technique which is used by aircraft as an integral part of the surveillance mission. It is also used as an input to a defensive aids system (Figure 5.5).

The task is to intercept, locate, record, and analyse radiated electro-magnetic energy for the purpose of obtaining tactical advantage, the ESM antennas are often mounted on the wing tips as a combination of antennas for improved detection performance to and obtain direction of arrival (DOA). The intention is to detect defensive or missile guidance radars at a greater distance than the target radar is capable of confirming detection.

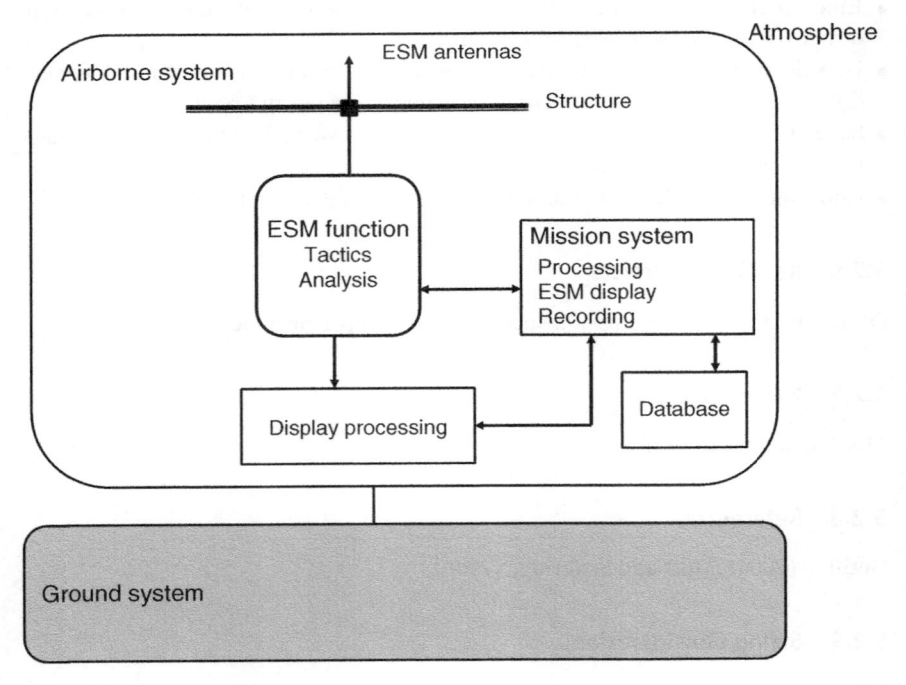

Figure 5.5 Illustration of a generic ESM system.

5.3 Electronic Support Measures (ESM)

ESM is used as a threat warning to detect transmissions that pose an immediate threat, known as a radar-warning receiver (RWR) which indicates an impending attack by fighter or ground to air missiles.

5.3.1 Purpose of System

To provide emitter information such as waveform, pulse repetition frequency (PRF), range, and bearing of hostile transmitters as a warning of threats and as a source of intelligence.

5.3.2 Description

A set of antennas to detect radar and RF transmissions with equipment to analyse the detected signals to determine their most likely source, together with the ability to detect the DOA of the signals. An on-board database allows the signals to be analysed to determine the type of transmitter, and the most likely platform carrying the transmitters. This is a passive sensor.

5.3.3 Safety/Integrity Aspects

Mission-critical.

5.3.4 Key Integration Aspects

Integration with mission computing and on-board database and data link for access to remote intelligence data bases.

5.3.5 Key Interfaces

- **Airframe/structure:** provision of apertures and suitable fasteners for sensitive antennas with maximum field of view and no blanking from fuselage components, protection of the pressure shell.
- Sensitive antenna co-axial cables must be installed with suitable radius of bends to avoid damage to the di-electric. Cables in the wings with long cable runs may require additional protective metal tube shielding.
- **Other systems:** exchange of information by direct wiring or data bus. It may be necessary to blank other transmitting systems for a short period to prevent interference.
- Flight deck and mission crew stations for controls and displays and human factors assessment.
- **Electrical power:** connection to appropriate bus bar with circuit protection devices, and most suitable harness runs.

- **Installation:** mounting tray for electronic LRUs and provision of suitable conditioning and cooling. Physical protection of sensitive cables.
- Security of attachment of all components.
- Consideration of EMC by radiation, induction, and susceptibility.
- Ground systems for download of data, provision of ESM data bus for mission system.

5.3.6 Key Design Drivers

Gathering of intelligence by passive means. Used as a means of providing protection of the host platform by speedy detection and identification of threats.

5.3.7 Modelling

Mission system test rig.

5.3.8 References

Adamy (2003), Bamford (2001), Moir and Seabridge (2006), Poisel (2003), Schleher (1999), van Brunt (1995).

5.3.9 Sizing Considerations

Analysis of the current and emerging threats, number and type of antennas, and workstation/displays.

5.3.10 Future Considerations

ESM will continue to development to match advancing developments of emitter technology. UAVs may provide a suitable solution.

5.4 Magnetic Anomaly Detection (MAD)

Magnetic anomaly detection (MAD) is a technique used by maritime forces to confirm the presence of a submarine before prosecuting an attack. This is primarily to ensure that previous search techniques have not accidently or mistakenly tracked a shoal of fish or detected a wreck. The detector is usually mounted at the end of a long boom projecting from the rear of the aircraft to ensure that the sensitive detector is not affected by stray magnetic fields or transmissions (Figure 5.6).

The aircraft will make a slow low pass over the designated contact area, if a magnetic deviation is found, then the aircraft will deploy a retro-fired smoke marker and then pull a high g turn to fly over the identified spot to deploy the appropriate weapon.

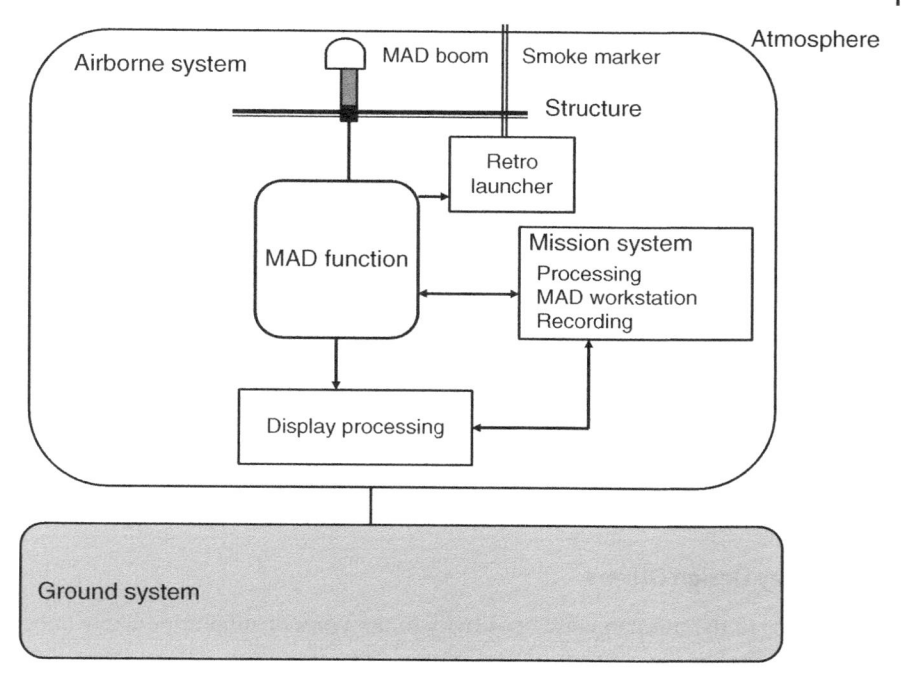

Figure 5.6 Illustration of a generic MAD system.

5.4.1 Purpose of System

To confirm the presence of large metallic objects under the sea (submarines) and to discriminate them from natural objects, prior to attack.

5.4.2 Description

A sensitive magnetic sensor mounted clear of any items of fuselage likely to cause interference or create a false signal. Used to confirm the presence of a submarine by maritime patrol aircraft.

5.4.3 Safety/Integrity Aspects

Mission-critical.

5.4.4 Key Integration Aspects

Integration with the mission computing system and the display philosophy to provide immediate and unambiguous confirmation of a submerged submarine. Control for retro-marker.

5.4.5 Key Interfaces

- Must be located so that there is no possibility of interference from any stray magnetic fields with the sensitive sensor.
- **Airframe/structure:** provision of apertures and suitable fasteners for MAD sensor on a boom at a defined distance from the rear of the aircraft with maximum field of view and minimum presence of magnetic materials. Protection of the pressure shell.
- Flight deck and mission crew stations for controls and displays and human factors assessment.
- **Electrical power:** connection to appropriate bus bar with circuit protection devices and most suitable harness runs.
- **Installation:** mounting tray for electronic LRUs and provision of suitable conditioning and cooling.
- Security of attachment of all components.
- Consideration of EMC by radiation, induction, and susceptibility.

5.4.6 Key Design Drivers

Ease of use to the mission crew, sensitivity to the types of submarine likely to be engaged.

5.4.7 Modelling

Mission system test rig.

5.4.8 References

A brief description can be found on Federation of American Scientists. Fundamentals of Naval Weapons Systems by the Weapons and Systems Engineering Department, US Naval Academy. Not dated.

5.4.9 Sizing Consideration

MAD sensor head, boom, display/chart recorder.

5.4.10 Future Considerations

Developments of sensor systems and indication to detect modern materials in new sub-surface systems. Unmanned small submarines using alternative non-magnetic materials may drive a need for new sensors.

5.5 Acoustic System

The acoustic system is a primary sensor system of the maritime patrol aircraft for anti-submarine operations. The system uses sonobuoys dropped in specific patterns to enable the mission crew to detect and track a submarine over a long period of time. Although the main aim of the system is to attack and destroy a submarine, the search action can be deemed successful if it results in the submarine remaining sub-surface and taking evasive action, thereby disrupting its original mission plan. A more localised search can be conducted by a helicopter using a dipping sonar detector (Figure 5.7).

The tactics to carry out a successful search will have been developed over many flying hours and are usually classified.

5.5.1 Purpose of System

To provide a means of detecting and tracking the passage of underwater objects such as submarines.

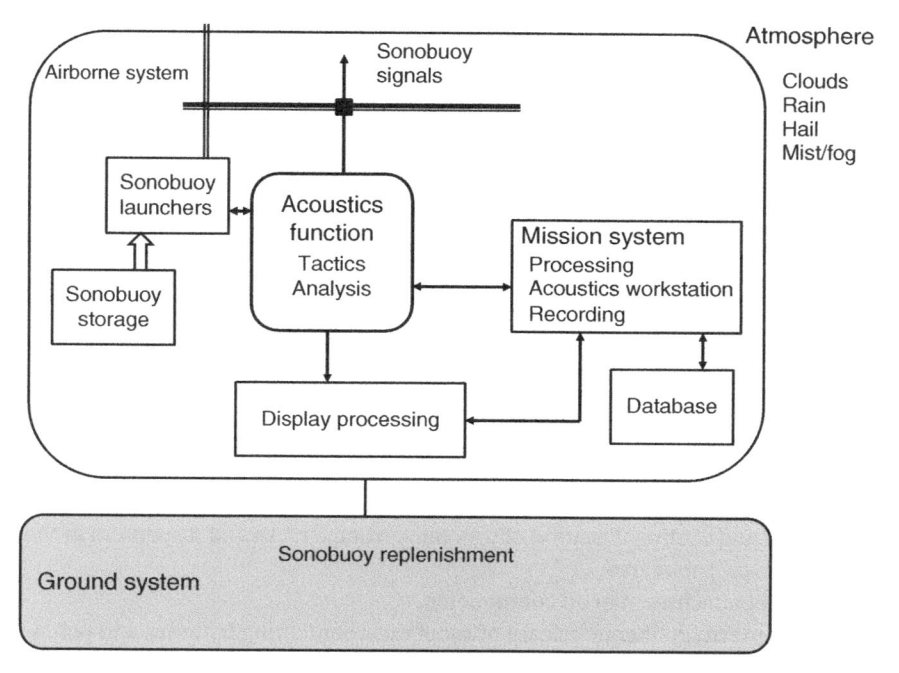

Figure 5.7 Illustration of a generic acoustic system.

5.5.2 Description

Passive and active sonobuoys are dispensed from the maritime patrol aircraft and provide a means of acoustic detection of submarines. Signals are transmitted back to the aircraft for analysis.

The overall system consists of the following major components:

- Single and multiple sonobuoy launchers
- Storage racks
- Different types of sonobuoy – active and passive
- RF receivers with high bandwidth
- Computing and analysis software with appropriate algorithms
- Workstation with large high definition screen
- Access to on-board and off-board intelligence data bases to aid recognition

5.5.3 Safety/Integrity Aspects

Mission-critical, sonobuoy dispensers in fuselage and potential depressurisation risk.

5.5.4 Key Integration Aspects

Integration with mission computing and displays.

5.5.5 Key Interfaces

- **Airframe/structure:** provision of apertures and suitable fasteners for installation of sonobuoy dispensers in the cabin with chutes for dispersal of the buoys into the sea. There is a potential depressurisation risk and suitable measures must be taken to prevent this. Wind tunnel tests will be required to confirm correct release and dispersal of sonobuoys.
- **Other systems:** exchange of information by direct wiring or data bus.
- Flight deck and mission crew stations (acoustic operators) for controls and displays and human factors assessment.
- **Electrical power:** connection to appropriate bus bar with circuit protection devices and most suitable harness runs.
- **Installation:** mounting tray for electronic LRUs and provision of suitable conditioning and cooling. Location of sonobuoy storage racks and dispensers in the mission crew cabin area.
- Security of attachment of all components.
- **Environment:** deliberate release of sonobuoys containing batteries and potentially hazardous materials into the sea.
- Consideration of EMC by radiation, induction, and susceptibility.

- Ground systems for download of data, provision of services on turn around, and replenishment of sonobuoys.

5.5.6 Key Design Drivers

Mission success, performance.

5.5.7 Modelling

Mission system test rig and acoustic test ranges.

5.5.8 References

Gardner (1996), Urick (1982), Urick (1983).

5.5.9 Sizing Considerations

Sonobuoy storage, types and number of sonobuoys (role fit), dispensers, workstation, and antennas. Duration of mission and number of sonobuoys required.

5.5.10 Future Considerations

Use of artificial intelligence (AI) and algorithms to aid recognition and confirmation of threats. Unmanned small sub-surface vessels will need development of tactics and analysis, as well as new sonobuoy sensors.

5.6 Mission Computing System

The mission computer will collate all sensor data and sensor operator inputs to provide a composite picture of the tactical situation. Access will be provided to remote databases to complete on-board analysis. The computer memory devices such as hard drive and all external sources of data must be capable of storage in a secure safe overnight and provision should be made for physical destruction in the event of a serious accident, forced landing in hostile territory, or capture of the aircraft (Figure 5.8).

5.6.1 Purpose of System

To collate the sensor information and to provide a fused data picture to the cockpit or mission crew stations.

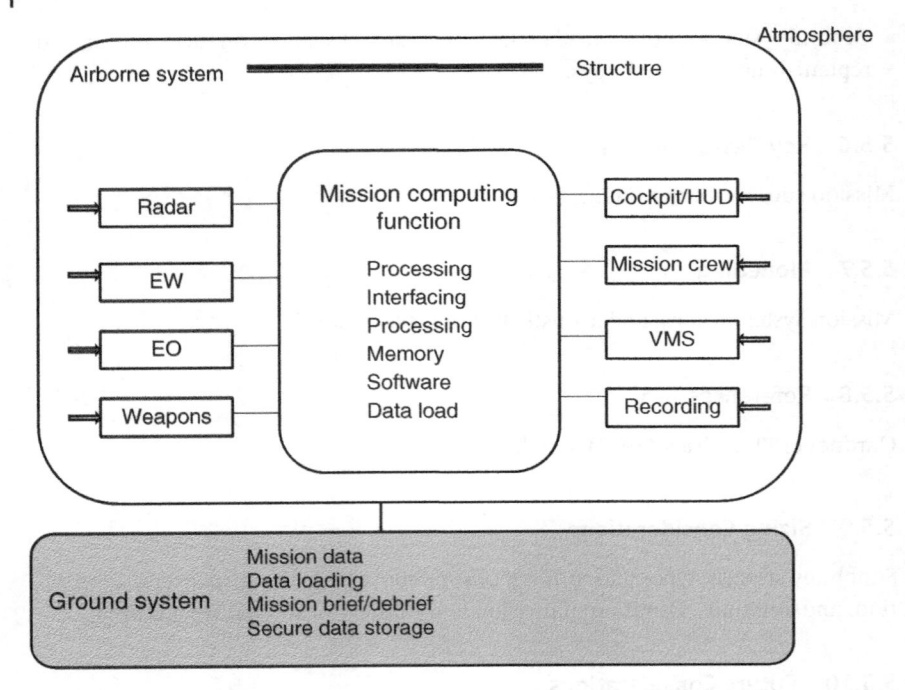

Figure 5.8 Illustration of a generic mission computing system (VMS, vehicle management system).

5.6.2 Description

Suitable architecture computing and interfacing system, appropriate data transmission systems, recording, and data loading.

5.6.3 Safety/Integrity Aspects

Mission-critical.

5.6.4 Key Integration Aspects

Integration with avionic systems, cockpit, and sensors; human factors.

5.6.5 Key Interfaces

Avionic and mission system data busses.
  Performs as integrating medium and control for all avionics and utility systems:
- **Other systems:** exchange of information by direct wiring or data bus. Interfaces to mission systems sensors and deployable components. Connection to the

avionics and vehicle systems will require an interface between commercial standard data abuses and military systems data bus.
- Flight deck and mission-crew workstations for controls and displays and human factors assessment.
- **Electrical power:** connection to appropriate bus bar with circuit protection devices, and most suitable harness runs.
- **Installation:** mounting tray or cabinet for electronic modules and provision of suitable conditioning and cooling.
- Security of attachment of all components.
- Consideration of EMC by radiation, induction, and susceptibility.
- Ground systems for download of data, provision of services on turn around, and provision of secure intelligence data.

5.6.6 Key Design Drivers

Mission success and performance.

5.6.7 Modelling

Operational analysis modelling and mission system test rig.

5.6.8 References

Jukes (2003), Moir and Seabridge (2013).

5.6.9 Sizing Considerations

Mission computer and recorders.

5.6.10 Future Considerations

Use of artificial intelligence (AI) and algorithms to aid recognition and confirmation of threats.

5.7 Defensive Aids

An aircraft operating in a hostile military environment needs to be equipped with measures for self-defence. Mission briefing will include a statement of threats on outward and return legs of the mission, as well as self-defence the pilot must be warned of real tactical threats to the mission and have the means to minimise their

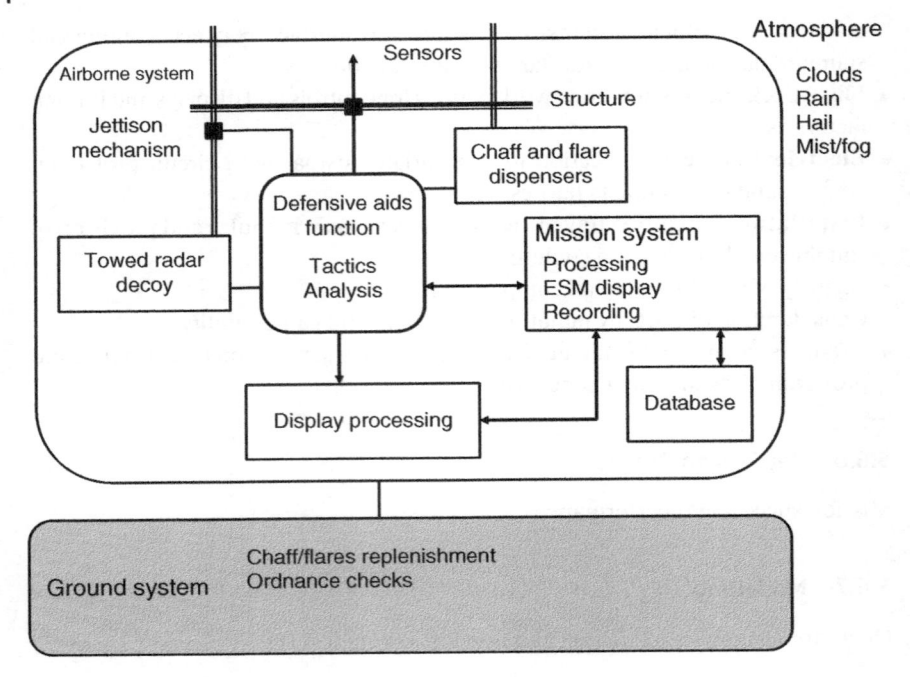

Figure 5.9 Illustration of a generic defensive aids system.

effectiveness (Figure 5.9). Most common threats to low-flying aircraft include the following:

- Small arms fire
- Radar guided anti-aircraft artillery (AAA)
- Shoulder launched surface to air missiles (SAM)
- SAM from ground sites, vehicles, or ships

The defensive aids system is a highly capable selection of measures to detect and deter attack by missiles. It is selective because the aircraft types have different vulnerability and because the threat changes rapidly.

To counter weapons or systems utilising some form of electronic system for guidance, a defensive aids system may include any or all of the following systems depending on the role and the intensity of the threat:

- RWR
- Missile warning receiver
- Laser warning receiver
- Countermeasures dispensing
- Towed decoy

5.7.1 Purpose of System

To provide a means of self-defence by detecting attack from offensive systems using electronic sensors and guidance systems and deploying countermeasures to deter the threat. Access to on-board and off-board intelligence databases to aid recognition is essential.

5.7.2 Description

A suite of sensors to detect missile approach, missile plume or missile homing radar, warning system and countermeasures such as chaff and flare, towed radar decoy, and active jamming.

- **RWR:** sensors placed at locations around the fuselage perimeter to provide full hemisphere horizontal coverage.
- **Missile warning receiver:** similar to RWR and including IR and UV detectors during a following missile launch.
- **Laser warning receiver:** full hemispherical coverage with sensors operating in the laser band.
- **Countermeasures dispensing:** deployment of packages of appropriate mixes of chaff and flare to confuse missile seekers or defence radars.
- **Towed decoy:** a heat source towed behind the aircraft to cause IR seekers to home in on the decoy rather than the aircraft jet pipes. The cable length is selected so that any missile head explosion is sufficiently remote from the aircraft to avoid damage.

5.7.3 Safety/Integrity Aspects

Mission-critical, loss of aircraft and crew is possible if the system fails to perform correctly.

5.7.4 Key Integration Aspects

Mission computing, cockpit, and countermeasures.

5.7.5 Key Interfaces

- **Airframe/structure:** provision of apertures and suitable fasteners for sensors and antennas and protection of the pressure shell. Stress for strength of attachment of the towed decoy cable to aircraft and suitable material, aerodynamic assessment for towed behaviour, and for dispersal of chaff and flares.
- **Other systems:** exchange of information by direct wiring or data bus.

- Flight deck and mission crew workstations for controls and displays and human factors assessment.
- **Electrical power:** connection to appropriate bus bar with circuit protection devices, and most suitable harness runs.
- **Installation:** mounting tray for electronic LRUs and provision of suitable conditioning and cooling. Location of antennas for maximum field of view and low drag.
- Security of attachment of all components.
- **Environment:** release of decoy materials into the atmosphere on trials and training ranges.
- Consideration of EMC by radiation, induction, and susceptibility.
- Ground systems for download of data, provision of services on turn around, and replenishment of fluids or provisions.

5.7.6 Key Design Drivers

Mission success and self-protection.

5.7.7 Modelling

Mission system test rig, EW ranges to test sensors, ranges to test chaff, and flare dispensing.

5.7.8 References

Moir and Seabridge (2006).

5.7.9 Sizing Considerations

Antennas, antenna pods, workstation/display, and countermeasures dispensers.

5.7.10 Future Considerations

There will be continuous development of systems sensors and countermeasures to match developing threats.

5.8 Station Keeping System

For military operation involving a number of aircraft flying in close formation in low visibility conditions, it is essential to provide suitable formation distances and recognition to avoid accidents or failure of the mission. A simple system of mutual separation is provided (Figure 5.10).

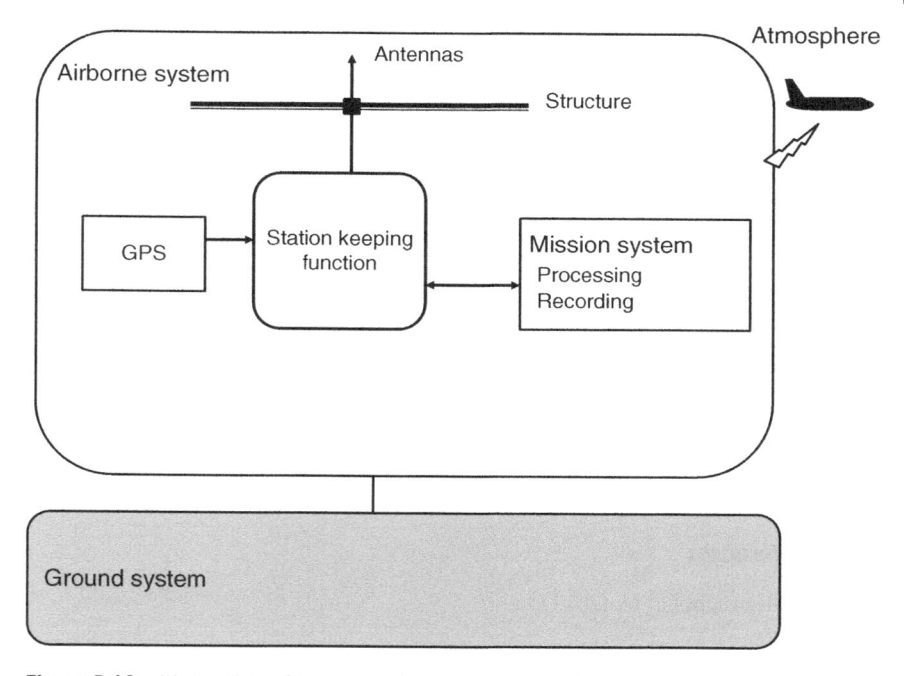

Figure 5.10 Illustration of a generic station keeping system.

5.8.1 Purpose of System

To provide a means of safely maintaining formation in conditions of poor visibility – especially for large transport aircraft. The system can allow up to 36 aircraft on four different frequency channels to fly instrumented formation in zero visibility. The system aircraft to operate within a 10 nautical mile radius of a selected participating master system on the same frequency, allowing for close contact between aircraft.

5.8.2 Description

Detection system and separation warning.

5.8.3 Safety/Integrity Aspects

Safety involved.

5.8.4 Key Integration Aspects

Communications.

5.8.5 Key Interfaces

- **Other systems:** exchange of information by direct wiring or data bus. Use of communications systems.
- Flight deck for controls and displays and human factors assessment.
- **Electrical power:** connection to appropriate bus bar with circuit protection devices, and most suitable harness runs.
- Consideration of EMC by radiation, induction, and susceptibility.

5.8.6 Key Design Drivers

Safety, safe operation of crew and aircraft, and mission success.

5.8.7 Modelling

Modelling by simulation and trials.

5.8.8 References

CAA (2014), Brachet et al. (2014).

5.8.9 Sizing Considerations

Display, number in formation, and scope of radio contacts.

5.8.10 Future Considerations

Number of aircraft and physical separation requirements. Development of Formation Flight System (FFS) in development by USAF.

5.9 Electronic Warfare System

The modern battlefield is rich in RF emissions. Some of these will be radio communications using speech or data; others will be sensors operating in the RF wave bands. Even under radio silence conditions, there will be transient transmission bursts and a rapid and sensitive system will be able to gain valuable intelligence (Figure 5.11).

Electronic warfare can be envisaged as a number of related disciplines, and example is shown below:

- Signals intelligence (SIGINT)
    - Communications intelligence (COMINT)
    - Electronic intelligence (ELINT)

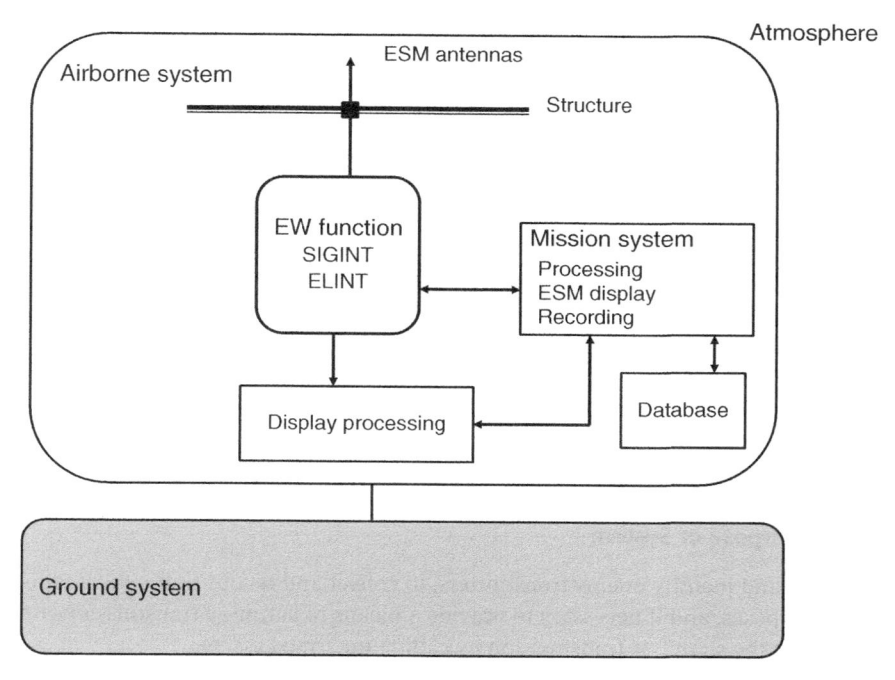

Figure 5.11 Illustration of a generic electronic warfare system.

- ESM
  - Threat warning and direction of arrival (DOA)
  - Target acquisition
  - Homing
- Electronic countermeasures (ECM)
  - Noise jamming
  - Deception jamming
- Electronic counter countermeasures (ECCM)
  - Anti-ESM
  - Anti-ECM

Monitoring of communication wavebands allows the COMINT operator to examine patterns of use that, even if the speech is unintelligible or encrypted, still reveals a lot of information. Patterns of communications traffic and changes in patterns over time can enable analysts to detect enemy intentions.

Monitoring of signals allows the ELINT operator to identify different emitters such as radio or radar. Analysis of the signal-frequency PRF and wave/pulse shape can be compared with known signals on a database. This will allow positive identification of particular emitter types. Using intelligence data, a good operator can even identify the type and identification of the platform hosting the emitter.

ESM is often used to gather intelligence and also as a means of detecting and identifying threat radars in order to initiate evasive action or countermeasures. Appropriately positioned sensitive antenna combinations, often contained in pods on the wing tips are used to gather signals from a wide band of frequency bands from ground or ship borne radars, defence radars, or missile guidance radar. This provides sufficient information to identify the threat and determine the DOA of the signal.

ECM can be initiated by the defensive aids system using jamming, deploying chaff and flare, or by deploying a towed radar decoy.

Jamming can be performed by an externally carried pod or by an accompanying specialist aircraft by emitting a high-power signal aimed at specific signals.

Counter-counter measures is aimed at finding ways of avoiding jamming or interference by enemy assets. Techniques include frequency hopping and LPI techniques. Systems need to be designed to enable such changes to be incorporated rapidly.

5.9.1 Purpose of System

To detect and identify enemy transmitters, to collect and record traffic for intelligence purposes, and if necessary to provide a means of jamming transmissions or destroying the source of transmission by calling up armed assets.

5.9.2 Description

Antennas are required to detect a wide spectrum of signals for COMINT and identification of radars for SIGINT. This applies to continuous or spurious transmissions. The system requires RF receiving and transmitting devices, recording, and access to analysis by operator, software with supporting databases.

5.9.3 Safety/Integrity Aspects

Mission-critical.

5.9.4 Key Integration Aspects

Antenna integration, mission computing, and on-board intelligence database.

5.9.5 Key Interfaces

Location of antennas on the structure for optimum field of regard and low drag. Data link and encrypted communications for connection to force commander:

- **Airframe/structure:** provision of apertures and suitable fasteners for sensors and antennas and protection of the pressure shell.

- **Other systems:** exchange of information by direct wiring or data bus.
- Flight deck and mission-crew workstations for controls and displays and human factors assessment.
- **Electrical power:** connection to appropriate bus bar with circuit protection devices and most suitable harness runs.
- Installation – mounting tray or workstation for electronic LRUs and provision of suitable conditioning and cooling.
- Security of attachment of all components.
- **Environment:** potential for high-energy RF transmissions on trials and training ranges.
- Consideration of EMC by radiation, induction, and susceptibility.
- Ground systems for download of data, provision of services on turn around, and provision of EW databases.

5.9.6 Key Design Drivers

Accuracy of detection and location and need to obtain intelligence on new emitters and current asset deployment.

5.9.7 Modelling

Mission system test rig.

5.9.8 References

Adamy (2003), Bamford (2001), Poisel (2003), Schleher (1999), van Brunt (1995), JASC Code (n.d.), ATA-100 (n.d.) Chapter 99.

5.9.9 Sizing Considerations

Antennas, antenna pods, and receivers.

5.9.10 Future Considerations

There will be continuous development of systems sensors and countermeasures to match developing threats and countermeasures.

5.10 Camera System

Photographic imagery image intelligence (IMINT) can be used to provide independent confirmation of SIGINT intelligence by providing a high-resolution

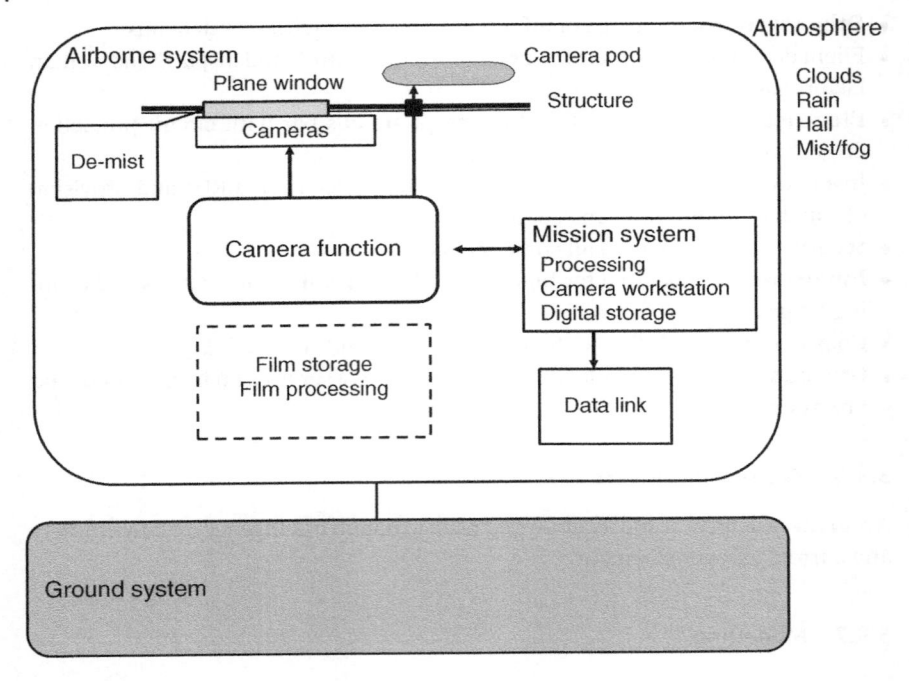

Figure 5.12 Illustration of a generic camera system.

permanent or digital image using ground mapping cameras. As well as confirming the types of enemy installations and mobile assets further intelligence can be obtained to identify numbers, groups, and battalion identification marks as well as deployment of troops. Although more often obtained from satellite imagery high or low altitude aircraft overflying the battlefield in time of conflict or peacetime can gather useful information. Cameras can also be used to provide real-time battle damage assessment (Figure 5.12).

5.10.1 Purpose of System

To record weapon effects, or to provide high-resolution images of the ground for intelligence purposes.

5.10.2 Description

Cameras installed in the fuselage or in fuselage/wing mounted pods to provide forward, rearward, downward, and oblique views. Surveillance cameras will be high resolution with mapping ability for high-quality images for intelligence purposes

(IMINT). Lenses will need appropriate anti-misting and any plane glass windows must be of high optical quality.

5.10.3 Safety/Integrity Aspects

Mission-critical.

5.10.4 Key Integration Aspects

Alignment with aircraft axis, structure, and mission system.

5.10.5 Key Interfaces

- **Airframe/structure:** provision of apertures and suitable fasteners for installation cameras and lenses with plane glass viewing window under fuselage or in under-wing pylons.
- **Other systems:** exchange of information by direct wiring or data bus.
- Flight deck and mission-crew workstations for controls and displays and human factors assessment.
- **Electrical power:** connection to appropriate bus bar with circuit protection devices, and most suitable harness runs.
- Mounting tray for electronic LRUs and cameras and provision of suitable conditioning and cooling.
- Security of attachment of all components.
- Consideration of EMC by radiation, induction, and susceptibility.
- Ground systems for download of data, provision of services on turn around, and replenishment of film or memory devices.

5.10.6 Key Design Drivers

Mission success, resolution of images used at high speed and high or low altitude. Vibration free mounting.

5.10.7 Modelling

Modelling by simulation and calculation.

5.10.8 References

Avery and Berlin (1985), Oxlee (1997).

5.10.9 Sizing Considerations

Area to be photographed or mapped, resolution required, day or night missions.

5.10.10 Future Considerations

Improved satellite imagery, use of UAVs may complement, or supersede this role.

5.11 Head Up Display (HUD)

The combat aircraft pilot cannot afford to spend a lot of time looking down at displays in the cockpit, especially during the combat phase of the mission. The head up display (HUD) is placed directly in front of the pilot and imagery projected so that it can be viewed head up. The HUD presents a virtual image of symbology or displays into the pilot's line of sight and focussed at infinity. To the pilot, the imagery appears to be in the same focal plane and fixed on the outside world scene. This allows the pilot to view information without having to re-focus their eyes to look inside the cockpit (Figure 5.13). Information associated with flight parameters and target parameters include the following:

- Flight path vector
- Horizon

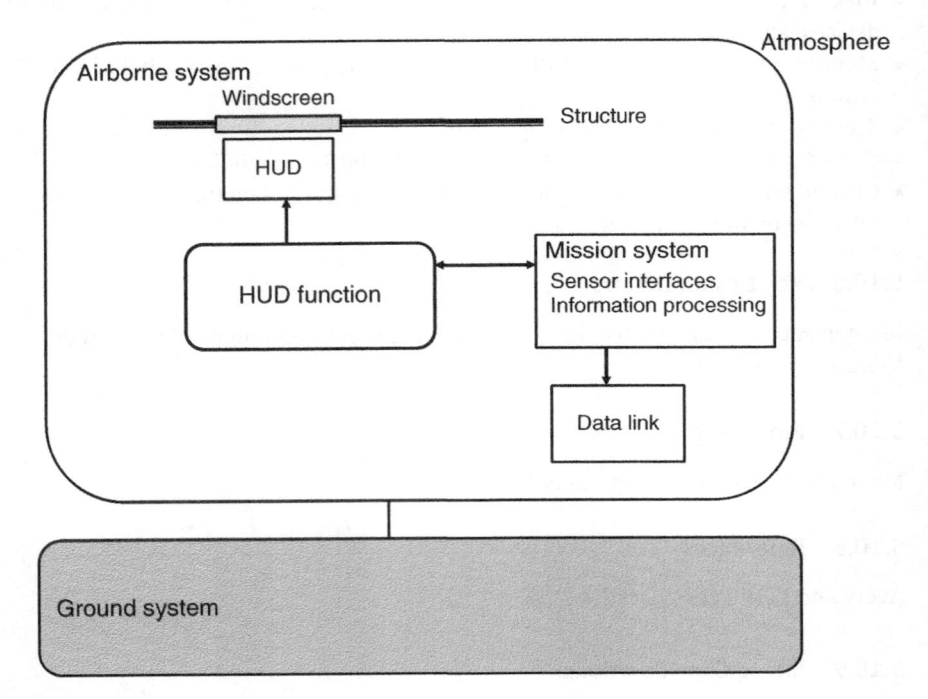

Figure 5.13 Illustration of a generic HUD system.

- Sightlines to targets
- Primary flight data – speed, height heading, etc.
- Flight guidance cues

5.11.1 Purpose of System

To provide the crew with primary flight information and weapon aiming information collimated to infinity, therefore, superimposed on the pilot's forward view. The HUD is usually procured as a complete item from a specialist company.

5.11.2 Description

Optical system to project the image focussed to infinity in the pilot's direct vision, connected to the avionic systems to obtain navigation and weapons data.

5.11.3 Safety/Integrity Aspects

Safety involved – safety critical if used for primary flight information.

5.11.4 Key Integration Aspects

The HUD must be integrated into the physical cockpit design to provide the correct information without obstructing the pilot's external field of view. The system must be clear of the ejection envelope. Human factors integration and cockpit display suite.

5.11.5 Key Interfaces

- **Airframe/structure:** installation of the HUD in the cockpit respecting ejection clearances.
- **Other systems:** exchange of information by direct wiring or data bus.
- Flight deck and mission crew workstations for controls and displays and human factors assessment.
- **Electrical power:** connection to appropriate bus bar with circuit protection devices, and most suitable harness runs.
- **Installation:** mounting tray for electronic LRUs and provision of suitable conditioning and cooling. Location of HUD for optimum visibility by the pilot and ejection clearances to be observed.
- Security of attachment of all components.
- Consideration of EMC by radiation, induction, and susceptibility.

5.11.6 Key Design Drivers

Combat performance may also be used as landing aid.

5.11.7 Modelling

Mission system test rig.

5.11.8 References

Jukes (2003), Moir and Seabridge (2006) Chapter 11.

5.11.9 Sizing Considerations

Required detail of imagery, viewing envelope, space in cockpit, ejection envelope.

5.11.10 Future Considerations

Used on civil aircraft types.

5.12 Helmet Mounted Systems

The field of view of the HUD is very limited when compared to the total hemisphere of regard afforded to the pilot in the bubble canopy of a modern fast jet fighter and limited when compared to the 120° field of view of a modern radar and the acquisition cone of a typical air-to-air missile. This limitation can be overcome by mounting displays into the helmet and allowing the image to be projected onto the visor and introducing eye tracking. This means that the pilot can be provided with stereoscopic, full colour image no matter in what direction the head is pointing. This makes it possible to cue, acquire, designate, track, and release weapons at targets significantly off-boresight without the need to manoeuvre the aircraft (Figure 5.14).

The helmet and visor were traditionally worn to give the pilot protection and certain life support and utility functions, such as

- Communications headset and microphone
- Oxygen
- Glare reducing sun visor
- Laser eye protection
- Head protection during ejection and crash

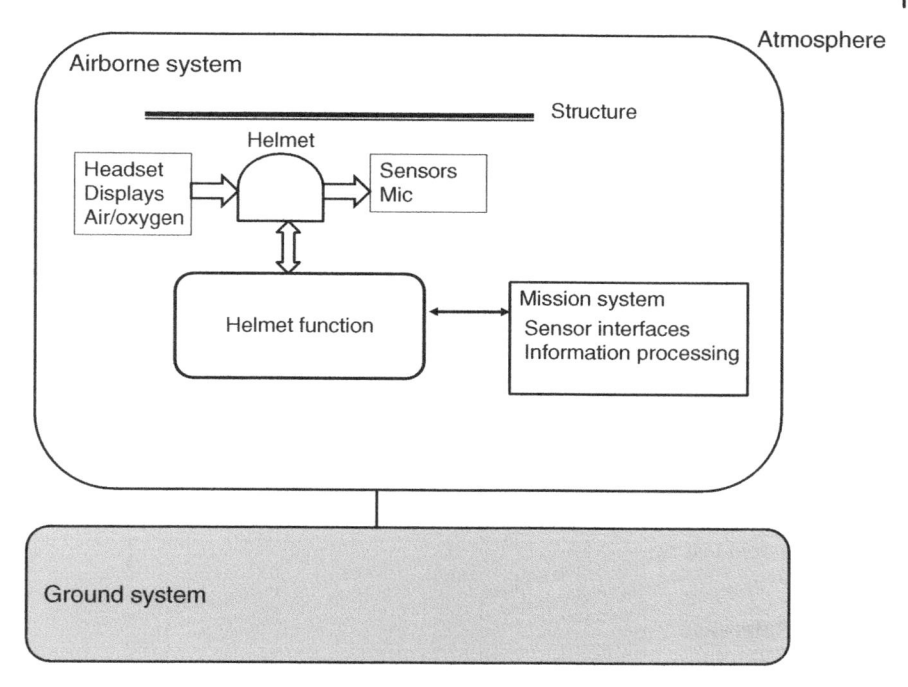

Figure 5.14 Illustration of a generic helmet system.

5.12.1 Purpose of System

To provide primary flight information and weapon information to the crew whilst allowing freedom of movement of the head, position of the head, and sighting information.

5.12.2 Description

Display surface mounted to the pilot's helmet may also contain a sighting mechanism.

5.12.3 Safety/Integrity Aspects

Mission-critical.

5.12.4 Key Integration Aspects

Integration with mission computing and avionics. Human factors integration with each pilot to ensure comfort and protection.

5.12.5 Key Interfaces

Interface with standard aircrew helmet.

- **Other systems:** exchange of information by direct wiring or data bus.
- Flight deck for controls and displays and human factors assessment.
- **Electrical power:** connection to appropriate bus bar with circuit protection devices, and most suitable harness runs.
- Security of attachment of all components to the helmet.
- Consideration of EMC by radiation, induction, and susceptibility.

5.12.6 Key Design Drivers

Combat performance, low workload, health, and safety (of user).

5.12.7 Modelling

Mission system test rig.

5.12.8 Reference

Jukes (2003).

5.12.9 Sizing Considerations

Amount of information to be processed and viewed, impact of helmet on pilot's neck and safety in high g conditions.

5.12.10 Future Considerations

The addition of gloves with sensors enables the pilot to perform virtual control actions by pointing, 'pressing', and gesturing. This combination of helmet displays and gloves makes the virtual cockpit a significant reality (see Section 5.15).

5.13 Data Link

Voice was originally used to transfer information between aircraft and ground centres; however, it is slow and prone to misunderstanding and transcription errors. In contrast, high bandwidth data links deliver more information, and they do it rapidly, often incorporating error detection and correction techniques, and they allow encryption to be used. Many data links are line of sight wavebands,

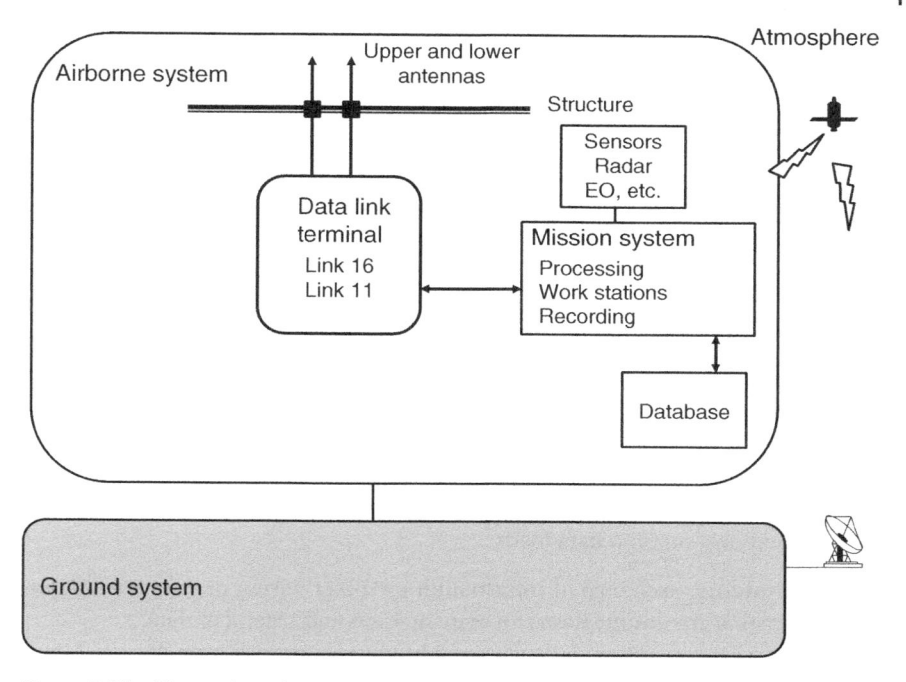

Figure 5.15 Illustration of a generic data link system.

but the use of Satcom permits transmission over the horizon to remotes centres (Figure 5.15).

Using a data link terminal connected to the mission computer allows the crew to send and receive encrypted voice or data messages. Two links are in common use:

- Link 16, often known as the joint tactical information distribution system (JTIDS) most commonly used on ultra high frequency (UHF) in the same frequency as identification friend/foe (IFF), distance measuring equipment (DME), and Tacan. The system transmissions are constrained to avoid any mutual interference.
- Link 11 used by forces engaged in joint maritime operations. NATO Link 11 (NILE) has been adopted as Link 22 (Stanag 5522).

5.13.1 Purpose of System

To provide transmission and receipt of messages under secure communications using data rather than voice.

5.13.2 Description

Terminal with encoding/decoding facility, mission data uploading capability, and encryption devices.

5.13.3 Safety/Integrity Aspects

Mission-critical.

5.13.4 Key Integration Aspects

Integration with suitable radio transmitters and data link protocol suitable for co-operative working.

5.13.5 Key Interfaces

Communications, mission data loads

- **Other systems:** exchange of information by direct wiring or data bus. Major use of existing communications for transmission and receipt of data.
- Flight deck for controls and displays and human factors assessment.
- **Electrical power:** connection to appropriate bus bar with circuit protection devices, and most suitable harness runs.
- **Installation:** mounting tray or workstation for electronic LRUs and provision of suitable conditioning and cooling.
- Security of attachment of all components.
- Consideration of EMC by radiation, induction, and susceptibility.
- Ground systems for download of data, provision of services on turn around, and mission system security information.

5.13.6 Key Design Drivers

Security of transmission.

5.13.7 Modelling

Mission system test rig.

5.13.8 References

Schleher (1999).

5.13.9 Sizing Considerations

Transmitter/receiver, message workstation, antenna.

5.13.10 Future Considerations

Alternative data links standards, improved encryption procedures and moves towards digital communication systems.

5.14 Weapon System

Many military aircraft are designed to carry and release weapons, either for self-defence purposes or to destroy enemy assets. Although some aircraft types carry weapons in a bomb bay, many types carry them under the wings or the fuselage. Guns are carried in the fuselage or in pods attached to the wings or fuselage. External pylons or bomb bays may also be used to dispense humanitarian aid packages such as air sea rescue equipment, aid packages (Figure 5.16).

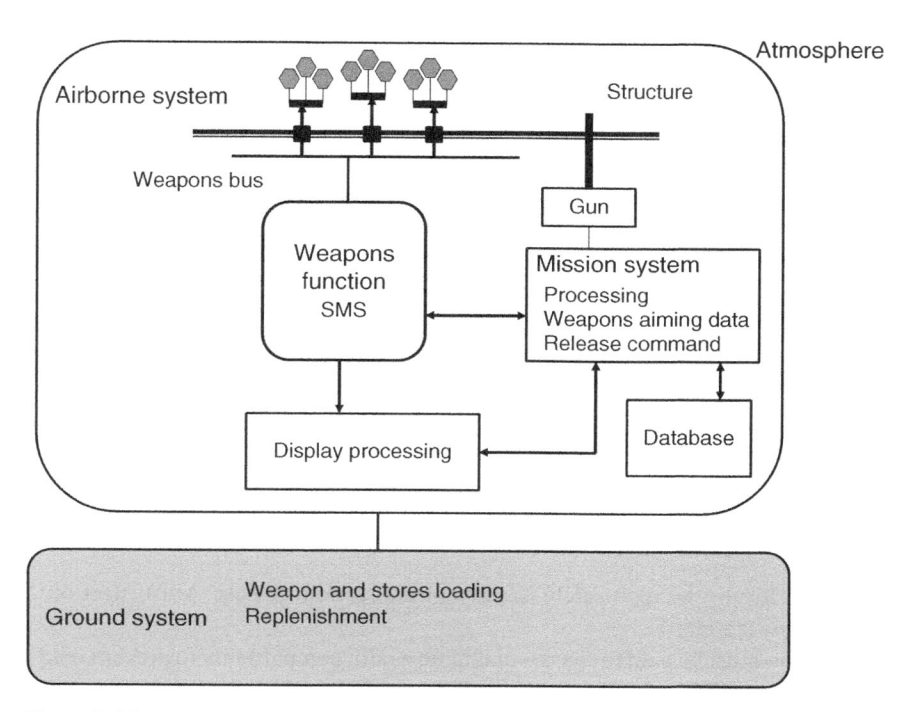

Figure 5.16 Illustration of a generic weapon system (SMS, stores management system).

The weapons carried include combinations of the following:

- Ballistic bombs
- Smart bombs
- Practice bombs
- Area denial weapons
- Air to ground weapons – short range and medium range
- Cruise missiles
- Air-to-air missiles
- Anti-radar missiles
- Anti-ship missiles
- Torpedoes
- Mines
- Guns

The weapons system includes

- The carriage and release mechanism
- Bomb bay carriage and doors
- Arming
- Targeting
- Emergency jettison

5.14.1 Purpose of System

To arm, direct, and release weapons from the aircraft weapon stations.

5.14.2 Description

System for management of external or internal stores, fuselage, wing or bomb bay carriers or pylons for weapons carriage, and safe methods of emergency release. Provision of targeting information from the mission computer. The system must include a mechanism for emergency jettison of all weapons.

5.14.3 Safety/Integrity Aspects

Mission-critical – weapon safety to prevent inadvertent release. Must meet ordnance safety standards.

Attention must be paid to electro-magnetic health compatibility to prevent inadvertent releases. Wiring must be designed to ordnance board requirements and separated from all other wires.

5.14.4 Key Integration Aspect

Navigation, mission computing, aerodynamics, separation of wiring from all other wiring, or sources of energy to prevent inadvertent release.

5.14.5 Key Interfaces

- **Airframe/structure:** provision of apertures and suitable fasteners for pylons, gun bays, and bomb bays. Each weapon release unit or pylon will require a strengthened 'hard point' on the structure. A similar hard point must also be provided for all external stores such as fuel tanks, sensor pods, instrumentation pods, and air sea rescue store. Use must be made of all standard stores interface standards to ensure compatibility with a wide range of weapons. Location of guns, consideration of gun firing vibration, and collection of links. Interface with engine to prevent smoke and pressure effects disturbing the engine intake flow.
- **Other systems:** exchange of information by direct wiring or data bus respecting the integrity and safety of the weapon system. Weapon system wiring will be identified by colour, type, identification, and must be separated from all other system harnesses to meet ordnance safety mandatory requirements. The wiring harnesses may need to be screened and protected.
- Flight deck and mission-crew workstations for clearly identified and guarded controls and displays and human factors assessment.
- **Electrical power:** connection to appropriate bus bar with circuit protection devices, and most suitable harness runs.
- **Installation:** mounting tray for electronic LRUs and provision of suitable conditioning and cooling. Mounting of guns and collection of expended links and cases. Equipment in the vicinity of guns may need anti-vibration protection.
- Security of attachment of all components.
- **Environment:** release of exhaust from missiles, and contamination from ordnance on trials and training ranges.
- Consideration of EMC by radiation, induction, and susceptibility.
- Ground systems for download of data, provision of services on turn around, and replenishment of weapons.

5.14.6 Key Design Drivers

Mission success, ordnance safety, and probability of kill. Ability to manage a wide variety of stores.

5.14.7 Modelling

Modelling of ballistics and weapons trajectories and modelling of probability of kill in different scenarios.

Wind tunnel testing for drag impact and safe separation, flight trials on live release ranges or gun-firing range.

5.14.8 References

Rigby (2013), Rigby (2010), MIL-STD-1760 (n.d.) (latest issue). JASC Code (n.d.), ATA-100 (n.d.) Chapter 94.

5.14.9 Sizing Considerations

Number and types of weapons to be carried to fulfil all roles and impact on pylons (wing, fuselage, or bomb bay), weapons (role fit), and cockpit controls.

5.14.10 Future Considerations

Weapon technology will move on to meet changes in the battlefield and to improve precision strikes. The system must be designed to accommodate a wide range of weapon types and must be readily adaptable to incorporate new weapons.

5.15 Mission System Displays and Controls

The key feature of the crew workplace, whether the fast jet cockpit or the surveillance aircraft mission stations are to provide access to components with which to input commands into the systems and to observe the behaviour of the systems by the display of graphics or text on the display screens or instruments (Figure 5.17a,b).

The pilot of a combat aircraft (often single seat) will make use of existing head down displays supplemented by HUD or helmet displays. Controls are provided by switches embedded into throttle levers and control column (hands on throttle and stick [HOTAS]) and panels designated to specific tasks such as communications, flight management systems.

The mission crew of a surveillance aircraft will use individual workstations with keyboard and roll ball. The data from the workstation are fused at the workstation of the tactical commander.

5.15 Mission System Displays and Controls | 231

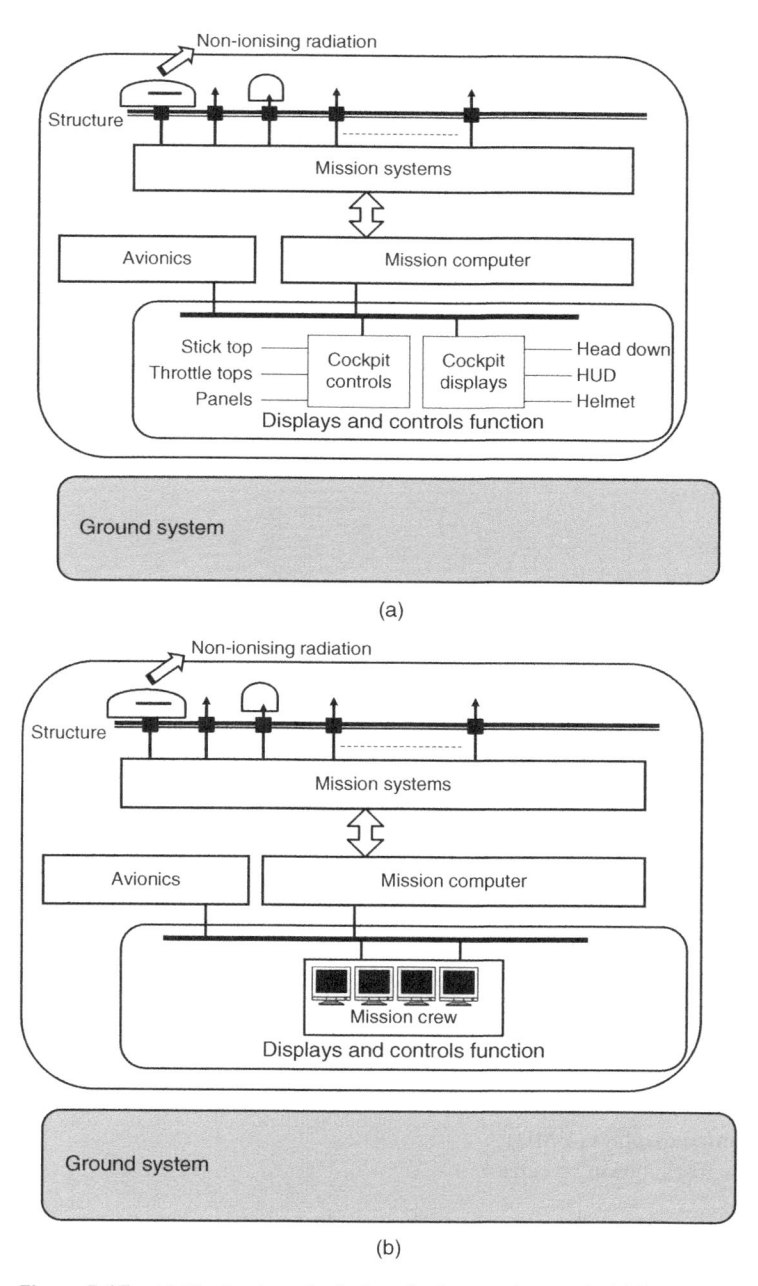

Figure 5.17 (a) Illustration of mission displays and controls (a) for combat aircraft and (b) for surveillance aircraft.

5.15.1 Purpose of System

To provide the pilot or mission crew with the means of controlling the mission systems and observing the consequences of system behaviour on a suitable set of displays.

5.15.2 Description

The controls provide an input to the systems either directly or by means of the aircraft computing architectures. Information on the performance of the systems is fed directly to the displays or from the computing architecture. This system is entirely contained in the cockpit or flight deck and is designed to meet regulatory and desirable aspects of comfort, reach and visibility in a safe environment. Human factors played an essential part of the design philosophy. Typical controls include the following:

- Control column
- Control column top (stick top)
- Throttle levers
- Communications panels
- Navigation control panel
- Auto-pilot flight director
- Discrete switches
- Soft keys or touch screens
- Mission-crew workstation key pad and roll ball

Control signals and data are provided by data bus and some discrete wiring to a set of display management computers. The redundancy of the computers and displays and their associated power supplies must be sufficient to support a safety critical system. The system must be capable of tolerating multiple failures without total loss of display. A separate and independent set of displays may be provided to provide essential information such as attitude, horizontal situation, altitude, and airspeed as a get-U-home display. Typical display surfaces include the following:

- HUD
- Helmet mounted displays (HMD)
- Mission-crew workstation screens

To supplement the visual displays, it is common to have aural cues such as audio tones to draw the pilot's or crew member's attention to warnings, and voice messages to add urgency to unusual conditions. Controls may be supplemented by direct voice input.

5.15.3 Safety/Integrity Aspects

The system is mission-critical and total loss of the system will result in mission abort. The design must avoid all common mode issues and should allow for the loss of one or more workstations without severe mission impairment. Weapon safety must be assured to prevent inadvertent release, preferably by ensuring one single crewmember being responsible for arming and release.

5.15.4 Key Integration Aspect

The display suite must be integrated with the aircraft navigation systems the flight deck and the weapon system. Aerodynamics must be involved in all aspects of antenna placement, turret design, sonobuoy launchers and weapon carriage, and bomb bay as well as safe separation of all weapon types.

5.15.5 Key Interfaces

- **Airframe/structure:** equipment mounting in a designated mission crew area or in the cockpit.
- **Other systems:** exchange of information by direct wiring or data bus.
- Mission crew of flight deck for installation of controls and displays on workstations and human factors assessment for reach, access, lines of sight, assessment of font sizes, and colours.
- **Electrical power:** connection to appropriate bus bar with circuit protection devices, and most suitable harness runs. Connection to bus bars to maintain power under all conditions to minimise total loss of displays.
- **Installation:** mounting trays and workstations for electronic LRUs and provision of suitable conditioning and cooling.
- Security of attachment of all components.
- Consideration of EMC by radiation, induction, and susceptibility.

5.15.6 Key Design Drivers

Robust computing, single failure tolerant system with the capability of handling large amounts of data, and to remain operational for very long periods of time.

5.15.7 Modelling

The graphic displays can be modelled by using VAPS for individual displays and slowly building up to a full set of screens. A mock-up of the mission crew area can be used with realistic mission scenarios to simulate a full mission. Wind tunnel for aerodynamics and weapon release behaviour.

5.15.8 References

Atkin (2010), Jukes (2003), Moir and Seabridge (2008) Chapter 12, Pallett (1987), Rankin and Matolak (2010).

5.15.9 Sizing Considerations

The duration of the mission and the information to be displayed will determine the number of mission crewmembers and their accommodation. Workstation must be designed for comfort, visibility, and for long-term usage in high vigilance conditions.

5.15.10 Future Considerations

Advances in sensor technology, display technology and miniaturisation combined with fast computing and display processing techniques combined in the helmet and visor have led to the 'virtual cockpit' becoming a reality. Sensors on the helmet allow sighting to be performed, and cameras around the aircraft allow a virtual outside view. Gloves equipped with sensors allow hand gestures to be converted into commands (see Section 5.12).

5.16 Mission System Antennas

This paragraph, as in Chapter 4, will consider the total set of antennas as a system. The reason for this is that although each antenna is carefully matched to its own system's needs in terms of impedance, frequency, power, and range, when it is installed on the fuselage exterior, it is operating in a dynamic and complex environment. It will be placed with other antennas each transmitting or receiving independently with little or no relationship to other antennas, unless this factor is incorporated into the design. Figure 5.18 shows the range of communication band antennas and radar antennas, their connection to some form of RF stage to allow receipt and transmission of signals and the use of the information by the mission crew.

Figure 5.19 shows an example of the antennas required by the mission systems described in this chapter. The locations shown are indicative only, and the exact location will depend on the size of the airframe and the nature of the transmissions from each antenna with relationship to its neighbours. It must be noted that it is unlikely that a single aircraft type would ever require all these antennas. These mission system antennas must be located and designed to operate in conjunction with the avionics antennas described in Chapter 4.

5.16 *Mission System Antennas* | 235

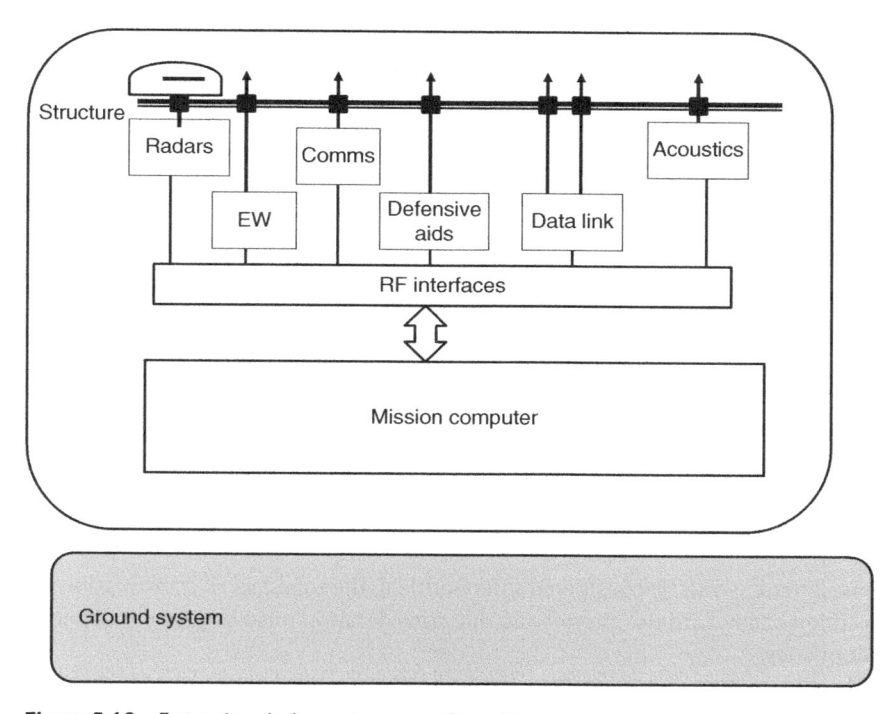

Figure 5.18 Example mission antenna configuration.

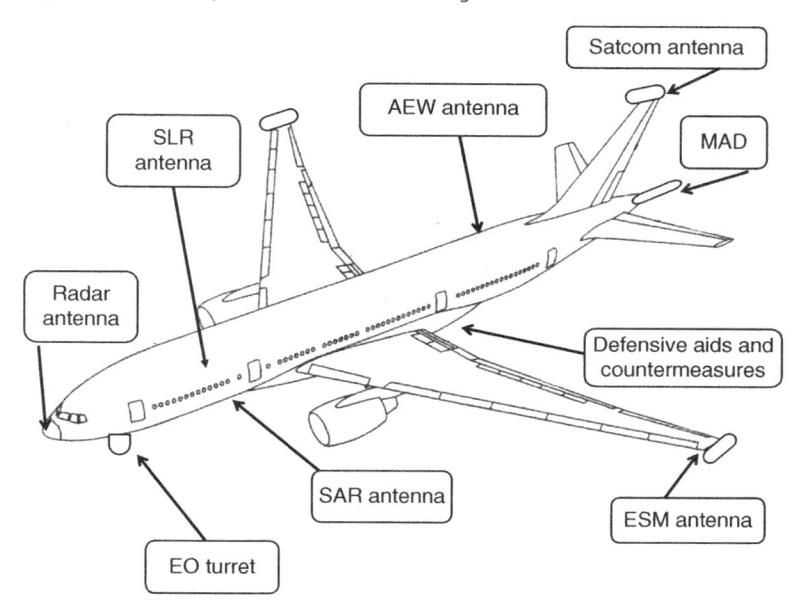

Figure 5.19 Example mission antenna layout.

5.16.1 Purpose of System

To provide a means of transmission and reception of RF signals for all relevant systems on the aircraft and to ensure interoperability of the systems throughout the flight with no mutual interference, despite weather or meteorological conditions. The system also provides ready access to internationally accepted emergency channels for transmission or for continuous monitoring.

5.16.2 Description

Many of the systems described above have a need to transmit information to other aircraft or ground systems and also to receive information from similar sources. The selection of wave band and radio type will be performed by the pilots directly by means of the flight management system matching waveband and frequency to the navigation aid selected.

5.16.3 Safety/Integrity Aspects

Some antennas will be considered safety-critical, the total loss of transmission or reception can hazard the aircraft and due consideration must be given to suitable redundancy.

Many of the signals are very sensitive and antennas need to be placed to provide the optimum field of regard. The signals are used to provide intelligence and in some cases to initiate immediate action. With this in mind, there must be no corruption or blanking of signals to generate non-trustworthy information that could lead to the use of weapons.

Some antenna cover and radome materials may be susceptible to long-term UV exposure at high altitude.

5.16.4 Key Integration Aspects

Integration with individual systems and with all other antennas to avoid mutual interference or blanking/disturbance of transmission.

5.16.5 Key Interfaces

Each antenna must be closely matched to its own system and the location of antennas on the fuselage must take into account the field of regard of each antenna. This particularly important for radar systems

- **Airframe/structure:** provision of apertures and suitable fasteners for antennas, optimum location of antennas, drag assessment, protection of the pressure shell.

- **Other systems:** exchange of information by direct wiring or data bus.
- Mission crew cabin or flight deck for controls and displays and human factors assessment.
- **Electrical power:** connection to appropriate bus bar with circuit protection devices, and most suitable harness runs.
- **Environment:** RF electromagnetic transmissions.
- Consideration of EMC by radiation, induction, and susceptibility.

5.16.6 Key Design Drivers

Compatibility of individual system and antenna, compatibility with complete antenna complement and interoperability, and low drag antenna designs.

5.16.7 Modelling

Mission system integration rig, antenna placement modelling, and realistic modelling of sub-scale mock-ups.

5.16.8 References

Macnamara (2010), Moir and Seabridge (2013) Chapter 9.

5.16.9 Sizing Considerations

Type of aircraft, complement of antennas, and wavebands.

5.16.10 Future Considerations

Conformal antennas and multiple user antennas to reduce drag.

References

Adamy, D.A. (2003). *EW 101 A first Course in Electronic Warfare*. London: Artech House.
Air Transport Association of America (ATA). Specification ATA100.
Avery, T.E. and Berlin, G.L. (1985). *Fundamentals of Remote Sensing and Airphoto Interpretation*. Hoboken NJ: Prentice Hall.
Atkin, E.M. (2010). Aerospace avionics systems. In: *Encyclopedia of Aerospace Engineering*, vol. 8: Chapter 391 (ed. R.H. Blockley and W. Shyy), 4787–4797. Wiley.
Bamford, J. (2001). *Body of Secrets*. Century.

Brachet, J.-B., Cleaz, R., Denis, A. et al. (2014). *Reference Material for a Proposed Formation Flight System*. Cambridge, MA: Department of Aeronautics and Astronautics Massachusetts Institute of Technology.

CAA (2014). *Safety and airspace regulation activity*. NO 2014-00-0110.

Gardner, W.J.R. (1996). *Anti-Submarine Warfare*. Sterling VA: Potomac Books.

Gething, M. (2004). *Jane's Electro-optic Systems*, 10the. London: Janes.

Hannen, P. (2013). *Radar and Electronic Warfare Principles*. Raleigh NJ: SciTech Publishing.

JASC Joint Aircraft System/Component Code. FAA and Joint Aviation Authority (European Civil Aviation Conference) code table for printed and electronic manuals.

Jukes, M. (2003). *Aircraft Display Systems*. Chichester UK: Wiley.

Lacomme, P., Marchais, J.-C., Hardange, J.-P., and Normant, E. (2001). *Air and Spaceborne Radar Systems*. Norwich: William Andrew.

Macnamara, T.M. (2010). *An Introduction to Antenna Placement and Installation*. Chichester UK: Wiley.

MIL-STD-1760. Department of Defense Interface standard: aircraft/store electrical interconnection system (consult latest issue).

Moir, I. and Seabridge, A. (2006). *Military Avionics Systems*. Chichester UK: Wiley.

Moir, I. and Seabridge, A. (2008). *Aircraft Systems*, 3e. Chichester UK: Wiley.

Moir, I. and Seabridge, A. (2013). *Civil Avionics*, 2nde. Chichester UK: Wiley.

Oxlee, G.J. (1997). *Aerospace Reconnaissance*. Sterling VA: Potomac Books.

Pallett, E.H.J. (1987). *Aircraft Electrical Systems*. Longmans Group Limited.

Poisel, R.A. (2003). *Introduction to Communication Electronic Warfare Systems*. London: Artech house.

Rankin, J.M. and Matolak, D. (2010). Aircraft communications and networking. In: *Encyclopedia of Aerospace Engineering*, vol. 8: Chapter 394 (ed. R.H. Blockley and W. Shyy), 4829–4852. Wiley.

Richards, M. (2014). *Fundamentals of Radar Signal Processing*, 2nde. London and NY: McGraw Hill.

Richards, M., Holm, W.A., and Schleer, J.A. (2014). *Principles of Modern Radar*. Raleigh NJ: SciTech Publishing.

Rigby, K. (2010). Weapons integration. In: *Encyclopedia of Aerospace Engineering*, Chap. 417, vol. 8 (ed. R.H. Blockley and W. Shyy), 5107–5116. Chichester UK: Wiley.

Rigby, K.A. (2013). *Aircraft Systems Integration of Air Launched Weapons*. Chichester UK: Wiley.

Schleher, C.D. (1978). *MTI Radar*. London: Artech House.

Schleher, C. (1999). *Electronic Warfare in the Information Age*. London: Artech House.

Skolnik, M.I. (1980). *Introduction to Radar Systems*. London and NY: McGraw-Hill.

Stendby, M. (ed.) (2012). *Jane's Radar and EW Systems 2011–2012*. London: Janes.

Stimson, G.W., Griffiths, H.D., Baker, C.J., and Adamy, D. (2014). *Introduction to Airborne Radar*. Raleigh NJ: SciTech Publishing.
Urick, R.J. (1982). *Principles of Underwater Sound*. Los Altos: Peninsula.
Urick, R.J. (1983). *Sound Propagation in the Sea*. Los Altos: Peninsula.
van Brunt, L.B. (1995). *Applied ECM*. Virginia: EW Engineering Inc.
Walton, J.D. (1970). *Radome Engineering Handbook*. NY: Marcel Dekker.

Further Reading

An interesting and informative history of the development of air intercept radar: *Looking forward – 60 years of fire control radar*. Richard Scott for Selex ES (A Finmeccanica company). Pub Terry O'Hare www.3-56media.com.

A historical paper describing the integration of the AI23 radar into the BAC Lightning leading to the integrated weapon system concept: Wilson, Tony. (2019). *Lightning Development Studies and Proposals*. Journal 72 of the Royal Air Force Historical Society, ISSN 1361 4231.

This book covers in detail the history and development of military flight helmets from the post-World War II era to the present, and includes over 120 different helmets and their associated equipment such as oxygen masks, boom microphonesWise, A.R. and Breuninger, M.S. (1997). *Jet Age Flight Helmets: Aviation Headgear in the Modern Age*. Schiffer Publishing.

Exercises

5.1 Propose a mission system and scrutinize the market and current research to look for new types of sensor. Examine how these new sensor types can be incorporated into your mission system architecture.

5.2 Can you design a basic mission system architecture to cater for new sensor types with minimum change to the system?

5.3 Consider a typical mission system or a mission system known to you to explore issues of obsolescence. Propose solutions to any issues that arise.

5.4 Are there any known or emerging changes to warfare that will demand new sensors or will force current sensors to become obsolescent?

5.5 Can you propose a modification to the system diagram to highlight any influences of links that are critical to mission success?

6

Supporting Ground Systems

It is important to recognise that the airborne systems will interact with a set of ground-based system as illustrated in Figure 6.1 (Seabridge and Moir, 2020). In most instances, these systems will be a normal element of the airborne systems and routinely collecting information and presenting it to a system on the ground for analysis and action. In other instances, the system will be for a particular part of the development process, for example the collection of flight test data. In the case of unmanned air vehicles, the system forms a closed-loop system to allow the aircraft to be piloted remotely. In all these cases, the requirements of the ground system must be considered in the design of the airborne system.

The collection and transfer of information from the aircraft to the ground system will vary according to the age and technology of the aircraft type. The wide scale use of high-speed data bus networks means that most of the data generated by the system is available on the data bus. The transfer in modern times is usually by data link or telemetry, but previous generations of aircraft have used magnetic tape reels or cassettes.

What is important is that collected data is uniquely identified and is in a format that is understood at the receiving station. Typical items for mutual agreement include the following:

- Data name
- Unique identifier
- Range
- Resolution
- Sampling rate

Typically the identification of data or parameters measured will in accordance with project standards, national or international standards, especially if the data is to be used by different parties.

A typical system for acquiring data from aircraft sensors, systems, and data buses is illustrated in Figure 6.2. This system will typically be used for flight test, maintenance management, and accident data recording systems.

Aircraft Systems Classifications: A Handbook of Characteristics and Design Guidelines, First Edition.
Allan Seabridge and Mohammad Radaei.
© 2022 John Wiley & Sons, Inc. Published 2022 by John Wiley & Sons, Inc.

242 | *6 Supporting Ground Systems*

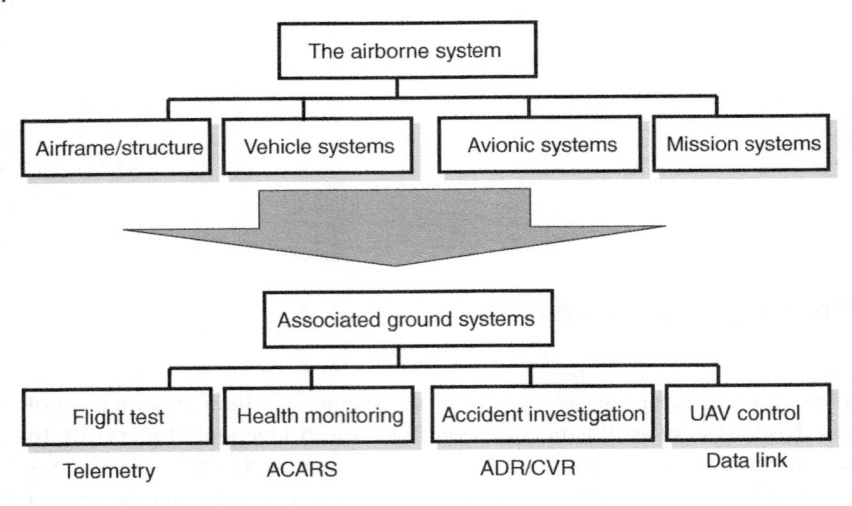

Figure 6.1 The integration of airborne and ground systems.

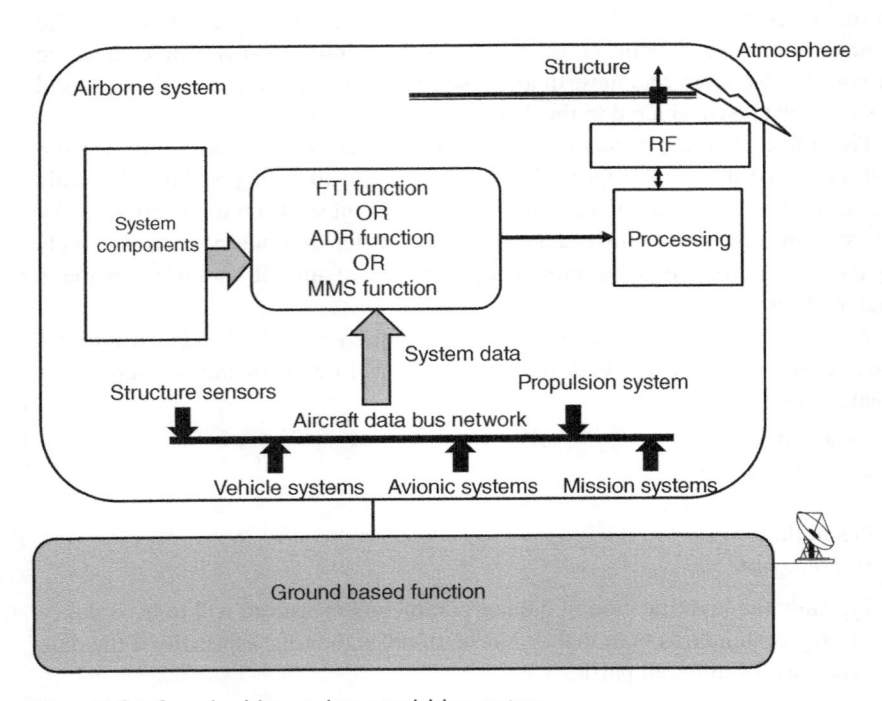

Figure 6.2 Generic airborne data acquisition system.

6.1 Flight Test Data Analysis

Flight-testing is an essential part of the process of gathering evidence to supplement ground test evidence to support the qualification of an aircraft type. Flight testing is required for the following types:

1. Experimental aircraft
2. Prototype aircraft
3. Production aircraft

Experimental aircraft have been developed at various stages in the history of aerospace to demonstrate the feasibility of new shapes, propulsion systems, new materials, and technology application. Often, only a single experimental aircraft is produced which becomes a valuable demonstration vehicle. It may have a short life because the need to reduce costs may lead to limited flight clearance for many components. Because of this flight testing needs to be focused, efficient, and effective. One example of a UK programme is the Experimental Aircraft Programme which was developed to demonstrate a number of technologies to reduce the risk to a follow-on production program (Seabridge and Skorczewski, 2016).

Prototype aircraft are a key tool in the test phase of aircraft development, where there is a need to collect information from the aircraft system for ground analysis, (Seabridge and Moir, 2020, Chapter 7). Most aircraft programmes include a prototype phase and each prototype will be equipped with the means of monitoring the behaviour of components of the systems and the performance of the aircraft. Full-scale prototypes are an extremely costly form of model, but the penalty of proceeding to series production without fully understanding the performance issues may be even more costly. The number of prototypes in any one programme is determined by the amount of testing to be done; the load of general and specialist systems tests is usually spread across the available prototypes to enable concurrent testing to take place. The results of the analysis inform the designers of the systems on the verification of their system design and will be used as part of the evidence of safe and correct operation. Most flight-testing is performed at the airfield at which the prototype was built; however, testing must be supported at specific trials at other sites. Examples are hot, cold and tropical trials, radar and EW ranges, and weapon release.

Production acceptance testing consists of a limited series of tests on each aircraft leaving the production line prior to delivery to a customer. The testing will be sufficient to confirm that the aircraft is safe to fly and forms the basis of handover to the customer.

The sensors and data bus connections that allow the collection of data from the aircraft and its systems are often referred to as flight test instrumentation (FTI).

244 | *6 Supporting Ground Systems*

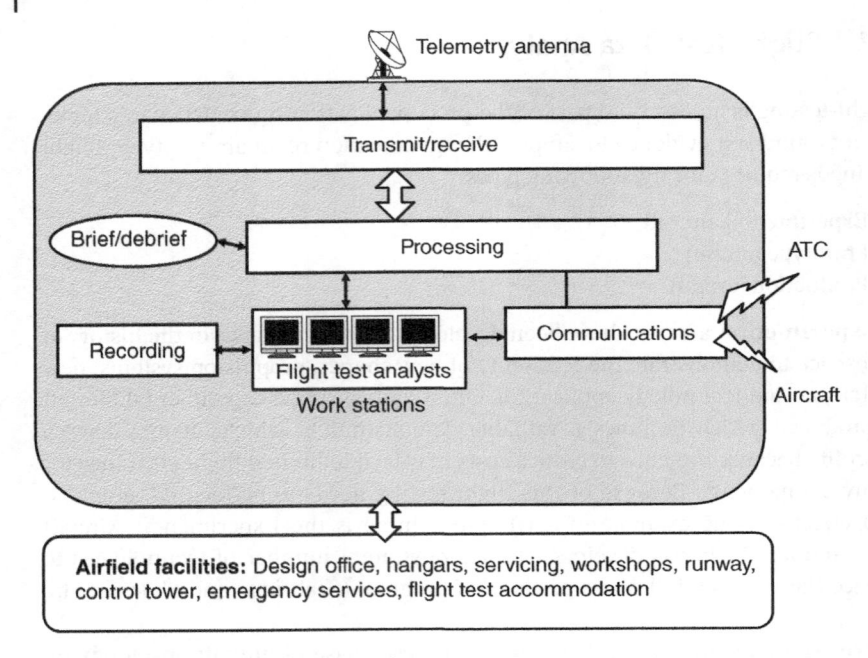

Figure 6.3 Flight-test ground-based system.

In the aircraft, data on the performance of the systems are collected from direct connection to system wiring or components. The design for this 'tapping' into the system wiring must be done with great care to avoid any interference with correct system operation from the connection of FTI. Alternatively, much information can be obtained from the aircraft data bus network. On legacy aircraft, the gathered data is stored in removable data media for removal after flight. On more modern aircraft, the data are transmitted to a ground station by telemetry to allow real time analysis of data. On large transport aircraft, the processing and workstations may be assembled in the passenger cabin so that the flight test specialist can be a part of the flight testing (Figure 6.3).

6.1.1 Purpose of System

To provide an environment for planning flight test missions, to manage tests and trials, to brief and debrief air crew, to collect and analyse data, to provide test reports to support qualification to provide safe management of all test activities, to design and manage the installation of FTI to collect the required aircraft data. At the flight-test centre, the telemetry signal from the aircraft is received and the data made ready for analysis. Specialist flight-test analysts perform the following tasks:

- Compile test procedures for each test flight.
- Conduct the pilot's briefing for the flight, communicate directly with the pilot and debrief the pilot immediately after the test.
- Analyse the data and provide detailed test results.
- Provide certification evidence for design and airworthiness.

6.1.2 Description

The system must provide a suitable office environment at the flight-test centre, provision of workstations, communications, and analysis tools for flight-test personnel. It will be necessary to support operations of the aircraft at remote bases such as air shows, hot and cold weather trials, range trials for military aircraft, flight-testing at partner airfields. This support must include facilities for personnel, normal and secure communications, and suitable analysis capability.

6.1.3 Safety/Integrity Aspects

The connection of FTI to the aircraft and its systems shall not degrade the performance of the systems. Any trials that would provide a hazard to structure, engines, and aircrew must be underwritten by design and airworthiness, examples include stall and departure tests, spinning, rejected take-off, and release of stores.

6.1.4 Key Integration Aspects

Integration and connection to the aircraft and its systems with no possible likelihood of interfering with the performance of the systems or of the aircraft. Provision of facilities for design and development of FTI, collection and analysis of data.

6.1.5 Key Interfaces

Access to aircraft to be tested, flight operations, ground personnel, aircraft design, and airworthiness.

6.1.6 Key Design Drivers

Focussed, efficient, and effective flight-test process to minimise time spent testing and to provide the correct results.

6.1.7 Modelling

Previous experience, laboratory experimentation, use of trials aircraft to test telemetry.

6.1.8 References

Gregory and Liu (2021).

6.1.9 Sizing Considerations

Number of aircraft to be tested simultaneously, duration of test programme.

6.2 Maintenance Management System

It is common for the health of airframe, engine, and aircraft systems to be monitored continuously to record observable failures, but more commonly to collect data to identify trends toward degraded performance so that more intelligent decisions can be made about equipment removal and replacement in a cost-effective manner. Systems are in use on many aircraft types to gather on-board data and relay it to ground facilities, such as engine health monitoring, structural health, and usage monitoring and prognostics/diagnostics systems are found in many types of aircraft. Data are relayed to the ground by data link and the ARINC communications and reporting system (ACARS).

Systems health data can be collected by monitoring of components or parameters in individual systems either directly or by extracting selective data from the aircraft data bus network. The type of information need to perform effective analysis must be included in the design of the individual system. Engine data can be collected from the engine control system (full authority digital engine control [FADEC]) and made available to the engine manufacturer as well as the aircraft systems. Structural health data can be collected by sensors or accelerometers incorporate in the airframe structure or skin materials (Figure 6.4).

Data is analysed and results stored in a data base that identifies the specific aircraft and fleet data. The results can be used to predict where and when to perform any maintenance actions and to order spares as appropriate.

6.2.1 Purpose of System

To collect real time data during routine flying in order to inform decisions about the most cost-effective manner to repair or service aircraft and to collect fleet-wide maintenance data.

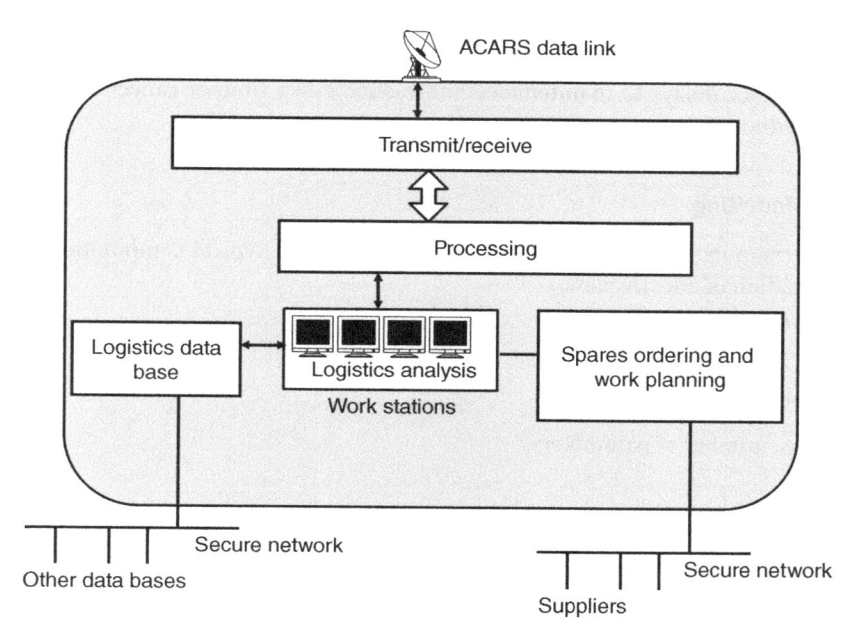

Figure 6.4 Logistics ground-based system.

6.2.2 Description

The system includes a suitable office environment, provision of workstations, communications, and analysis tools for flight test personnel.

6.2.3 Safety/Integrity Aspects

The connection of maintenance management system data collection devices to the aircraft and its systems shall not degrade the performance of the systems. The decision about the deliberate carry-over of failures in redundant systems to another flight must be performed with great care to reduce risk.

6.2.4 Key Integration Aspects

Integration with national and international logistics systems and suppliers to ensure rapid reporting and corrective actions in identical types.

6.2.5 Key Interfaces

Interface to aircraft wiring, initially by design and then by physical connections.

6.2.6 Key Design Drivers

Need to reduce delays to maintenance and reduce down time or cancellations. Need to reduce maintenance costs in a safe and effective manner.

6.2.7 Modelling

Computer simulation of system complexity, modelling of typical computations, and estimation of effectiveness.
No references are available.

6.2.8 Sizing Considerations

Fleet sizes, number of parameters.

6.3 Accident Data Recording

By regulation, newly manufactured aircraft must monitor at least 88 important parameters such as time, altitude, airspeed, heading, and aircraft attitude. To aid in an investigation following an accident or incident. Data are collected from direct connections to aircraft systems and from aircraft data bus networks on a continuous basis in order to assist in determining the cause of accidents. The data must meet customer and Civil Aviation Authority/International Civil Aviation Organisation (CAA/ICAO) requirements for the parameter and for data formatting.

The data are commonly stored in an accident data recorder designed to withstand the rigours of high-speed impact crash, fire, and submersion in seawater. Systems data is complemented by cockpit voice recordings and video recording. The recorder is painted bright orange and is equipped with a locator beacon capable of operation for at least 30 days. The recorder will be recovered after an accident and the data analysed to replay critical moments in the flight prior to the accident.

The systems data can be supplemented by a cockpit voice recorder used to record the audio environment in the flight deck. This is typically achieved by recording the signals of the microphones and earphones of the pilots' headsets and of an area microphone in the roof of the cockpit, including communications with air traffic control.

Cockpit image recorders provide information that supplements the cockpit voice recorder (CVR). The rationale is that what is seen on an instrument by the pilots of an aircraft is not necessarily the same as the data sent to the display device. This is particularly true of aircraft equipped with electronic displays. A mechanical instrument panel is likely to preserve its last indications, but this is not the case with an electronic display.

6.3.1 Purpose of System

To provide an environment for the recovery of the recorder and removing the memory devices and recovering all the data contained to aid with an investigation. The facility must also have the capability to deal with data recorders that have been mechanically damaged, subject to fire and intense heat, and exposed to prolonged submersion in seawater. It must be possible to read data media recovered from damaged recorders. This is a specialist activity and is often the preserve of government or national air safety agencies.

6.3.2 Description

The system includes the collection of mandatory parameters from the aircraft systems and structure which is stored in a rugged accident data recorder mounted in the aircraft.

6.3.3 Safety/Integrity Aspects

The connection of accident data recorder (ADR) wiring to the aircraft and its systems shall not degrade the performance of the systems.

6.3.4 Key Integration Aspects

Integration of ADR connections with the aircraft wiring and data bus design.

6.3.5 Key Interfaces

Interface with aircraft wiring by design and physical implementation.

6.3.6 Key Design Drivers

Efficient collection of data, rugged design to withstand fire, shock, and immersion in seawater. Rapid interrogation of recorded data when required.

6.3.7 Modelling

Design based on previous implementations and best practice.
   No references are available.

6.3.8 Sizing Considerations

Number of parameters, duration of flights, period for retention of data during extra long flights.

6.4 Mission Data Management (Mission Support System)

Specialist mission system analysts perform the following tasks:
- Compile plans for each mission.
- Prepare intelligence, codes and mission plans.
- Provide these to the aircrew or upload into the aircraft mission system.
- Conduct the pilot's briefing for the flight, communicate directly with the pilot, and debrief the pilot immediately after the mission.
- Analyse the data and provide detailed mission success results.

Data is acquired and stored by the mission computer from all sensors and the navigation system to provide a complete record of the mission. This is used by the mission crew to enable them to make decisions and to complete the mission using their track and location and sensor data. All of this is required by the tactical commander to command the mission and all the data can be stored on portable media (cassette, CD, hard drive) or it can be down loaded on a secure data link. It is then available for analysis of the mission, supported by the crew debrief, to analyse the mission, and to provide information to brief the next mission. A generic system diagram is shown in Figure 6.5.

6.4.1 Purpose of System

To provide an environment for the conduct of military missions which include briefing the mission, providing on-line data analysis and mission navigation, comparison of sensor data with on-board or off-board data bases to advise the mission commander, recording of sensor data and all decisions and providing the mission debrief.

6.4.2 Description

The system includes a suitable collection and collation of sensor data and presentation on operator workstations, appropriate long-range and tactical area navigation and data recording.

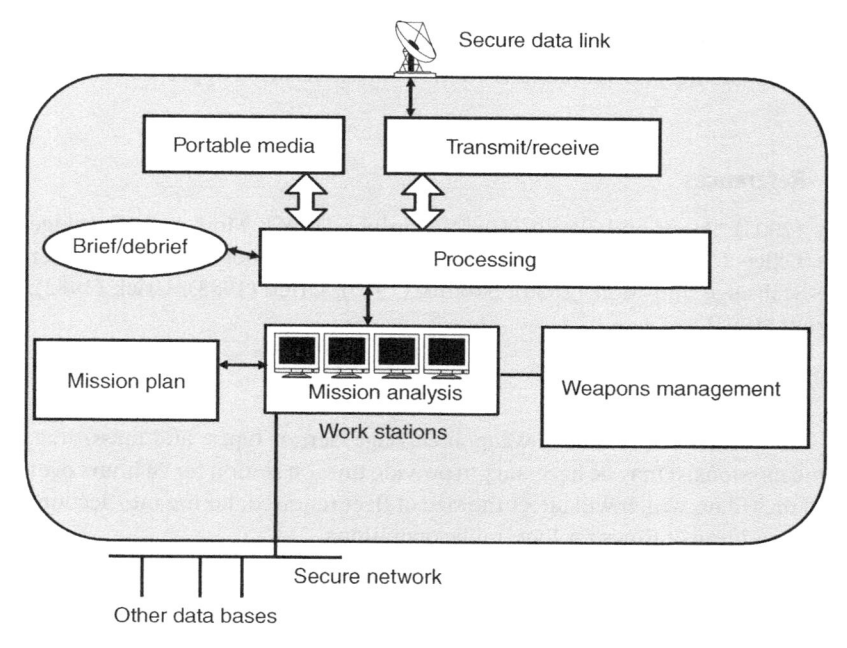

6.4 *Mission Data Management (Mission Support System)* | 251

Figure 6.5 Mission management ground-based system.

6.4.3 Safety/Integrity Aspects

The aircraft shall remain safe under all mission conditions which may require evasive action and reaction to enemy defence threats.

6.4.4 Key Integration Aspects

Integration with the mission computer to enable download of all relevant data. Secure communications and data links. Integration with national and internal agency communications networks.

6.4.5 Key Interfaces

Communications interface with main operating bases, with other co-operating forces and with the intelligence community.

6.4.6 Key Design Drivers

To make a significant contribution to the performance on any single mission and to enable collation of intelligence to support other missions.

6.4.7 Modelling

Computer modelling and simulation of various missions using appropriate sensor models.

6.4.8 References

Adamy (2003); Airey and Berlin (1985); Gardner (1996); Moir and Seabridge (2006); Oxlee (1997); Poisel (2003); Rigby (2013); Schleher (1978); Schleher (1999); Seabridge and Moir (2020); Skolnik (1980); Urick (1983); Urick (1982); Van Brunt (1995).

6.4.9 Sizing Considerations

Range and duration of missions, weapon carriage, sensor types, and mass/drag. For some missions, it may be necessary to provide time on station for 24 hours over a period of 30 days which will affect the size of fleet required, taking into account reliability and transit times for long-range operations.

6.5 UAV Control

Unmanned air systems are being used to collect information and to conduct military action, usually under control of a human command structure. Even in the event of such vehicles acquiring more autonomy, there will still be a need for information to be gathered on the ground for analysis and for commands to be sent to the vehicle. This will require the vehicle to be designed with telemetry and with communication paths to download information and upload commands.

The human machine interface portion of the unmanned air vehicle (UAV) system is the ground control system, also known as ground control system (GCS), command and control station (CCS), mission control ground station (MCS), mission planning, and control ground station (MPCGS) (Figure 6.6).

6.5.1 Purpose of System

To provide an environment for a human operator to control the unmanned vehicle and to receive, store, and process data received from the UAV sensors.

6.5.2 Description

The control environment includes one or more of the following:

- A simple hand-held controller in a ruggedised case and line of sight communications antenna.

6.5 UAV Control

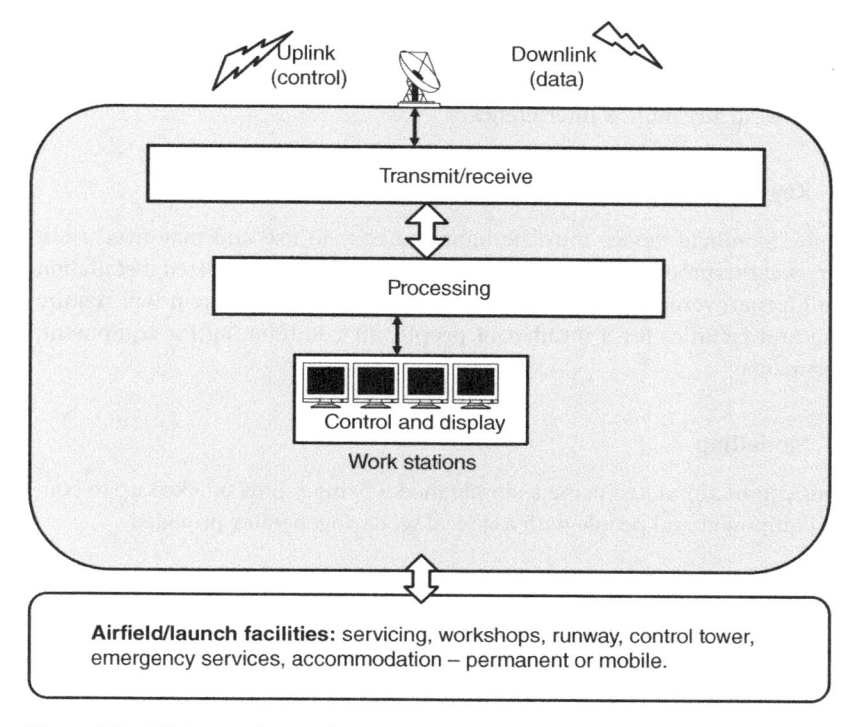

Figure 6.6 UAV ground control system.

- One or more multi-terrain land vehicle(s) to include transportation of the UAV, the launch and recovery equipment, communications and space for one or more workstations.
- A full-scale command and control centre, either a building or transportable containerised accommodation to house multiple operators and suitable analysis and control modules.

6.5.3 Safety/Integrity Aspects

Segregation of the flying crew from sensor operators to minimise operator error. May require command destruction of the vehicle to minimise risk to people and property on the ground in the event of an unrecoverable failure.

6.5.4 Key Integration Aspects

Integration with a mission management system and network access to an overall command and control structure. Provision of an integrated launch facility with fuel, power sources, accommodation, and facilities.

6.5.5 Key Interfaces

Interface with the control and data links with the appropriate bandwidth and separation to avoid any mutual interference.

6.5.6 Key Design Drivers

A mobile, handheld device must be light and easy to use and may need rudimentary weatherproofing. A mobile system may require a ruggedized installation in an all-terrain vehicle. A full-scale command and control system will require housing and facilities for a number of people, air conditioning for equipment, and occupants.

6.5.7 Modelling

Missions can be simulated using a simple model using rooms mocked up to contain all equipment and people with a typical scenariogenerator provided.

6.5.8 References

Sadrey (2020).

6.5.9 Sizing Considerations

Number of vehicles to be operated simultaneously, suitable bandwidth for command, and high-resolution sensor download. Speed of command link to maintain suitable vehicle stability and control over extreme long ranges.

References

Adamy, D.A. (2003). *EW 101 A first Course in Electronic Warfare*. Artech House.
Airey, T.E. and Berlin, G.L. (1985). *Fundamentals of Remote Sensing and Airphoto Interpretation*. Prentice Hall.
Gardner, W.J.R. (1996). *Anti-Submarine Warfare*. Brassey's Publication.
Gregory, J.W. and Liu, T. (2021). *Introduction to Flight Testing*. Wiley.
Moir, I. and Seabridge, A. (2006). *Military Avionics Systems*. Wiley.
Oxlee, G.J. (1997). *Aerospace Reconnaissance*. Brassey's Publication.
Poisel, R.A. (2003). *Introduction to Communication Electronic Warfare Systems*. Artech House.
Rigby, K.A. (2013). *Aircraft Systems Integration of Air Launched Weapons*. Wiley.
Sadrey, M.H. (2020). *Design of Unmanned Aerial Systems*. Wiley.

Schleher, C.D. (1978). *MTI Radar*. Artech House.
Schleher, C. (1999). *Electronic Warfare in the Information Age*. Artech House.
Seabridge, A. and Moir, I. (2020). *Design and development of aircraft systems*, 3rde. Wiley.
Seabridge, A. and Skorczewski (2016). *EAP: The Experimental Aircraft Programme*. BAE Systems Heritage Department. Available from Amazon.
Skolnik, M.I. (1980). *Introduction to Radar Systems*. McGraw-Hill.
Urick, R.J. (1982). *Principles of Underwater Sound*. Peninsula Publishers.
Urick, R.J. (1983). *Sound Propagation in the Sea*. Peninsula Publishers.
Van Brunt, L.B. (1995). *Applied ECM*. EW Engineering Inc.

Exercises

6.1 Propose a set of requirements for the processing the results of photographic or image surveillance covering all aspects of image capture from a wide range of cameras/sensors, including the dissemination of results to all users.

6.2 From the requirements describe a suitable ground based system operation.

6.3 Examine the security issues associated with your implementations.

6.4 If a commercial organisation wished to provide such a facility as a bought out service, what security arrangements would have to be in place to satisfy the needs of military, law enforcement and civilian agencies.

7

Modelling of Systems Architectures

7.1 Introduction

The systems described in Chapters 3–5 form the basis of architectures that will be used in a completed project. At the early stages of the design, it is most likely that a number of different architectures will emerge. A mechanism is required for assessing the suitability of these architectures and arriving at an optimal solution based on measures such as mass, volume, power consumption, cost, and fitness for purpose. This can be a time-consuming task and an automated model driven methodology is to be preferred (Butz 2007). This chapter looks at some methods currently being examined and proposes an example methodology.

The system architecture and the proposed methodology studied in this chapter are relevant to avionics integration and architecture since the avionics system architectures have reached to a level of complexity that designing a new architecture and/or upgrading legacy architecture manually does not work well. In other words, it is crucial to exploit the full potential concept of integrated modular architectures (IMA) and/or distributed integrated modular architectures (DIMA) architectures which seems impossible through designing architectures by hand. Therefore, in recent years, a trend to create computer-aided design and model-based design tools is growing. The main goal of these tools is to provide an automatic and/or semi-automatic avionics architecture optimisation tool. The input to these tools is a model that represents the system and allows for formal analysis, verification, evaluation as well as simulation. On the other hand, avionics integration (functional and physical) is also an important task which tries to make the best use of software and hardware resources. Particularly, avionics integration is to utilise a mechanism for using resources, management processes to identify the sharing of resource capabilities, the potentiality of reuse of resources in order to improve equipment utilisation, operating efficiency, and availability. In what follows the very basic principle of avionics integration is defined.

Aircraft Systems Classifications: A Handbook of Characteristics and Design Guidelines, First Edition.
Allan Seabridge and Mohammad Radaei.
© 2022 John Wiley & Sons, Inc. Published 2022 by John Wiley & Sons, Inc.

7.1.1 Principle of Avionics Integration

In general, systems architecting is intended to create and build systems. It tries to find trade-off, balance, and compromise among the customers' needs and existing resources and technologies as well as technical and/or operational requirements (NASA 2007). Avionics system architecting, in particular, is an important and challenging task that can help engineers to visualise concepts by enabling requirements to be mapped, decomposing functions, and determining functional links to physical mapping of the functions into an aircraft structure. In other words, the system architecture is a means of visualising concepts, in the early design stage, to discuss things and agree upon interfaces, functional integration and allocation, and the usage of commercial off the shelf systems (COTS) components and standards independently of physical implementation.

System integration, on the other hand, is the design process in which decisions are made to integrate sub-systems and functions into a total system architecture irrespective of how the system will be fabricated. Finally, avionics architecture is referred to a general arrangement of systems, sub-systems and equipment that together perform a set of functions defined as avionics architecture considering top-level requirements, functional allocation, logical connections, and physical interfaces.

The term integration is widely used in aerospace industry including integrated team, integrated products, and integrated solutions which invoke a variety of definitions. It therefore seems that the meaning of integration becomes ambiguous in many cases. The dictionary definition of integrated is 'made up of parts'. Integration thus is the process of bringing these parts together to make a whole. Consequently, to define integration, the parts that are going to be combined and merged need to be defined. In aircraft system design, the integration can happen in different levels like component, function, system, process, and information level (Wang and Zhao 2020). Integration happens because engineers want to investigate solutions that are more efficient in operation and in their use of equipment. To do so, designers often incorporate many functions into one hardware component and/or line replaceable item (LRI) (Seabridge and Moir 2020).

Therefore, the first step in avionics system integration is functional integration which brings together functions. In other words, functional integration is the level of integration that defines how the functions are partitioned and how they interrelate. In functional integration, the distinctions between previously separate boundaries are the principle safety issue. On the other hand, physical integration is the level of integration that defines how the functions are implemented by hardware and software components. It is therefore the task of deciding how the system is to be implemented in real world in terms of its geographical location, electrical isolation, logical view of its environment, and hardware modularity.

It should be noted that, if a system contains 'n' functions, the physical integration of that system may not yield a system containing 'n' physical elements. In other words, there may not be a one-to-one mapping of functional to physical elements (Nickum et al. 2002). In the process of physical integration, the system functional partitioning may become unclear. There is no formal link between physical and functional integration; however, there is a trend, and that is, as the level of functional integration increases, the necessary co-operation between functions becomes more complex and demanding. This is true about integrated modular avionics (IMA). While IMA architecture reduces constraints like size, weight, and power consumption, the level of system integration has increased. To overcome the complexity of systems integration and optimise avionics architecture, researchers are striving to provide tools and/or methods to ease the process and exploit the full potential of the integrated architecture. To do so, the first step is to model the avionics architecture.

7.2 Literature Survey of Methods

Aircraft systems design in general and avionics systems design in particular are complex tasks as a number of factors need to be taken into account including size, weight, power consumption, and cost of equipment as well as safety constraints and environmental conditions. This also leads to a high interaction between systems, and engineering domains that necessitates the need for a well-established systems engineering approach to handle it.

7.2.1 Model-Based Development

A model is a simplified presentation of a real system which represents the important aspects of the system with regard to a particular purpose (Weilkiens et al. 2015). Models have been used in engineering in various forms including construction of physical models/prototypes, symbolic models like mathematics and computer aided engineering (CAE) tools. In what follows, model-based development (MBD) and model-based system engineering (MBSE) and their relations are defined.

MBSE: It is a formalised engineering approach which uses a central model to capture requirements, and supports systems engineering activities like design, analysis, verification, and validation.

MBD: It is a mathematical and visual method based on abstract representations with predefined semantic and syntax, which address complex systems design.

Relation: The principle of MBSE is used to execute MBD.

7.2.2 Modelling and Analysis Techniques

What is meant by modelling here is a system model that represents a system, in particular, an aircraft system like a fuel system and/or an avionics system which are made up of hardware and software as well as human interaction. The model can be a model of models known as meta-model (Andersson 2009). An aircraft model, for instance, is composed of several models which is still a system model although at a higher complexity level. At the conceptual level, avionics system, particularly, defines the functions and the relationship among them. In further detail designs, the functions' interaction, flow, and physical equation can also be added to predict performance and dynamic behaviour of a model. One engineering approach to model aircraft systems is called MBSE which uses a central model to capture requirements, architecture, and design to help system architects in different levels of aircraft system design (Andersson and Sundkvist 2006).

There are a number of modelling techniques in aircraft system and avionics system design, each of which is appropriate for a set of problems. The main goal of modelling is to provide a tool for concept generation, automatic architecture optimisation, and trade-off studies of generated architectures. Architecture modelling determines the system boundary, its components and subsystems as well as its interfaces. Architecture modelling is also known as systems architecting. The main task of systems architecting is to provide a balance between requirements and actual products. Different various design techniques are being used to help systems developers to understand, decompose, analyse, and document the problems including axiomatic design, design structure matrix (DSM), and function/means tree.

Axiomatic and DSM are design matrices that can systematically analyse and document the relations in the design process. The axiomatic method is based on two rules/axioms which use a matrix to visualise, analyse, and transform the customer requirements to functional requirements (FRs), design parameters, and process variables (PVs). Figure 7.1 shows the process in axiomatic design. It includes mapping customer requirements into functional requirements (FRs). To satisfy the FRs, design parameters are defined in the physical domain. Finally, process variables (PVs) are defined to satisfy design parameters. This mapping process is shown linearly in a design matrix.

DSM is another tool and technique in systems engineering for system decomposition and integration. It provides a visual representation of a complex system as well as design parameters with their interdependencies and flow information which support innovative solutions to decomposition and integration problem. Any changes that may affect the system can also be analysed by DSM i.e. if a component needs to be changed, all the dependencies and interfaces can be quickly identified. It can also help the design team to optimise the flow of information

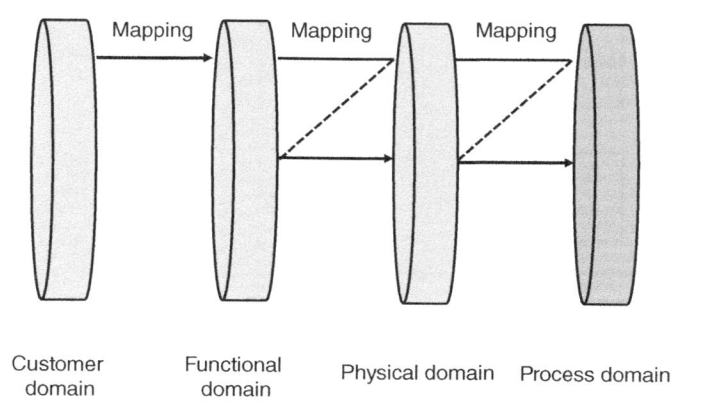

Figure 7.1 The fundamental concept of axiomatic design.

between different interdependent components. The House of Quality (HoQ) is another matrix-based method that combines the analysis of functional decomposition to component dependencies. More matrix-based methods can be found in Gavel (2007).

In a large-scale problem, matrix-based methods are difficult and time-consuming to handle, but function/means tree is more suitable. This method is also used for functional decomposition, allocation to means i.e. components to fulfil the requirement and concept generation. The function/means tree has hierarchical structure functions and means on various levels. It can also be used to represent alternative solutions, from which a final candidate can be determined (Johansson and Krus 2006). Since integrated modular avionics architecture has many software applications, and hardware as well as dependencies among them, this allocation-based design technique is more appropriate. Figure 7.2 depicts a generalised inheritance mechanism which is called generic object inheritance enabling a quick reuse and modification of conceptual product and/or solution models at any level in their hierarchical break down structures (Johansson et al. 2008).

The developed methodology, explained in Section 7.3, used this technique for avionics functional breakdown and allocation of avionics line replaceable units (LRUs) for each function. Each function, in some cases, a combination of two or three functions is allocated to LRUs (means/solutions) from different vendors. This has created a set of data which is used for investigating various architectures as well as trade-off studies. The best allocation is based on physical and operational requirements needed. In other words, the avionics functions are assigned to components (LRUs) and components are assigned to installation locations.

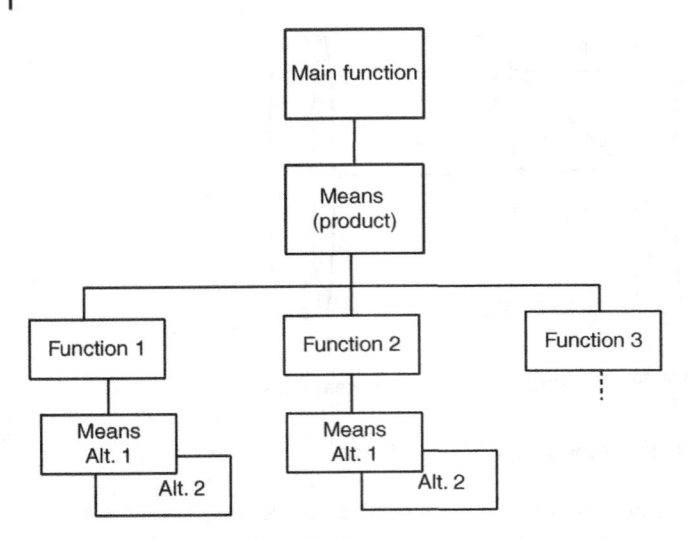

Figure 7.2 Function/means tree.

7.2.3 Integrated Development Environment

The advancement in domain specific modelling has created powerful tools for their main purposes. Many tools, however, have a limited scope of modelling and/or analysis which necessitates the need to integrate information/models from various modelling domains. To analyse and simulate a system in a large scope i.e. aircraft level, models and tools from various modelling techniques need to be integrated. For instance, an integrated system model could be models from hardware, models of embedded software, and models of the environment of the system e.g. other subsystems. In aircraft systems development, particularly avionics systems development, a set of engineering tools that are designed to work together are known as 'integrated'. These tools will decrease the non-productive time which is spent on user interaction with separate tools as well as transferring data among various engineering tools. Some of the most mature tools that are being used in aircraft systems development are DOORS, Focal Point, Modelica, Dymola, SCADE, MATLAB Simulink, Unified Modelling Language (UML), and Systems Modelling Language (SysML).

7.2.4 Model Management

For a complex systems design, when moving from document-based method to a model-based one, it is crucial to understand how to manage the set of models that have to be explained and planned. In other words, model management is to ensure that model-based methods and tools fit into the existing development environment for aircraft and/or avionics systems. Moreover, the modelling techniques and models as well as the supporting tools need to be managed during the whole of system's life cycle i.e. every change in a model has to be validated.

7.2.5 State-of-the-Art Avionics Architecture Modelling

A couple of modelling approaches are being used for avionics architectures including static modelling, safety modelling, and dynamic timing analysis. Here the emphasis is on models used in the context of IMA and DIMA architectures, and their pros and cons are addressed. Some general systems modelling approaches like UML and SysML can also be used to model some aspects of IMA architectures; however, UML profiles and their lack of discipline makes it difficult to verify, evaluate, and optimise IMA architecture automatically.

7.2.6 Key Aspects to be Modelled

The main goal in avionics system architecture modelling is to enhance the planning and the design process as well as improving the operational capability of the desired architecture. Thus, it is critical to fully understand all aspects of the avionics system architectures including the driving requirements, IMA architecture components and the process of function allocations to hardware, and hardware mapping of the components into aircraft structure. Figure 7.3 represents an overview of the IMA/DIMA architecture elements from software mapping to hardware mapping and their associated constraints and qualitative/quantitative measures.

The **system layer** is used to describe the functional breakdown and/or decomposition i.e. each subsystem has its dedicated technical specifications like resources (computation and storage), connectivity and capacity as well as other properties associated to IMA/DIMA architectures such as design assurance level (DAL). The aim of this layer is to identify the system tasks, logic, and connections as well as required sensors/actuators and interfaces for functional subsystems. For instance, a high lift system task/function is the extension and retraction of flaps and slats triggered by a flap lever in the cockpit. This may include a sensor acquisition, a monitoring, and actuator modules. These modules are enabled by a set of hardware components in hardware layer including common processing modules (CPMs), remote data concentrators (RDCs), and sensors.

The **hardware layer** is concerned with hardware devices and their configurations as well as network topology that support the functions in the system layer. The typical devices in IMA/DIMA architecture are CPMs, RDCs, input/output (I/O) modules, sensors, actuators, switches, and cables.

264 | *7 Modelling of Systems Architectures*

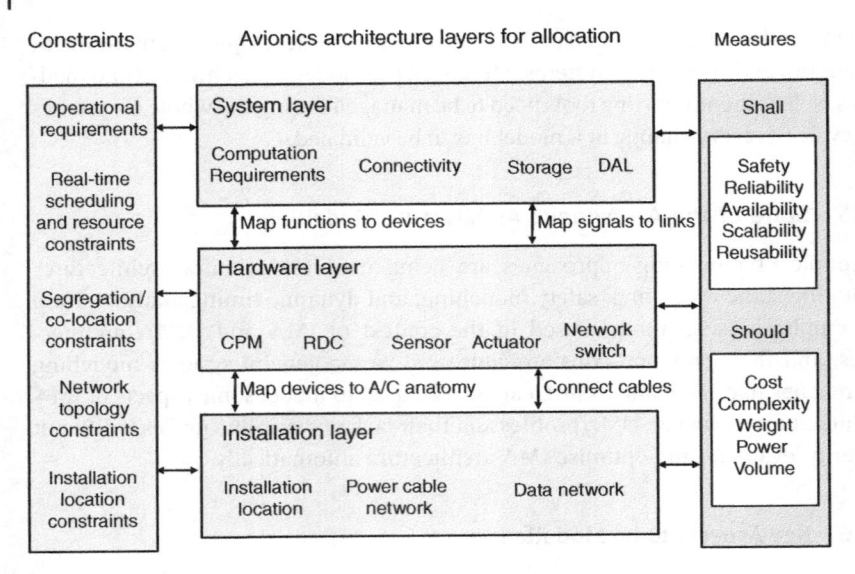

Figure 7.3 IMA/DIMA system architecture elements and design layers.

The **installation layer** allocates hardware to aircraft anatomy. It is indicated by installation locations and power network. The devices can be placed at different installation locations like avionics bay, cockpit, and centre and tail of the aircraft. These locations have their own constraints including mass, volume, cooling capability, number of slots, and power supply. These criteria must be taken into account for the optimisation of architecture while placing the devices in their installation locations. It can affect both the cable length and safety requirements.

All the three layers must be carefully modelled and designed by taking into account the required measuring criteria to achieve an optimised avionics architecture. In what follows some modelling techniques and/or tools are reviewed.

7.2.7 Architecture Analysis and Design Language (AADL)

Architecture analysis and design language (AADL) is one of the most popular tools in the avionics domain. It is a modelling language to describe software and hardware of critical and embedded systems. The hardware in the AADL is modelled as a composition of processors, memory, buses, and devices. Software is comprised of processes, threads, subprograms, and data. The system is comprised of an arbitrary set of software and hardware components. Several systems can be implemented. Further, components can be linked together by either signal relations or by child relations (hierarchical dependencies), for instance, assigning a process to a processor. AADL is standardized by Society of Automotive Engineers (SAE), including an extensible markup language (XML) data format and graphical representation (Feiler et al. 2006).

7.2.8 AADL-Based Modelling of IMA

The different layers shown in Figure 7.3 have been modelled by AADL. Fraboul and Martin proposed an IMA domain model consisting of four layers including application, architecture, mapping, and execution (Fraboul and Martin 1999). The application model includes Air Radio Inc. (ARINC) 653 partitions and ports as well as their connections. The architecture layer comprises CPM and buses. The application layer is bound to the architecture in the mapping layer. Finally, it is viable to automatically derive discrete even simulations by combining layers with the execution layer, which holds run-time information on hardware and software.

Delange et al. also proposed an AADL scheme to model ARINC 653 partitioned systems (Delange et al. 2002). The scheme defines how ARINC 653 partitions, processes, communication ports, health monitoring as well as ARINC 653 hardware is modelled. It defines the AADL classes to use and set of attributes. The proposed rigid modelling provides a derivation of executable dynamic schedule models. The approach is used for model-based verification of system and ARINC 653 OS layers. Further, a simulation approach of configured IMA system down to hardware is given by Lafaye et al. (2011). This paper shows how to transform AADL models in SystemC models. SystemC models are executable and used for hardware software interaction simulations. A unique one-to-one mapping between AADL classes to SystemC modules is given. Moreover, with the rigid modelling rules, transaction level SystemC models (TLM) can automatically be derived. Also, a demonstration shows how to verify processor and memory load limits on an IMA device loaded with system software over time.

7.2.9 Mathematical-Based Modelling of IMA

Förster et al. proposed π-calculus as a textual modelling language for avionics and IMA systems specification and verification (Förster et al. 2003). π-calculus is a formal mathematically motivated process specification language. The modelling elements are processes and their communication. The execution of π-calculus allows formal verification of the logical correctness of the modelled systems. The proposed model covers functional and communication specifications as well as hardware and network topology. The automatic derivation of specification is enabled by specifying software–hardware bindings. The simulation of this implementation is used to verify complex software/hardware systems.

In addition to IMA devices, avionics networking i.e. Aircraft Data Communication Network (ADCN) modelling has also been modelled in some literatures.

Avionics full duplex switched Ethernet (AFDX) end-to-end delays are analysed by Charara et al. where a comparison is made of network calculus, the queuing network modelling approach, and model checking (Charara et al. 2006). Network calculus is a mathematical model expressing each network node by queuing capability and queue size. For given message sizes and arrival rates, this can be propagated through the network automatically. The queuing approach builds up networks from basic building blocks like links, buffers, and multiplexers. Underlying behaviour and configurable properties of each block allow a simulation of network traffic to obtain delay information. Model checking uses timed automata to express network nodes. Comparing the three approaches, the precision goes up from network calculus to model checking as the calculation effort does.

Lauer and Ermont also proposed a modelling and simulation approach to verify latency and freshness requirements in AFDX networks based on the tagged signals model (Lauer and Ermont 2011). The loaded IMA systems are modelled as processes and signals transmitted over timed channels. IMA devices with applications are expressed as processes. AFDX links are timed channels. The transforming tagged signal models are expressed in mathematical optimisation problems in which end-to-end latency and freshness parameters can be calculated for different network setups. The proposed approach is applied on a flight management system (FMS) composed of a switched AFDX network, RDCs, and the cockpit panels.

7.2.10 General Assignment and Mapping Problem

The assignment problem answers the question of how to assign m (jobs, students) items to n other items (machines, tasks). Since there are many possible ways to perform this assignment, it is then defined as an optimisation problem to achieve the best suited assignment for the problem (Burkard et al. 2009). The mathematical model is defined as follows:

Let $x_{ij} = 1$ if task i is assigned to person j and 0 otherwise. Let C_{ij} be the cost of assigning i to j. Then, the objective function is to minimise the total cost of the allocation which is to

$$\min \sum_{i=1}^{n}\sum_{j=1}^{n} C_{ij} x_{ij} \tag{7.1}$$

Each task goes exactly to one person and each person gets only one job. These are given by the following constraints:

$$\sum_{j=1}^{n} x_{ij} = 1 \quad \forall i \tag{7.2}$$

$$\sum_{i=1}^{n} x_{ij} = 1 \quad \forall j \tag{7.3}$$

The decision variable is defined as a binary integer programming (BIP), i.e. $x_{ij} \in \{0, 1\}$.

One of the applications of assignment and/or mapping problem is the allocation of software applications to processors in distributed computing modules (Bokhari 1987). The objective in these problems is that a set of tasks is assigned to processors for the shortest time execution. The assignment can be either static or dynamic. The processes are restricted by resources like calculation time, memory, or bandwidth. The network can also be considered. In other words, the communication topology plays a significant role when optimising the overall execution time. The assignment problem classes differ in multiplicity of assignment, the number and type of resources, and the number of assignment layers, constraints, and the objective functions. This problem is well-known to be NP-hard in most cases, i.e. it cannot be solved by a deterministic algorithm in certain amount of time.

In other words, the optimal solutions can be found by an exhaustive search, yet there are n^m ways to assign m tasks to n processors, for instance, the search is often impossible. Therefore, optimal solution algorithms only exist for small problems. This combinatorial optimisation problem can be solved by exact and heuristics methods which results in optimal and suboptimal solutions respectively. The best solution algorithm depends upon the problem properties and the problem size. The exact methods usually work for small instances, yet for large instances, heuristics are created. Two examples of distributing computing systems defined as static assignment problem and solved by heuristics are given below.

Lo created a three step heuristics method for assigning tasks to processors to optimise overall execution time and communication cost. First, tasks are assigned in an artificial two processor network, which is optimal but not complete. Thus, it is repeated iteratively with updated processor and network capacities. Finally, the unassigned tasks are assigned by trial to the first capable processor. They managed to get great objective values in short runtimes for a group of 34 tasks and six processors (Lo 1988). Kafil and Ahmad used the A^* algorithm to solve the same problem. The algorithm is a non-deterministic global search algorithm using a search tree to calculate the costs of visited parts. The method further was extended by reducing the number of nodes, which happened by an initial guess. The results for 20 tasks and four nodes have been calculated in very short runtimes (Kafil and Ahmad 1997).

The distributed systems assignment problems are basically similar to the issues in IMA and DIMA architectures design. For example, assigning the avionics functions to IMA devices is an assignment problem. Also, software mapping is a static assignment problem. Moreover, IMA task assignment depends strongly on resources like power, memory, and I/O types. The safety requirements are also very important. Similar examples of these types of problem can be found in operations research. In what follows the IMA related works are overviewed.

7.2.11 An Overview of IMA Architecture Optimisation

The aim in avionics architecture optimisation is to support the IMA design process with automated and/or semi-automated architecture generation. To achieve this goal, the IMA design problem must be expressed formally and mathematically. A mathematical algorithm is then applied to the model to perform an exhaustive design space exploration. The input and output of the model is IMA architecture models. This enables designers and/or decision makers to investigate various possible architectures as well as the optimal one.

The automated design and sizing of IMA architecture is a relatively young field of research. However, the idea of how to make the best use of resources, resource allocation problem, in computer science and general assignment problems in operations research are very old. In addition, automated design and optimisation are also carried out in other industries and disciplines like space, automotive, economics, and infrastructure planning. In what follows an overview of literature on general distributed computing system optimisation as well as IMA architecture optimisation is given.

Over the past decade, the implementation of IMA architecture has been widely used by aircraft manufacturers. IMA architecture optimisation emerged with the creation of the first IMA systems. In early years, the priority was on validation rather than optimality. The main issues in IMA architectures are on processor scheduling and finding safe and reliable mappings of functions and signals in networks. Many approaches previously have been studied for model-based design and verification of IMA architectures. Most of them are related to one hardware device or a single avionics system. Only a few approaches exist that handle the whole software and hardware architectures, multi-devices and multi-applications, as well as spatial distribution.

Salomon and Reichel developed a static model for the automatic design of IMA architecture i.e. the allocation of avionics functions to IMA devices and placing the devices in aircraft structure (Salomon and Reichel 2013). The model provides some automation in IMA design process by applying an optimisation algorithm and focuses on safety evaluation. The model expresses logical, physical, and structural aspects of avionics systems. It includes propagation failure layers, function instantiation, hardware, hardware types, and geometry. The failure propagation model expresses system logic in terms of functions and connections. The failure propagation is dealt within the three stages including 'ok', 'passive', and 'out of control'. Signals and components are 'passive' if an error is detectable and 'out of control' if not.

The hardware types define all the components with their technical specifications like failure probability and cost. These components are then allocated to a geometry model. The function instantiation expresses the operational system which comprises multiple instances of the functions and connections from the failure propagation model. The redundancy level is also defined in a function instance model which represents the functions of the failure propagation model. The complete model provides the automatic derivation of fault trees and the computation of reliabilities with some simplification like excluding failures canceling each other (Salomon and Reichel 2011). The optimisation method used is a conventional genetic algorithm (GA), and is applied to flight control system architecture. The method and the algorithm do work well for small and medium architecture like flap control system; however, for large-scale architecture e.g. the whole flight control system with a high redundancy level, it does need an improvement to produce optimal results.

Dougherty et al. in Vanderbilt University in collaboration with Lockheed Martin developed a tool called ScatterD to optimise embedded flight avionics systems (Dougherty et al. 2011). The domain studied is mission computer and flight control systems. The deployment problem is expressed as a multi-resources bin-packing problem. In other words, the computer aided design tool deploys software applications to hardware while satisfying a number of complex constraints including processing time and real-time scheduling like processor time, memory size, and bandwidth. The optimisation algorithm implemented is a hybrid heuristics, genetic algorithm (GA), and particle swarm optimisation (PSO). The tool manages to reduce the required processors and network bandwidth consumption. The installation location of hardware and the operational capability of the system architecture are not studied.

In another literature, Manolios et al. from Georgia Institute of Technology in collaboration with Boeing for B787 project developed a general framework called component-based system assembly (CoBaSA) that implements a constrained component assembly technique (Manolios et al. 2007). The main objective of the tool is to create an environment for construction of large industrial systems by integrating components, particularly COTS components. CoBaSA software includes an expressive language for component interfaces, properties, and system-level and component-level constraints. Further, it uses a pseudo-Boolean solver to solve the constraints using SAT-based method. The tool enables automatic solving of system assembly problem directly from a system requirement. The tool mainly contributes to a greater reliability, lower cost development, shorter development cycles as well as less testing and validation in system design and integration.

CoBaSA is further used for the integration of IMA avionics architecture components for the Boeing 787 Dreamliner. The assembly problem of IMA architecture involves mapping of avionics function to line replaceable modules (LRM), mapping of LRM to cabinets, mapping of RDCs to switches as well as sensors and actuators. Moreover, system architects have to consider some constraints including worst case execution time (WCET), I/O timing, memory, latency, network jitter, and so on. The implementation of CoBaSA has hugely reduced the assembly time for Boeing compared to their current methods.

Shi and Zhang also developed a tool for avionics integration optimisation using mathematical programming (Shi and Zhang 2016). The tool is created in three steps called system organization, system integration, and requirement analysis. First, the system organization is based on linear programming (LP) and is used to select the best vendor products according to the system performance requirements like minimising and/or maximising a particular cost function. In this step, suitable devices are selected for the architecture. The properties considered for the devices to be selected are performance, weight, size, power, processor unit, and DAL. Then, in the system integration phase, designers are able to form the optimal architecture with all the constraints defined by applying optimisation algorithms.

The algorithm used in this work is GA and PSO where users can select the algorithm for their problems. The requirement analysis further verifies the integration design by Boolean logic. The major requirement assessed in the paper is safety. Since the safety requirements are defined as Boolean, a satisfiability modulo theory (SMT) solver is used for requirement verification. The tool is basically developed in Java, and is eclipse-based. In all steps, the user needs to manually choose the safety objectives and evaluation algorithm. Spatial distribution of avionics modules in aircraft structure is not studied in this work, and the human machine system is not taken into account in the model for safety assessment.

Zhang and Xiao modelled the DIMA as a cyber physical system (CPS) integration scheme with physical layer and function layer. The physical layer focuses on mapping of CPM and RDC into predefined locations including avionics bay, cockpit, centre, and tail (Zhang and Xiao 2013). The function layer is about the allocation of tasks/functions to CPM, RDC as well as sensors and actuators. The integration constraints defined are influenced by the maximum resources available in each location including mass, slot, and cooling capability for physical layer. For the functional layer, constrains are related to calculation resources like memory and segregation constraints of functions.

To implement the scheme and apply multi-objective optimisation for the DIMA system, software based on MATLAB was developed. The multi-objective optimisation problem is solved by using an improved lexicographic optimisation technique for minimising the weight and maximising the reliability. The model represents that the reliability measure in optimisation is improved. Moreover, the comparison between IMA and DIMA shows that DIMA proves a better performance in reliability.

Salzwedel and Fischer presented a new methodology for optimising avionics architecture at aircraft level (Salzwedel and Fischer 2008). The proposed method is developed to deal with bounded and statistical uncertainties of early design stages related to the components of an architecture. To handle these complex tasks, two techniques have been employed. One technique is used to develop an automated design at system level that tackle the uncertainties, and the other one is developed for system level optimisation of avionics architecture and function. The automated design provides an environment for simulation by connecting to a database of architecture components and their connection information. The Monte Carlo method is also used to handle components uncertainties. The process comprises modelling of the architectural components and network system, automated mapping of avionics function into architecture, developing an XML database format, and architectural optimisation at aircraft level. The approach managed to minimise cost, weight, and cable length, and maximise availability and generates an XML description of the optimised IMA architecture. The optimisation results in reducing wiring by 68%, cost by over 78%, and increasing availability by several orders of magnitude.

Annighöfer et al. developed a model-based methodology for architectural design of a DIMA architecture. A meta-model is defined for DIMA architecture design from system requirements and aircraft anatomy up to functional allocation to IMA hardware devices, networking, and physical installation locations. As aircraft systems require calculation power to execute their logic, and I/O interfaces to connect their peripherals (sensors and actuators), they must meet safety and performance constraints like reliability and signal latencies which are also modelled. A number of cost functions are also defined to evaluate the system architecture. These are mass, ship set cost (SSC), operational interruption cost (OIC), initial provision cost (IPC). This work is of one a few works that is suitable for aircraft modelling. In Annighöfer et al. (2011) the IMA platform which is the combination of hardware, the system applications, signals, and peripherals are modelled in different layers including software mapping and hardware mapping. The software mapping is the allocation of avionics functions to IMA hardware, and the hardware mapping is the allocation of hardware to installation locations in aircraft structure. The system constraints defined are peripheral, segregation, atomic, latency, devices, installation locations, and power constraints.

Moreover, they proposed a novel method for automated device type selection, sizing and mapping of IMA architecture based on binary programming (BP) (Annighöfer et al. 2013). The tool is developed in eclipse modelling framework (EMF). The optimisation is implemented in MATLAB. The model employs a combinational multi-objective solver including Pareto front, branch-and-cut algorithm, and a genetic algorithm (NSGA-II). Pareto front shows a better performance in optimality. For single objective, solving a COTS solver like CPLEX and GUROBI is used as well. As a case study, the method is applied to four aircraft systems including bleed air system (BAS), pneumatic system (PS), ventilation control system (VCS), and over heat detection system (OHDS). The optimisation problem is solved for mass and OIC as objective functions. The results show a great improvement in mass and OIC compared to manual design.

They managed to extend their model further to signal routing and network topology. An AFDX topology is described by devices and links, and presented as an undirected graph (Annighöfer et al. 2014). To achieve the optimal topology, the problem is formulated as binary programming (BP) which is a combinatorial optimisation problem. The network topology evaluated by costs like mass contribution while satisfying a number of constraints. The topology mass is calculated from the mass of all switches used and the cable mass of all links. The optimisation problem is solved for mass and OIC as objective functions. The results show an improvement in signal routing and network topology. The method is used to optimise full or sub-parts of avionics architectures for certain objectives such as mass and cost while considering system requirement.

Lee et al. developed a scheduling tool and algorithm for optimisation of airplane information management system (AIMS) Boeing 777 cabinets (Lee et al. 2000). AIMS is a time synchronised distributed computing system which includes several processors and I/O boards. The proposed algorithm calculates the time scheduling for all partitions, tasks, and bus messages. The objective is that all partition and task deadlines are held and the capacity of processors and buses is not exceeded. They came up with a two-level algorithm creating the processor's schedules first and then calculating the bus schedule.

Zhang used mixed integer linear programming (MILP) to find virtual link (VL) trees in an AFDX network (Zhang 2008). This work illustrates how to formulate linear constraints and binary variables such that a consistent routing tree from each source node to each destination node is retrieved. The objective is an overall low bandwidth utilisation of AFDX links. It is implemented for a topology of eight nodes that up to 1600 VLs can be routed in four hours. However, global optima are not retrieved.

Bauer et al. in Airbus Toulouse developed a decision-analysis tool for the optimisation of fly-by-wire flight control system architecture of airliners (Bauer et al. 2007). The tool for fly-by-wire flight control system architecture postulates a combination of actuators, power circuits, and flight computers for each control surface. Different actuator technologies are considered including servo-control (S/C), electro hydrostatic actuator (EHA), and electrical-backup hydraulic actuator (EBHA). The objective is to select the system architecture that keeps the weight as low as possible while fulfilling safety and technological constraints. They chose branch-and-bound algorithm to solve this discrete optimisation problem for Airbus A340 roll control system architecture. Two scenarios are solved including 3H in which three hydraulic circuits power the flight control actuators and 2H-2E where two hydraulic and two electrical circuits power the flight control actuators. The resulting architecture weighs 3.1 kg less than the reference architecture.

Literatures reviewed above were the most relevant published works related to automated IMA architecture design and optimisation. The scope of each approach, however, is limited to a certain aspect. The baseline for all of them include problem formulation, constraint definitions, and solving techniques like integer programming (IP), BIP, and mixed integer programming (MIP). Nevertheless, examples can be found in other industries as well.

7.2.12 Architecture Design and Optimisation in Other Industries

Similar topics from other industries like space and satellite systems are also popular. Since determining the optimal placement of avionics boxes on the spacecraft is a difficult task which is normally performed manually, Jackson and Norgard developed a tool to optimise avionics box placement (Jackson and Norgard 2002). This has been defined as a multi-objective optimisation problem for optimising the placement of avionics boxes on the exterior panels of a spacecraft in which multiple cost functions and constraints must be satisfied. The objectives are to minimise the amount of harness wiring and the length of radio frequency (RF) cable runs while keeping the thermal loading and mass distribution across panel within acceptable limits. The input information into the problem are avionics boxes dimension, masses, power dissipation, mounting location of fixed components, connectivity between boxes and interconnection priority weighting. Further, a simulated annealing (SA) algorithm is proposed to solve the problem. SA reaches the optimal solution by perturbing the current best solution to explore more of the solution space.

Fabiano and his colleagues also developed a tool for the spacecraft equipment layout since the decision-making of where to place electrical equipment is a difficult task while taking into account many factors like position of centre of mass, moment of inertia, heat dissipation, and electro-magnetic interference (EMI) and integration issues at the same time. This task is usually done by a group of system engineers which takes time and as soon as a feasible solution is found, it becomes the baseline. In other words, all possibilities and layouts are not explored completely and the solution is not necessarily the optimal one. In this paper, a tool based on Excel is developed that employed optimisation techniques which provides the system engineering team and easier way to explore the layout conceptual design space. The algorithm used is M-GEO which is a multi-objective algorithm. Finally, decision-making criteria are used to select solutions from the Pareto frontier. Works with similar concepts can be found in literatures (Yang et al. 2016; Dayama et al. 2020; Sousa et al. 2013).

Annighöfer also developed a formal mathematical-based model for the European space launcher ARIANE 5 to upgrade its avionics systems as well as future launchers (Annighöfer et al. 2015a; Annighöfer et al. 2015b). Since the ARIANE 5 launcher was first developed in early 1990s, it has a federated avionics architecture, and its avionics systems may not be necessarily optimal with respect to the current technological advancement and requirements. The proposed architecture has used the IMA architecture concept. Moreover, the avionics architecture design is formulised as a binary programming which includes function allocation to IMA devices and mapping the device into an installation location while satisfying a number of constraints like power, mass, and segregation. Twenty installation locations are available for IMA devices and 158 locations for peripherals (sensors and actuators). MIP and COTS solvers like CPLEX and GUROBI are used to solve the optimisation problem. The results led to a huge optimisation in mass, power consumption, and reduction of cable length.

A similar concept from automotive industry is a commercial tool called PREEVision. It is a model-based design of electrical and electronic (E/E) architectures for the automotive domain (PREEvision 2020). The main goal of the tool is to provide a component database which supports automatic design. The model consists of functional, component, and installation layers as well as the evaluation of requirements. The components are automobiles' electronic control unit (ECU) and data buses like controller area network (CAN). The functions are then assigned to an ECU, and the ECU is assigned to installation locations. Moreover, the function signals that are assigned to wires or buses can be automatically routed. The complete model and tool enable different instantiations of the same architecture.

Hardung in his PhD dissertation developed a framework for the optimisation of the allocation of functions in automotive networks (Hardung 2006). Multiple objectives were defined including costs and busload for optimisation, while a number of constraints like memory consumption, I/O-pins consumption, timer consumption, and bandwidth availability are met within acceptable limits. In other words, the hardware is assumed fixed, and the software components are to be allocated to ECUs. A database model in structured query language (SQL) is built to store all relevant information including ECUs' technical specifications like the weight of ECU and wiring harness, cables, cost, and their functional links. The model is then implemented for central door locking and keyless entry where the details of the architecture is shown which include network topology, resources, costs, weight, busload as well as supplier complexity. The supplier complexity is defined by the minimum number of suppliers involved in the development of the ECU. This is transferred into a set-covering problem. The optimisation algorithm proposed is ant colony optimisation (ACO) which is also compared to evolutionary algorithm (EA) and is shown that ACO is faster than EA at the beginning. However, after a while, ACO cannot find better solutions. While EA is slow at the beginning, it is continuously improving in later phases. Due to better performance, at the end the ACO is selected as the final approach.

7.2.13 Comparison of Modelling Approaches

All the above approaches can be utilised to express IMA architecture in a certain scope. All of them are driven by mathematical programming which can formulate the IMA design problem and speed up the development process automatically and/or semi-automatically through simulation or formal verification. What is common among models is their formulation i.e. the overall definition of the problem mathematically. This includes decision variables, objective functions, and constraints. The decision variables are defined in various forms like IP, binary programming (BP), and MIP. Also, the objective functions defined are different, for instance in Salomon and Reichel (2011) and Zhang and Xiao (2013), the objectives are related to improving reliability and safety of the architecture whereas in Dougherty et al. (2011), the objective is to minimise the hardware (processor) and the required network bandwidth. The main objective functions in Annighöfer et al. (2014) and Annighöfer et al. (2013) are to minimise weight and costs.

Moreover, models in Salomon and Reichel (2013), Manolios et al. (2007), Annighöfer et al. (2011), and Salzwedel and Fischer (2008) are capable of expressing aircraft-level architecture which is within the scope of this research as well. The other issues are related to the separation of software and hardware as well as the level of automation. Software mapping and/or hardware mapping alone cannot express the complete model of an architecture. Automatic design is only addressed in Annighöfer et al. (2013) and Salomon and Reichel (2013). Salomon and Reichel (2013) mainly focuses on finding redundancy structures and lacks resources for installation location. Further, the types and the number of constraints in different models vary. A thorough modelling of IMA architecture which proposes various cost functions and constraints can be found in a patent by Airbus (Minot 2010) in which constraints are formalised by a set of linear inequalities. However, the most common constraints studied in literatures can be classified as peripheral, segregation, latency, power, and installation location constraints.

In conclusion, linear programming which is a widely used technique to express real-world problems into mathematical forms as used in literatures is selected in this research as well. The problem of IMA architecture optimisation is defined as an assignment problem in that it is to assign the best avionics LRUs/LRMs from database to the proposed integrated architectures. A set of costs functions are defined and constraints are expressed in inequalities forms. The decision variables are defined as binary variables. The contribution of this research into modelling is adding the volume and weight constraints to the architecture as well as introducing the operational capability as a new cost function.

7.2.14 Comparison of Optimisation Methods

Linear programming (LP) problems, in general, can be solved by many algorithms. However, in IMA avionics architecture as well as many other aircraft systems architectures, the variables are usually defined as either IP and/or binary programming (BP). IP problems are not easy to solve and are known as NP-hard discrete combinatorial optimisation problems. In other words, they cannot be solved in polynomial time due to the vastness of the solution space (Galli 2018; Jensen 2002). IP problems can be solved by three different algorithms:

- The exact algorithms guarantee optimal solutions, but they may need a huge number of iterations. They are branch-and-cut, branch-and-bound, dynamic programming algorithms, Boolean satisfiability, and decomposition.
- The heuristic algorithms guarantee sub-optimal solutions, but the quality is not guaranteed. While the running time may be polynomial in some cases, they may find a good solution fast. They are like greedy algorithms, local search, meta-heuristics (PSO), Tabu search, and simulated annealing.
- The approximation algorithms provide suboptimal solutions in polynomial time and the sub-optimality has a bound. They are like linear programming retaliation and Lagrangian relaxation.

The exact methods are usually suitable for small-scale problems, yet for large-scale problems heuristics are developed. The assignment problem, in particular, which has been used to define software/hardware allocation, installation location, and task assignment is a branch of LP and IP problems. One particular example of using IP in aircraft avionics fleet upgrade optimisation can be found in Guerra et al. (2016). Most of the other task assignment problems in IMA architecture (e.g. Lo 1988; Kafil and Ahmad 1997; Salomon and Reichel 2013; Dougherty et al. 2011) are combined with time constraints for scheduling and network transmission. They all used heuristic methods like evolutionary algorithms (EV) including GA, NSGA-II, PSO, and ACO for these discrete optimisation problems. There is no literature that has compared these methods in IMA architecture optimisation; however, the performance of ACO and PSO are reported to be faster than the GA (Hardung 2006).

The other categories use COTS solvers like SAT-solver, CPLEX, GUROBI, and GAMS. In Sagaspe and Bieber (2007), Carta et al. (2012), Manolios et al. (2007), and Annighöfer et al. (2014), COTS solver is used and results are compared to heuristic methods. In many cases, the COTS solver results are more accurate than heuristics.

In conclusion, the problem, in this study, is defined as an assignment problem which is a combinatorial optimisation problem i.e. very large feasible solutions exist. To solve this problem, the branch and bound algorithm which is a widely used algorithm for solving large scale NP-hard combinatorial optimisation problems is selected. Moreover, the PSO algorithm is also implemented to compare the results using a weighted sum method (WSM) as well as solving the multi-objective optimisation problems for finding Pareto-frontier.

7.3 Avionics Integration Architecture Methodology

The established framework and method presented here is realised based on industrial processes and Cranfield University's internal projects in Aerospace Vehicle Design (AVD) group as well as avionics network and architecture lecture notes (Seabridge 2017) and other aircraft flight deck and systems documents (Airbus 1998; Airbus 1999; Airbus 2011; Airbus 2006; ATR 2015; Cessna 2000; BAE Systems 2007). Figure 7.4 represents an overview of the framework.

Based on aircraft level avionics requirements, the avionics system integration and architecting starts from a top-level functional decomposition to provide the framework for the avionics systems design and integration. This leads to the equipment specifications with every requirement being satisfied by the equipment performance parameters and/or operational capabilities. In other words, for each avionics function, at least three avionics LRUs are investigated from various vendors. The operational capability of each LRU is evaluated separately against a set of criteria using multi criteria decision-making (MCDM) method, simple additive weighting (SAW). Further all the technical specifications of avionics LRUs as well as their manufacturers are recorded in a database. Finally, the proposed initial system architecture for the automatic flight control system (AFCS) architecture and the whole avionics system architecture are modelled using mathematical programming. The problem is defined as an assignment problem i.e. it is to assign the best avionics LRUs from the database while satisfying a number of design constraints including mass and volume in installation locations.

The optimisation problem is then solved by applying branch and bound algorithm for single objective cost functions and PSO for multi-objective cost functions. The cost functions defined here in this case study include minimising the overall weight of the architecture, minimising volume, minimising power consumption, maximising the reliability and operational capability as well as trade-offs studied between these cost functions. The proposed method is not developed for a specific aircraft type, and it is meant to be a general method that can be used for any aircraft type and/or architecture. However, the proposed avionics architecture is similar to a short to medium haul civil aircraft architecture

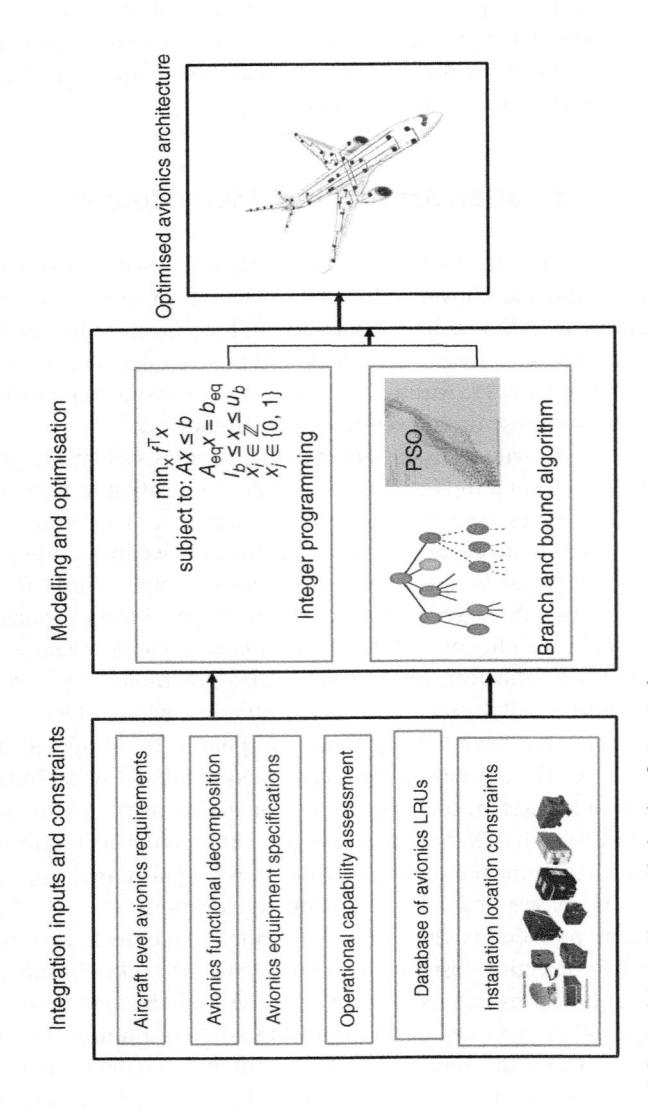

Figure 7.4 Avionics integration optimisation framework.

7.3 Avionics Integration Architecture Methodology

which is used here as a case study. The proposed architectures have used the concept of IMA architecture although they are not fully integrated.

7.3.1 Aircraft Level Avionics Requirements

Requirements engineering is essential for creating an acceptable avionics system. The designed avionics system cannot perform as expected by the customer unless the customer requirements as well as other stakeholders' requirements and related regulations and standards are well documented and understood by designers. The major driving requirements are from safety, mission, cost, and certification perspectives. In other words, a variety of technical and functional requirements for on-board avionics equipment have to be captured which are comprised of the following generic documents:

- Airworthiness requirements including the requirements related to aircraft equipment and systems as well as international airworthiness standards like CS-25, FAR-25, CAR-525, etc.
- Functional requirements for the on-board aircraft equipment
- Safety requirements
- Reliability requirements
- Avionics network and interface requirements
- Installation and environmental condition requirements

Here, the focus is on the communication, navigation, surveillance (CNS)/air traffic management (ATM) functions required to be on-board the aircraft. In other words, the requirements captured are mainly from regulatory documents. The functions that the avionics systems are expected to fulfil are shown in Figure 7.5 and they are meant to meet all CNS/ATM requirements as well as FANS-1/A. It should be noted that all the future air navigation system (FANS) and/or CNS/ATM requirements here are derived from an LRU perspective. The functional breakdown shown can be generally attributed to any civil aircraft.

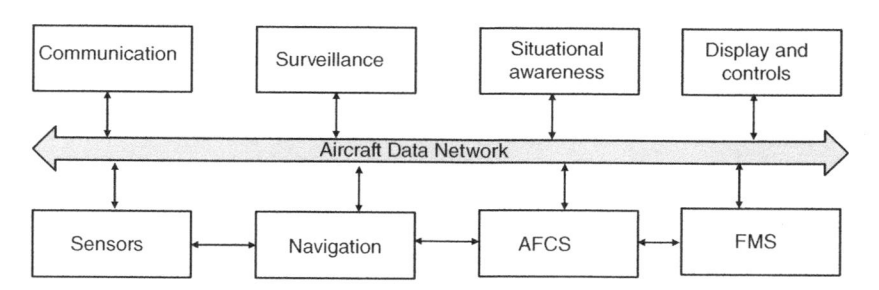

Figure 7.5 Avionics datum functional architecture.

The avionics requirements are concerned with the functional capability of avionics equipment on the aircraft. It is only after this requirement statement that the avionics design team can justifiably determine an equipment list. Then, for each item in equipment list, equipment technical specifications can be prepared. The operational capability required within each function is outlined further in detail in avionics technical specifications and operational capability assessment phase. The following requirements are obtained from the operational requirement of civil aircraft (EUROCONTROL 2017; Airbus 2007). The specific requirements are related to CNS and ATM systems as well as data transmission. The proposed architecture is to meet basic CNS/ATM requirements as well as improving operational capabilities by implementing new technologies like head-up display (HUD) and enhanced vision system (EVS) among others.

Table 7.1 shows the capabilities that are required for the proposed avionics architectures, and further these capabilities are evaluated from an LRU perspective. In other words, some capabilities are embedded as software within a particular LRU e.g. required navigation performance (RNP) and required time of arrival (RTA) are loaded in FMS and/or multi-function display unit (MCDU). Moreover, the accuracy and performance of some of these capabilities are further classified that determine the operational capability of avionics systems at aircraft level. The definition and classification is taken from a range of aviation international rules and regulations including International Civil Aviation Organization (ICAO) Doc 4444: PANS-ATM, ICAO Annex 2, 10, and 14 among others (ICAO 2007; ICAO 2013; ICAO 2005). The avionics systems are classified in three main groups including communication, navigation, and surveillance.

The next column in Table 7.1 is related to avionics capabilities. These capabilities are defined from an LRU perspective. However, in some cases, some avionics capabilities are loaded into an LRU as software. Also, some LRUs are investigated that are capable of doing more than just one avionics function (e.g. traffic/terrain/transponder collision avoidance system [T3CAS]) which is an advanced integrated surveillance system featuring a traffic alert and collision avoidance system (TCAS), terrain awareness warning system (TAWS), and Mode S transponder with automatic dependent surveillance-broadcast (ADS-B).

7.3.2 Avionics Functional Decomposition

Functional decomposition is a technique used for describing an avionics system in very general terms. The avionics requirements are divided into a set of 'top level' functions. The process seems to be straight-forward, however, there is no clear-cut way of doing this. Figure 7.6 shows the functional breakdown adopted as the datum functional architecture from an LRU perspective. This decomposition evolves from the operational requirements of the aircraft which determines the functions needed.

Table 7.1 Aircraft level avionics systems requirements.

Avionics systems	Avionics capabilities	Explanations
Communication	8.3 kHz VHF	VHF transceiver voice channel spacing
	SATCOM	SATCOM Airborne radio telephone communication via a satellite
	CPDLC	CPDLC Controller Pilot Data Link Communications (CPDLC) is a means of communication between controller and pilot, using data link for ATC communications
	VDL-M2	VDL-M2 VHF Data Link mode 2 is a means of communication between aircraft and ground stations
Navigation	GBAS CAT I/II/III	Ground Based Augmentation System CAT I is a landing system capability known as GLS whose performance is equivalent to ILS CAT I (down to 200 ft). CAT II/III are under development
	SBAS APV CAT I/II	SBAS APV CAT I/II Satellite Based Augmentation System CAT I is a landing system capability whose performance is equivalent to LPV minima (down to 200 ft)
	RTA (FMS)	RTA (FMS) Required Time of Arrival enables the pilot to define a waypoint with a specific arrival time ± 30 s. If the aircraft cannot meet the 30 s requirement, the flight crew will be notified
	RVSM	RVSM Reduced Vertical Separation Minimum for above FL290 from 2000 to 1000 ft
	Autopilot/Flight Director	Autopilot/Flight Director software, which is integrated with the navigation systems, is capable of providing control of the aircraft throughout each phase of flight
	RNAV 10	RNAV 10, which is designated as RNP 10 in the ICAO's PBN Manual, is an RNAV specification for oceanic and remote continental navigation applications
	RNAV 5 (B-RNAV)	RNAV 5 (B-RNAV), also referred to as Basic Area Navigation (B-RNAV), has been in use in Europe since 1998 and is mandated for aircraft using higher level airspace. It requires a minimum navigational accuracy of ± 5 nm for 95% of the time and is not approved for use below MSA
	RNAV 2	RNAV 2 supports navigation in en-route continental airspace in the United States
	RNAV 1 (P-RNAV)	RNAV 1 (P-RNAV) is the RNAV specification for Precision Area Navigation (P-RNAV). It requires a minimum navigational accuracy of $\pm$ iron for 95% of the time

(Continued)

Table 7.1 (Continued)

Avionics systems	Avionics capabilities	Explanations
	RNP 4	RNP 4 is for oceanic and remote continental navigation applications
	RNP 2	RNP 2 is for en-route oceanic remote and en-route continental navigation applications
	RNP 1	RNP 1 is for arrival and initial, intermediate and missed approach as well as departure navigation applications
	Advanced RNP	Advanced RNP is for navigation in all phases of flight
	RNP APCH	RNP APCH and RNP AR (authorisation required) APCH are for navigation applications during the approach phase of flight
	RNP AR APCH	RNP APCH and RNP AR (authorisation required) APCH are for navigation applications during the approach phase of flight
	RNP 0.3	RNP 0.3 is for the en-route continental, the arrival, the departure and the approach (excluding final approach) phases of flight and is specific to helicopter operations
Surveillance	ADS-B	A means by which aircraft, aerodrome vehicles and other objects can automatically transmit and/or receive data such as identification, position and additional data, as appropriate, in a broadcast mode via a data link
	Mode S Transponder	Mode S Transponder is a Secondary Surveillance Radar (SSR) process that allows selective interrogation of aircraft according to the unique 24-bit address assigned to each aircraft. Recent developments have enhanced the value of Mode S by introducing Mode S EHS (Enhanced Surveillance)
	TCAS/T2CAS/T3CAS	TCAS/T2CAS/T3CAS Currently, TCAS II version 7.1 is mandated in European airspace

Based on this functional decomposition which is derived from aircraft level avionics requirements, an equipment list, in this case LRU, is prepared. For each avionics LRU, at least three different examples are investigated from various vendors. The recorded LRUs are different in their physical specifications as well as their operational capabilities. This then leads to the problem of choosing the best LRU in order to optimise architecture in some criteria like weight and also improve the operational capability. The initial system architecture based of these equipment lists is drawn below.

7.3 Avionics Integration Architecture Methodology

Communication	Surveillance	Situational awareness	Displays and control
• VHF • HF • SATCOM • Datalink	• TCAS • ADS-B • TAWS • CVR/FDR	• EFB • HUD/EVS • FILR/SVR	• PFD • ND • MFD • HOTAS

⟵──────────── Aircraft data network ────────────⟶

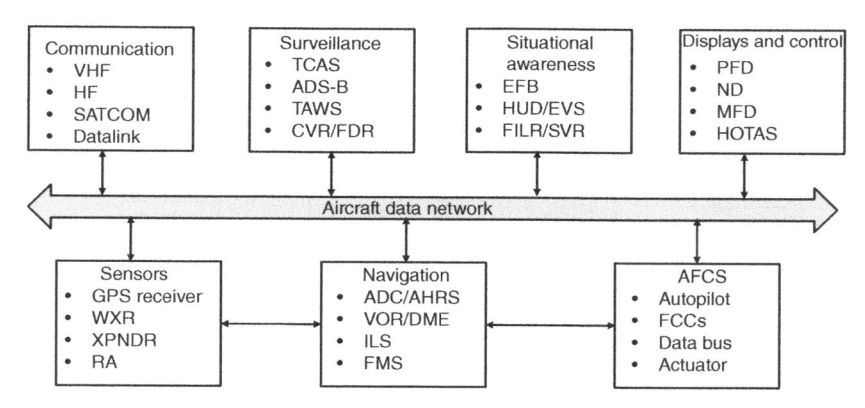

Sensors	Navigation	AFCS
• GPS receiver • WXR • XPNDR • RA	• ADC/AHRS • VOR/DME • ILS • FMS	• Autopilot • FCCs • Data bus • Actuator

Figure 7.6 Avionics functional decomposition from an LRU perspective.

7.3.3 Automatic Flight Control System (AFCS) Architecture

Avionics system architecting is the determination of the necessary interconnections and functional interrelationships between the components of an avionics system which is a complex task. Typically, this task is shared between the airframe manufacturer and the avionics supplier to ensure that all relevant factors and implications are taken into account. The initial system architecture is adopted directly from the integrated datum functional architecture and functional decomposition. Figure 7.7 illustrates the proposed system architecture for AFCS.

The proposed AFCS guarantees three functions including autopilot (AP), flight director (FD), and yaw damper (YD). The main components are two integrated avionics cabinets (IAC) which host automatic fight control application (AFCA) and exchange data with two air data computers (ADC), two attitude and heading reference systems (AHRS), two flight control computers (FCC), and seven display units. The navigation sensors are very high frequency omnidirectional range (VOR)/instrument landing system (ILS), global positioning system (GPS) receivers, and radio altimeters (RA). This architecture in modelling is referred as the small-scale architecture i.e. each avionics function and/or LRU will be defined as an avionics node in a network and/or architecture. Further, the AFDX/ARINC664 data bus is selected as the main data transmission system for the proposed architecture. The idea of IMA has also been implemented in two IACs. The IAC supplies resources for avionics applications including memory, I/O, and computations which are shared.

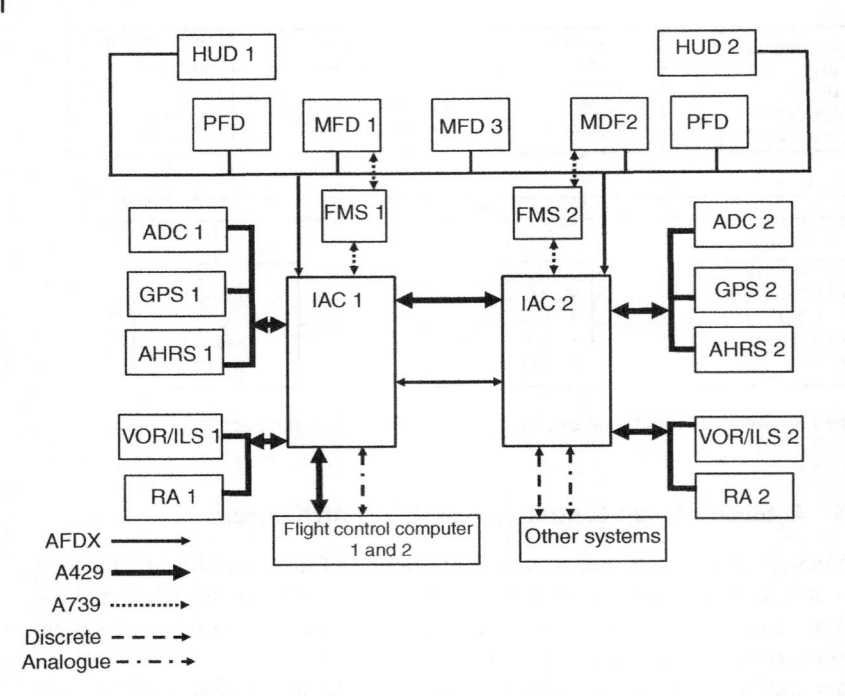

Figure 7.7 Automatic flight control system architecture.

The IACs host the following avionics applications:

- Flight warning application (FWA)
- Auto-flight control system (AFCS)
- Centralised maintenance application (CMA)
- Data concentration application (DCA)
- Switch module application (for AFDX)

Moreover, each IAC is composed of LRMs including CPM (core processing module), different I/O types, and switch module for AFDX. CPM can host avionics applications and provide the connection to avionics data network (ADN). The various I/O types provide connections for conventional avionics that cannot be directly connected to ADN. One IAC interfaces with other aircraft systems by different means of communications like discrete, analogue, and A429. The IMA information is shown in three main systems including two primary flight displays (PFDs), three multi-function displays (MFDs), one of which is for engine and warning display. This architecture further extended to the whole avionics system architecture as below.

7.3.4 Avionics Systems Architecture

Based on functional decomposition and the AFCS architecture, the architecture is extended to the whole avionics systems including navigation, communication, and indicating and recording systems as well as terrain and avoidance systems. Figure 7.8 illustrates the proposed avionics system architecture.

The IMA part is the same as explained in AFCS architecture. Some functional integrations are applied in that two or three functions can be performed by an LRU. For instance, TCAS, transponder mode S, and ADS-B are integrated in one LRU called T3CAS. Furthermore, the cockpit voice recorder (CVR) and flight data recorder (FDR) are also integrated in one LRU. This means that LRUs with these capabilities are found while investigating technologies for avionics functions. This architecture in the modelling section is referred as the large-scale architecture. The same as the small scale, the avionics functions and/or LRUs will be defined as avionics nodes in a network and/or architecture.

7.3.5 Avionics Equipment List and Technical Specifications

In order to optimise the proposed architectures and trade-off studies, it is necessary to quantify avionics LRU physical parameters as well as their performances and operational capabilities. Based on the proposed AFCS and avionics system architectures, the avionics equipment required are as follows. For AFCS architecture, ADC, AHRS, GPS receiver, VOR/ILS receiver, RA, FCC, HUD, PFD, MFD, FMS, and two IACs. These 11 avionics LRUs are used for the small-scale architecture optimisation.

Furthermore, for the large-scale architecture, some other LRUs are also added including weather radar (WXR), T3CAS, electronic flight bag (EFB), cockpit and flight data recorder (CVR/FDR), satellite communications (SATCOM), and very high frequency (VHF). These 17 LRUs are considered as the large-scale architecture optimisation problem. The technical specifications and performance description of the avionics LRUs are taken from Jane's Avionics as well as their companies' data sheet. The manufacturers are chosen according to the Flight International civil avionics directory. The physical specifications of avionics LRUs recorded are mass, power consumption, volume, and mean time between failure (MTBF). The performance and/or the operational capability of each LRU is evaluated separately to be quantified.

7.3.6 Avionics LRUs Operational Capability Assessment

The operational capability of each avionics LRU is defined as the performance parameters and capabilities that an LRU can perform. Since each avionics LRU

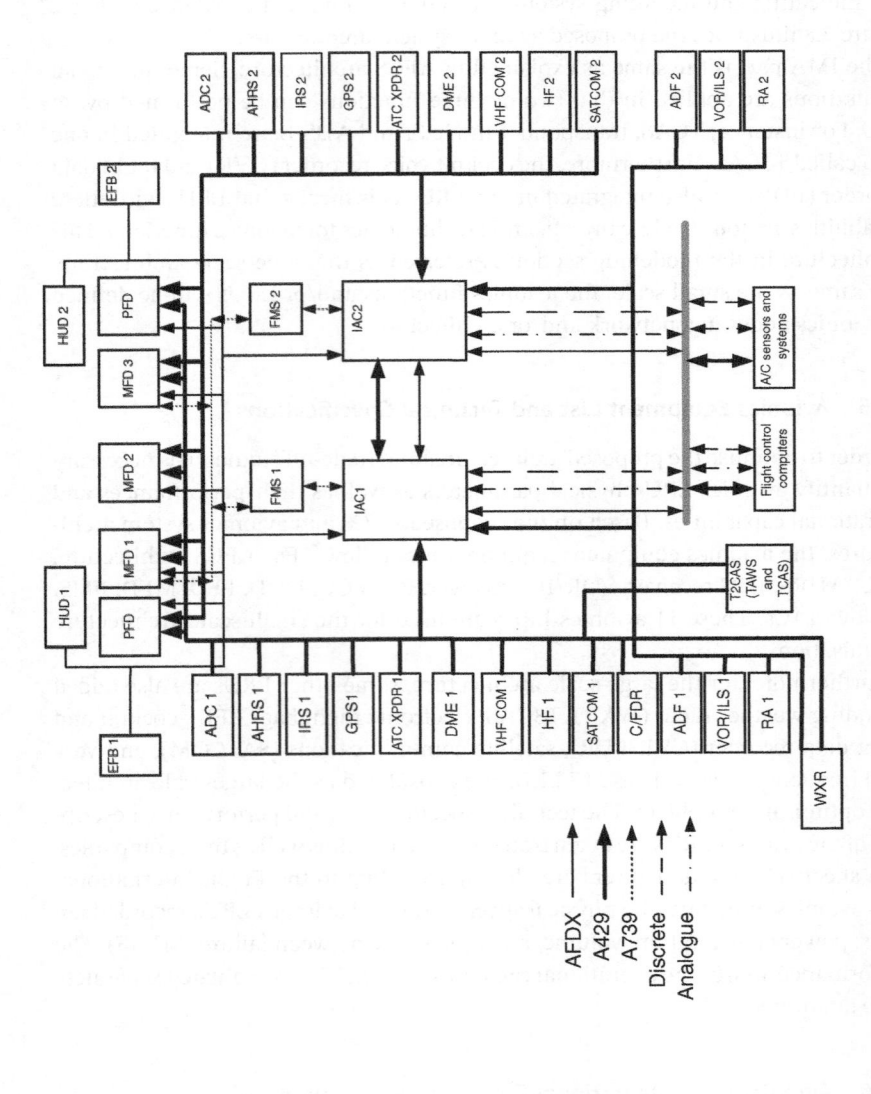

Figure 7.8 Avionics system architecture.

performs different functions that provide a different capability, they need to be evaluated separately against a set of criteria related to their functions. It is very challenging to evaluate new technologies, in general, and avionics LRU, in particular as a number of criteria involved in each assessment. Generally, technology assessment steps are technology identification, selection, and evaluation. The important task in technology selection and evaluation is to establish a set of evaluation criteria. Many big companies and departments have their own method to assess technology readiness level (TRL) and selection including DOE (DOE 2015), DOD (DOD 2019), and NASA (NASA 2016). It is also common to develop a tool to do technology assessment, a good example of this can be found in literature from Georgia Institute of Technology (Kirby 2002). The main focus in these reports is how to identify the TRL; however, here the author used existing technologies (LRUs) for integration and further optimisation of avionics architecture. Therefore, TRL is not in the scope of this literature although other established processes like identification and scoping of new technologies were applied according to those very established frameworks. In particular, the selection process is the main focus as it is to select the LRUs with the maximum operational capabilities.

Since in most cases, technology evaluation and selection criteria are more than one task, a solution is to use MCDM methods to handle this complexity (Triantaphyllou 2000). The methods help decision makers in the presence of multiple and/or conflicting criteria. What is common among these techniques is a set of technology alternatives, multiple decision criteria, and the attitude of the decision makers in favour of one criterion over the other as well as the preference of the technology alternatives. MCDM techniques help decision-makers to assess the overall performance of technology alternatives which will be further help for the optimisation of design solutions (Ching-Lai and Kwangsun 2011).

7.3.7 Simple Additive Weighting (SAW) Method

The SAW technique is a simple and widely used method for multi-attribute decision problems. The method is based on the weighted average i.e. weighting factors $[w_1, w_2, \ldots, w_n]$ are assigned to the criteria by the decision makers. The multi-criteria values with their weighting factors are summed into a single performance metric. SAW then selects the most preferred alternative A^* which has the maximum weighted outcome as it is represented in equation below:

$$A^* = \left\{ A_i \mid \max \sum_{j=1}^{n} w_j x_{ij} \right\}, i = 1, 2, \ldots, m, j = 1, 2, \ldots, n \qquad (7.4)$$

Table 7.2 Scales for technology alternative comparison.

Intensity of importance	Definition	Explanation
1	Equal importance	Two alternatives equally contribute to the objective
3	Moderate importance	Judgement slightly favoured one alternative over another
5	Strong importance	Judgement strongly favoured one alternative over another
7	Very strong	One alternative is strongly favoured over another
9	Extreme importance	The evidence of favouring one alternative over another is of the highest possible order

The SAW method process consists three steps:

Step 1: Building a decision matrix for technology alternatives (avionics LRUs) regarding to objectives and the relevant criteria by using a similar approach to Saaty 1–9 scale of pair-wise comparison (Saaty 1980) presented in Table 7.2.

Step 2: Calculate the normalised decision matrix for positive criteria:

For positive criteria:
$$r_{ij} = \frac{x_{ij}}{\max(x_{ij})} \quad (7.5)$$

For negative criteria:
$$r_{ij} = \frac{\min(x_{ij})}{x_{ij}} \quad (7.6)$$

Note: In this research, all the criteria defined and considered are positive.

Step 3: Calculate the normalised weighted matrix and evaluate each alternative by the following formula:

$$A_i = \sum_{j=1}^{n} w_j x_{ij} \quad (7.7)$$

where x_{ij} is the score of the ith alternative with respect to the jth criteria, and w_j is the weighted criteria.

7.3.8 Operational Capability Assessment

As mentioned before, for each avionics LRU, at least three different units from various vendors with different technical specifications are investigated. In this section, the operational capability of each LRU is evaluated by using SAW method. Since each avionics LRU performs a particular function/functions, the criteria that are defined for evaluation are almost different for each of them. Sabatini also developed a set of criteria for avionics LRUs in (2000). However, he did not use any quantitative method to evaluate these criteria. For instance, the AHRS is assessed against a set of criteria including attitude range, pitch/roll accuracy, heading accuracy, and RNP capability. For AHRS, four different LRUs are recorded that can be distinguished by their identifications (ID).

All the steps mentioned in the SAW method are applied to AHRS operational capability assessment including establishing a decision matrix, normalised matrix, and normalised weighted matrix which is led to a ranked and preferred choices. Table 7.3 shows all the criteria defined for each avionics LRU and went on the same processes to be evaluated. The selected AHRS guarantees the following performance for instance:

Attitude accuracy: 0.1° for straight and level flight, 0.2° in dynamic situations
Pitch/roll accuracy: 0.1°
Heading accuracy: 0.1°
Capability: Maintaining RNP envelop in loss of satellite.

Each assessment is done in a separate Excel sheet similar to the ARHS operational capability assessment. In this way, avionics LRUs operational capabilities are quantified as well as ranked.

7.3.9 Avionics Equipment Database

Avionics LRUs for each function/functions with their technical specifications and operational capabilities as well as the manufacturers are recorded in an Excel database. The technical specifications recorded are the ID of each LRU taken from their manufacturer identifications. The technical specifications recorded include physical characteristics of each LRU including mass, volume, and power consumption. The reliability of each LRU is also recorded by the MTBF which is used for components reliability assessment. The operational capability of each LRU which is calculated separately in operational capability assessment is also recorded.

7.3.10 Avionics Integration Optimisation Software Architecture

Based on the proposed methodology and framework for optimisation of IMA architecture a tool and/or software has been developed. Figure 7.9 represents a top-level design of avionics integration optimisation software system (AIOSS). The tool has three main parts including a database of avionics LRUs, avionics

Table 7.3 Operational capability criteria assessment of avionics LRUs.

Avionics LRU	Criteria assessment					
ADC	PA accuracy	IAS accuracy	Temperature accuracy	Mach number accuracy	RVSM compliant	
VOR/ILS	Deviation accuracy		Number of channels		Capability (VOR/ILS/MB/DGPS)	
RA	Height accuracy		Altitude range		Altitude range	
FMS	LNAV/VNAV capability	RNP/RNAV capability		RTA capability	DataLink/CPDLC	
PFD	Display area	Resolution		Viewing angle	Brightness	
MFD	Display area	Resolution		Viewing angle	Brightness	
HUD	Resolution	Field of view		Approach capability	Function integration (EVS, SVS)	
FCC	CPU	Memory		Interfaces	Application software	
IAC	CPU	Memory		Interfaces	Application software	
WXR	Max detection range	Azimuth coverage		Elevation coverage	Capabilities	
T3CAS	Bearing accuracy	Range capability		Operating altitude	No. of functions integrated	
EFB	Resolution	Viewing angle		CPU and memory	Functions	
C/FDR	Recording time	Impact shock		Penetration resistance	Deep sea pressure	
SATCOM	Data rate	Service coverage		Functions	Capability (FANS and ACARS)	
VHF	Pre-set channels	Channel change time		VDL Mode2 availability	Channel spacing (25, 50, 8.33 capability)	
AHRS	Attitude range	Pitch/roll accuracy		Heading accuracy	RNP capability	
GPS receiver	Altitude accuracy	Velocity accuracy		Position accuracy	Approach capability (SBAS/LPV or GBAS/GLS)	

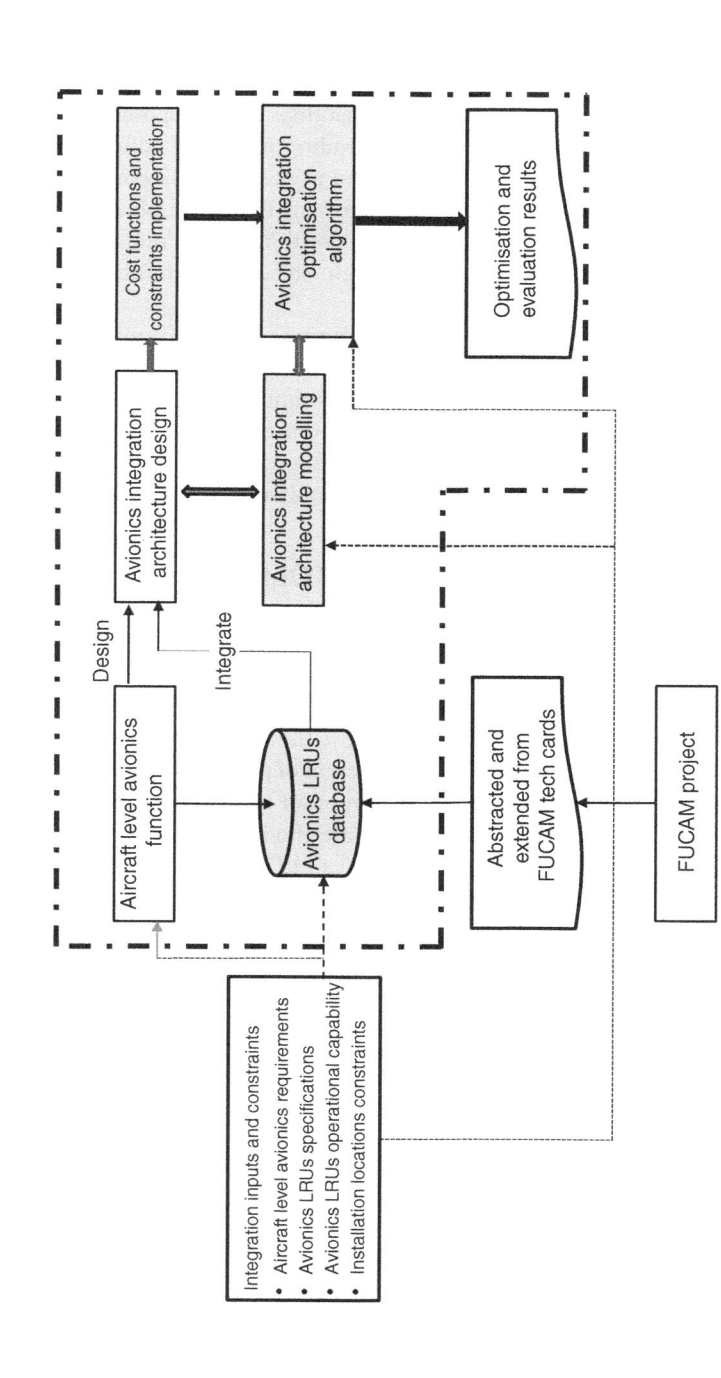

Figure 7.9 Top-level AIOSS architecture.

integration architecture modelling, and optimisation algorithms. The integration constraints and inputs to the software are the avionics LRUs mass, volume, power consumption, MTBF, and the operational capability. The proposed avionics architectures have some design constraints including mass and volume in the installation locations which are also implemented in the software. Furthermore, the modelled architectures are also implemented. For the single-objective optimisation, the software used branch-and-bound algorithm and for multi-objective optimisation, PSO is used although some single-objective cost functions are solved by both methods for comparison.

Finally, the avionics integration optimisation software provides a semi-automatic optimisation and evaluation of avionics architecture by reporting various possible architectures including minimum weight, minimum power consumption, minimum volume, maximum MTBF, and maximum operational capability. It also provides the trade-off architectures for minimum weight and maximum operational capability as well as minimum weight and maximum MTBF.

7.4 Avionics Integration Modelling of Optimisation

The mathematical programming and optimisation algorithms have been used to model avionics architectures and provide some level of design automation in many cases. The nature of the optimisation problems in these areas of study is understood as combinatorial discrete optimisation. Due to the vastness of solution space in these types of optimisation problems, the selection of problem formulation and solving algorithms must be carefully planned in order to maintain aircraft level applications. Here, the proposed architectures are modelled and defined as an assignment problem i.e. to assign the best avionics LRUs to the architecture and installation locations while satisfying some design constraints based on some cost functions required.

All the proposed optimisation algorithms are expressed as an integer and/or binary programming. Integer and/or binary programming algorithms can be solved by COTS solvers as well as MATLAB functions. Moreover, contradicting objectives, for trade-off studies, require multi-objective optimisation algorithms to calculate global optimal Pareto frontier. Here, PSO is proposed for multi-objective optimisation. The binary programming problems subject to single-objective optimisation are solved by 'Intlinprog' function of MATLAB which uses branch and bound algorithm to solve the problems.

7.4.1 IMA Architecture Layers for Allocation Modelling

As earlier in Section 7.2.5 and shown in Figure 7.3, the IMA architecture has three major layers for allocation problems including function mapping to software,

software mapping to hardware, and hardware mapping to installation locations. In each layer, different various constraints need to be kept. These constraints can be classified as follows:

Resource constraints: They are related to the computing resources in software mapping to hardware and/or devices like central processing unit (CPU), memory, and power.

Physical constraints: These are mainly related to installation locations when mapping hardware to installation locations including weight, volume, cooling capability, and the number of slots.

Segregation/collocation constraints: These are required when two functions and/or hardware cannot be assigned to the same device and/or location due to safety or any other concerns.

Connectivity constraints: This shows the connections between IMA devices as well as virtual links.

Performance constraints: They describe specifications that a design must satisfy including real-time scheduling and bandwidth.

The type of the allocation problem in this research concerns hardware mapping to installation location while keeping the installation location constraints within acceptable limits. For each function and/or functions, at least three avionics LRUs are recorded and the optimisation problem is defined to allocate the best possible LRUs to avionics architecture to optimise as well as investigate various architectures from an LRU perspective. The problem is formulated by integer linear programming (ILP) as described below.

7.4.2 Integer Programming

IP is a discrete extension of linear programming (Wolsey and Nemhauser 1999) where some or all of the variables are integer. It is also referred as ILP in which the constraints and objective functions defined are linear. A further restriction on IP is when the decision variables defined {0, 1} which is called binary programming (BP). BP in matrix notation is to optimise a linear cost function f

$$f^T x \tag{7.8}$$

Subject to

$$A_x \leq b$$
$$A_{eq} x = b_{eq}$$
$$\mathbf{x} \in \{0,1\}^n \tag{7.9}$$

where f^T refers to vector transpose and $\mathbf{x}$ is vector of decision variables. A, b, A_{eq}, and b_{eq} are matrices. Integer and binary programming have the same complexity and are solved with the same algorithms like branch-and-bound, LP-relaxation, cutting-planes, branch-and-cut, and branch-and-price.

7.4.3 Avionics LRUs Assignment Problem

The proposed architectures in this chapter are modelled here. The AFCS architecture is defined as the small-scale architecture and the avionics architecture is defined as the large-scale architecture. In both architectures, due to the similarity of the avionics LRUs between Captain's side and the First Officer's side, only half of the LRUs are considered. In other words, they are meant to be symmetric architectures. Therefore, the binary programming variables are defined as follows:

$$y_{il} = \begin{cases} 1 & \text{if node } i \text{ is assigned to location } l \\ 0 & \text{otherwise} \end{cases} \tag{7.10}$$

and

$$x_{ijl} = \begin{cases} 1 & \text{if equipment } j \text{ is assigned to node } i \text{ to location } l \\ 0 & \text{otherwise} \end{cases} \tag{7.11}$$

The objective function defined is to minimise the weight of the proposed architecture

$$\min f = \sum_i \sum_j \sum_l m_{ij} x_{ijl} \tag{7.12}$$

Further for the optimisation problem, the technical specifications and operational capabilities of each LRU are also defined as follows:

m_{ij}: The mass of each avionics LRU in node i with its associated equipment j

p_{ij}: The power consumption of each avionics LRU in node i with its associated equipment j

v_{ij}: The volume of each avionics LRU in node i with its associated equipment j

mtbf_{ij}: The reliability of each avionics LRU in node i with its associated equipment j in terms of MTBF

OC_{ij}: The operational capability of each avionics LRU in node i with its associated equipment j

Other objective functions defined by the author can be found in Radaei (2021).

7.4.4 Avionics Integration Constraints

The limited mass and volume available in installation location is very critical as explained earlier. These two constraints have been considered in modelling. In other words, for allocation of avionics LRUs to their installation locations, these two limitations must be maintained. For mass, in particular, there is also a defined target mass limitation imposed by designers in that the chosen architecture based on selected LRUs must not exceed that value. In other words, based on the defined objective functions, the selected LRUs must satisfy these constraints as well.

Mass: The mass of each avionics LRU is recorded in capturing technical specifications. Each installation location has a mass constraint which is implemented in architecture modelling for allocation of each LRU into installation locations.

Volume: The actual box dimension of each LRU is recorded in capturing technical specifications. Each installation location has volume constraint which is implemented in architecture modelling for allocation of each LRU into installation locations.

Here, three installation locations are taken into account including flight deck, avionics bay, and centre fuselage shown in Figure 7.10. Each location has mass and volume constraints and are defined as follows in avionics architecture modelling.

M: The overall maximum allowable weight of the chosen architecture

$M_{cockpit}$: The maximum allowable weight of avionics LRUs in flight deck installation location

M_{bay}: The maximum allowable weight of avionics LRUs in avionics bay installation location

M_m: The maximum allowable weight of avionics LRUs in centre installation location

$V_{cockpit}$: The maximum allowable volume of avionics LRUs in flight deck installation location

V_{bay}: The maximum allowable volume of avionics LRUs in avionics bay installation location

V_m: The maximum allowable volume of avionics LRUs in centre installation location

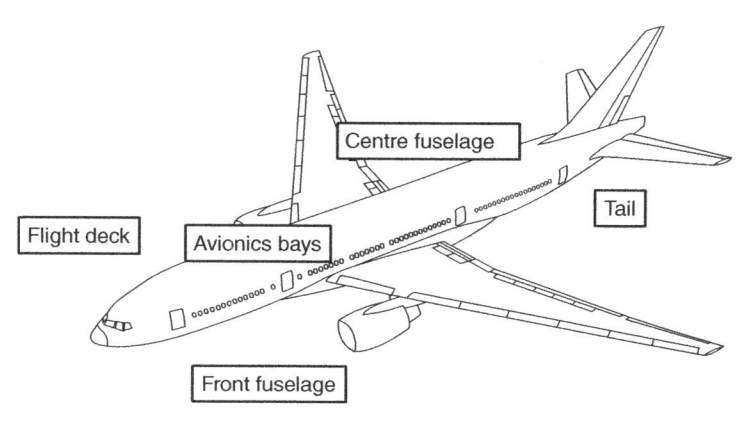

Figure 7.10 Avionics LRUs installation locations.

7 Modelling of Systems Architectures

The assignment of avionics nodes to installation location problem has to satisfy the following constraints:

a. Assignment constraint: each avionics node can only be assigned to one installation location i.e.

$$\sum_{l} x_{ijl} = 1 \tag{7.13}$$

b. Segregation and/or co-location constraint: this constraint guarantees that two avionics LRU have to be installed separately, not in the same place. As the variables defined as binary the constraint is defined as below i.e.

$$\sum_{j} \sum_{i \in (PFD,MFD,HUD,FMS)} x_{ijl} = 0, \ \forall l \in L - \{\text{Flight Deck}\} \tag{7.14}$$

$$\sum_{j} \sum_{i \in (ADC,AHRS,GPS,RA,VORILS,IAC)} x_{ijl} = 0, \ \forall l \in L - \{\text{Avionics Bay}\} \tag{7.15}$$

$$\sum_{j} \sum_{i \in (ADC,AHRS,GPS,RA,VORILS)} y_{il} = 5, \ \forall l \in \{\text{Avionics Bay}\} \tag{7.16}$$

$$x_{FCC,j,l} = 0 \qquad\qquad l \in L - \{\text{Center}\} \tag{7.17}$$

Equation (7.14) ensures that avionics LRUs like PFD, MFD, HUD, and FMS can only be installed in the flight deck, whereas Eq. (7.15) guarantees that avionics LRUs like ADC, AHRS, GPS, RA, VOR/ILS, and IAC can only be installed in the avionics bay. Equation (7.16) ensures that ADC, AHRS, GPS, RA, and VORILS are installed next to each other and separate from the IAC. Equation (7.17) states that FCC avionics LRU can only be installed in the centre installation location.

c. The mapping of avionics LRUs to installation locations has to satisfy the maximum allowable weight for each installation locations including flight deck, avionics bay, and centre i.e.

$$\sum_{i} \sum_{j} m_{il} x_{ijl} \leq M_{\text{flight deck}} \tag{7.18}$$

$$\sum_{i} \sum_{j} m_{il} x_{ijl} \leq M_{\text{bay}} \tag{7.19}$$

$$\sum_{i} \sum_{j} m_{ij} x_{ijl} \leq M_{\text{m}} \tag{7.20}$$

d. The mapping of avionics LRUs to installation locations has to satisfy the maximum allowable volume for each installation locations including flight deck, avionics bay, and centre i.e.

$$\sum_{i} \sum_{j} v_{ij} x_{ijl} \leq V_{\text{flight deck}} \tag{7.21}$$

$$\sum_i \sum_j v_{ij} x_{ijl} \leq V_{\text{bay}} \qquad (7.22)$$

$$\sum_i \sum_j v_{ij} x_{ijl} \leq V_{\text{m}} \qquad (7.23)$$

In brief, the mapping of avionics LRUs to installation locations is to minimise the weight of the chosen architecture according to the cost function defined in Eq. (7.12) while satisfying constrains equations from (7.13)–(7.23).

7.5 Simulations and Results for a Sample Architecture

For demonstration and evaluation of the proposed method, the developed avionics architectures are considered as a case study. The proposed architectures are artificial since there was no real aircraft as a reference for this study. However, the system functions, peripherals and the aircraft anatomy considered is like a single-aisle aircraft with two engines. For the installation of avionics LRUs, three installation locations are considered including flight deck, avionics bay, and centre. Table 7.4 shows the installation locations and their related physical constraints.

Each installation location has mass and volume limits e.g. each location has a maximum allowable mass and volume to install avionics LRUs. In what follows, avionics architecture optimisation problem is solved using GAMS and MATLAB.

7.5.1 GAMS

In this section, the assignment problem of avionics LRUs i.e. the allocation of the best avionics LRUs from database to avionics architecture and mapping of avionics LRUs into their installation locations while keeping the mass and volume constraints within limits is modelled and solved using branch-and-bound algorithm in GAMS. Based on the cost function defined in Eq. (7.12), the objective is to select and map the LRUs into their installation locations in order to keep the weight of the selected architecture minimum. Figure 7.11 represents the results in GAMS environment.

Table 7.4 Installation location constraints.

Installation locations	Mass (kg)	Volume (in. m^3)
Flight Deck	50	3100
Avionics bay	40	2500
Centre	30	1000

298 | 7 Modelling of Systems Architectures

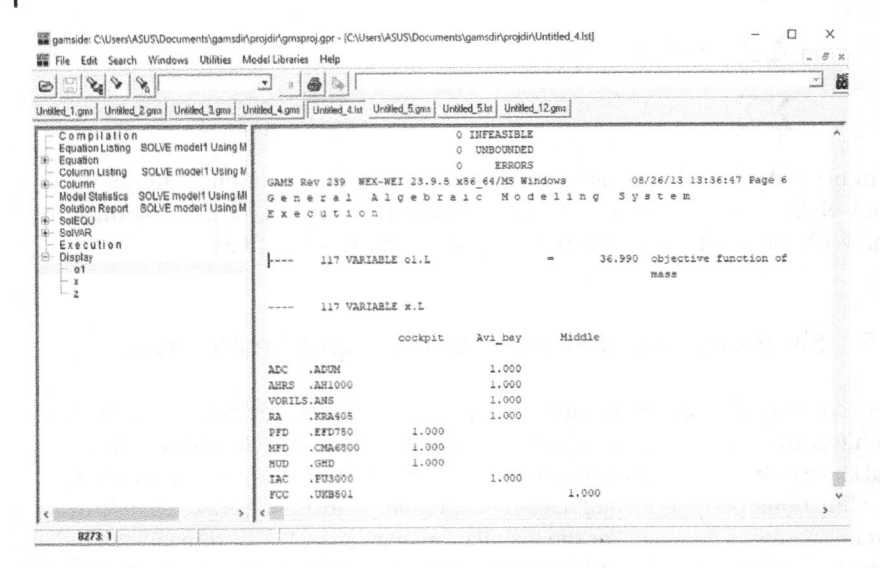

Figure 7.11 Solving the optimisation problem in GAMS.

As can be observed, the cost function value for minimum weight architecture is $z = 36.99$ kg for the AFCS architecture. Both avionics LRU selection and mapping are carried out in one run. The number 1 shows that the avionics LRU is assigned to that location. The PFD, MFD, HUD, and FMS are located in the flight deck. The FCC is located in the centre location, and the others including ADC, AHRS, GPS receiver, VORILS, RA, and IAC are located in the avionics bay. The associated ID of each assigned avionics LRU for weight minimisation is also reported on the left side.

7.5.2 MATLAB Function and PSO

In this section, the defined optimisation problem is solved by MATLAB function, Intlinprog, and PSO algorithm. In other words, the minimum weight architecture for AFCS architecture is simulated based on the cost function defined in Eq. (7.12) while satisfying constraints (7.13)–(7.23). The weight of the architecture and/or the amount of cost function for 11 avionics nodes and/or LRUs is $z = 36.9$ kg which is exactly the same value calculated by GAMS. In other words, the selected LRUs lead to an architecture with minimum weight. The allocated LRUs from the database for minimum weight for ADC, AHRS, GPS, VOR/ILS, RA, FMS, PFD, MFD, FCC, and IAC are as follows respectively: ADUM, AH1000, GPSWAAS, ANS, KRA405, UNS1FW, EFD750, CMA6800, GHD, UKB501, and PU3000, which are also the same as the LRUs selected by GAMS solver.

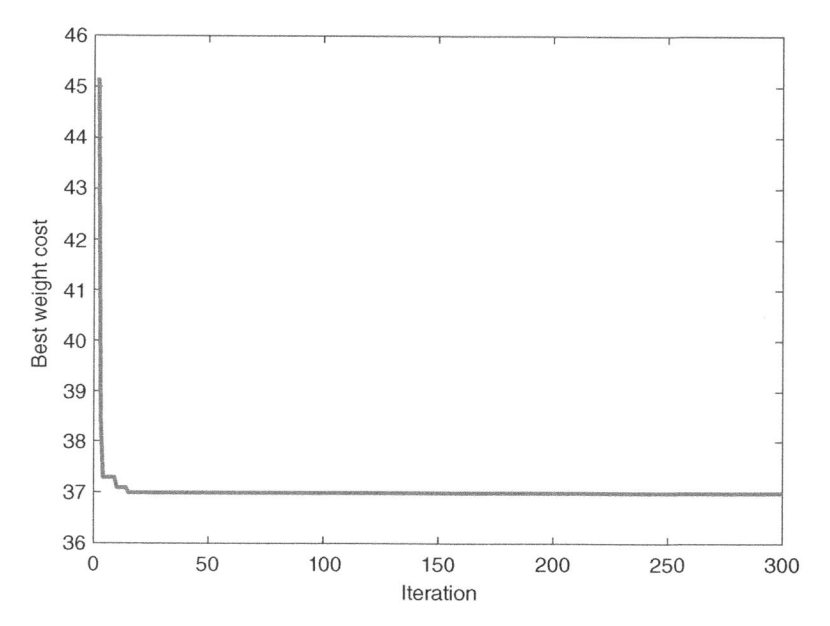

7.5 Simulations and Results for a Sample Architecture | 299

Figure 7.12 Best weight cost for AFCS architecture.

The above problem has also been solved by PSO algorithm as a single-objective optimisation. Figure 7.12 shows the value of weight cost function using PSO algorithm which is calculated $z = 36.99$ kg. The PSO parameters used in this simulation are shown in Table 7.5.

The problem has previously been solved by an exact method using MATLAB Intlinprog. The parameters, particularly, c_1, c_2 are chosen in a way that the algorithm converge to $z = 36.99$ kg. Three different runs were performed and each reached the desired value. In other words, they are chosen to result in the optimal solutions already calculated by the exact method. There is a general rule for setting these two learning factors that is $c_1 + c_2 \leq 4$. In general, having a higher value of ω, c_1, and c_2 causes to explore and create newer solutions, however, the lower value of which exploit the current solutions and make them better.

Table 7.5 PSO parameters for minimum weight architecture.

Iteration	Population	ω	c_1	c_2
300	50	1	0.3	0.3

7.6 Conclusion

A general method and tool have been developed to automate as well as optimise avionics integration architecture. The proposed avionics architecture has been assigned by the best possible avionics LRUs to achieve minimum weight architecture and mapping the selected LRUs to their installation locations while keeping the mass and volume constraints within limits.

References

Airbus (1998). A319/A320/A321 Flight Deck and Systems Briefing for Pilots.
Airbus (1999). A330/A340 Flight Deck and Systems Briefing for Pilots.
Airbus (2006). A380 Flight Deck and Systems Briefing for Pilots.
Airbus (2007). Getting to Grips with FANS (Future Air Navigation System).
Airbus (2011). A350 Flight Deck and Systems Briefing for Pilots.
Andersson, H. (2009). Model based systems engineeeing in avionics design and aircraft simulation. Linkoping University.
Andersson, H. and Sundkvist, B.-G. (2006). Method and integrated tools for efficient design of aircraft control systems. *25th International Congress of the Aeronautical Sciences*.
Annighöfer, B., Kleemann, E., and Thielecke, F. (2011). Model-based development of integrated modular avionics architecture on aircraft level. *Deutscher Luft- und Raumfahrtkongress*.
Annighöfer, B., Kleemann, E., and Thielecke, F. (2013). Automated selection, sizing, and mapping of integrated modular avionics modules. *IEEE/AIAA 32nd Digital Avionics Systems Conference (DASC)*.
Annighöfer, B., Reif, C., and Thielecke, F. (2014). Network topology optimisation for distributed integrated modular avionics. *Digital Avionics Systems Conference*, Colorado Springs, CO, USA.
Annighöfer, B., Nil, C., Sebald, J., and Thielecke, F. (2015a). Structured and symmetric IMA architecture optimization: use case ARIANE launcher. *34th Digital Avionics Systems Conference (DASC)*.
Annighöfer, B., Nil, Ç., Sebald, J., and F. Thielecke (2015b). ARIANE-5-based studies on optimal integrated modular avionics architectures for future launchers. *6th European Conference for Aeronautics and Space Sciences*.
ATR (2015). ATR 72-600 Maintenance Training Manual.
BAE Systems (2007). BAE 146/AVRO 146 RJ Seriess, Aircraft Flight Manual.
Bauer, C., Lagadec, K., Bès, C., and Mongeau, M. (2007). Flight control system architecture optimisation for fly-by-wire airliners. *J. Guidance Control Dyn.* 30 (4): 1023–1029.

Bokhari, S.H. (1987). *Assignment Problems in Parallel and Distributed Computing.* Norwell, MA, United States: Kluwer Academic Publishers.

Burkard, R., Dell'Amico, M., and Martello, S. (2009). *Assignment Problems.* Philadelphia, USA: Society for Industrial and Applied Mathematics (SIAM).

Butz, H. (2007). Open integrated modular avionics (IMA): state of art and future development road map at Airbus Deutschland. *Workshop on Aircraft Systems Technologies (AST)*, Hamburg, Germany.

Carta, D., Oliveira, J. D., and Starr, R. (2012). Allocation of avionics communication using Boolean satisfiability. *31st Digital Avionics Systems Conference.*

Cessna (2000). Cessna Citation XLS Instrumentation and Avionics.

Charara, H., Scharbarg, J.-L., Ermont, J., and Fraboul, C. (2006). Methods for bounding end-to-end delays on an AFDX network. *Real-time Systems, Euromicro Conference.*

Ching-Lai, H. and Kwangsun, Y. (2011). *Multiple Attribute Decision Making: Methods and Applications.* Berlin Heidelberg: Springer-Verlag.

Dayama, N. R., Todi, K., Saarelainen, T., and Oulasvirta, A. (2020). GRIDS: interactive layout design with integer programming. *Association for Computing Machinery*, Honolulu, HI, USA.

Delange, J., Pautet, L., and Plantec, A. (2002). Validate, simulate and implement ARINC653 systems using the AADL. *Proceedings of the ACM SIGAda International Conference*, New York.

DOD (2019). *Technology Readiness Assessment Guide.* DOD.

DOE (2015). *Technology Readiness Assessment Guide.* DOE.

Dougherty, B., Schmidt, D., White, J., and Kegley, R. (2011). *Deployment optimisation for embedded flight avionics systems. CrossTalk*, Nov–Dec 2011, 31–36.

EUROCONTROL (2017). *Avionics Requirements for Civil Aircraft.* EUROCONTROL.

Feiler, P. H., Gluch, D. P., and Hudak, J. J. (2006). The architecture analysis & design language (AADL): an introduction. Carnegie Mellon University.

Förster, S., Fischer, M., Windisch, A., and Balser, B. (2003). A new specification methodology for embedded systems based on the π-calculus process algebra. *Proceedings of the 14th IEEE International Workshop on Rapid Systems Prototyping.*

Fraboul, C. and Martin, F. (1999). Modeling advanced modular avionics architectures for early real-time performance analysis. *Proceedings of the Seventh Euromicro Workshop on Parallel and Distributed Processing.*

Galli, L. (2018). Algorithms for integer programming. IEEE.

Gavel, H. (2007). On aircraft fuel systems conceptual design and modeling. PhD Thesis. Linkoping University.

Guerra, C. J., Carmichael, N. E., and Nielson, J. T. (2016). Aircraft avionics strategic fleet update using optimal methods. *IEEE Aerospace Conference*, Big Sky, MT, USA.

Hardung, B. (2006). Optimisation of the allocation of functions in vehicle networks. PhD thesis. Erlangen: Erlangen University.

ICAO (2005). *Annex 2, Rules of the Air*, 10the. ICAO.

ICAO (2007). *ICAO Doc 4444: PANS-ATM (Procedures for Air Navigation Services)*, 15the. ICAO.

ICAO (2013). *Annex 14, Aerodromes, Volume 1 Aerodrome and Operations*, 6the. ICAO.

Jackson, B. and J. Norgard (2002). A stochastic optimisation tool for determing spacecraft avionics box placement. *IEEE Aerospace Conference*, Big Sky, MT, USA.

Jensen, P.A. (2002). *Operation Research Models and Methods*. Wiley.

Johansson, O. and Krus, P. (2006). Configurable design matrixes for system engineering applications. *The Proceeding of IDETC/CIE ASME*, USA.

Johansson, O., Andersson, H., and Krus, P. (2008). Conceptual design using generic object inheritence. *Proceedings of the ASME International Design Engineering Technical Conference*, New York.

Kafil, M. and Ahmad, I. (1997). Optimal task assignment in hetrogeneous computing systems. *Hetrogeneous Computing Workshop*.

Kirby, M. R. (2002). Technology identification, evaluation and selection. *Aerospace Systems Design Laboratory, Georgia Institute of Technology*.

Lafaye, M., Gatti, M., Faura, D., and Pautet, L., (2011). Model driven early exploration of IMA execution platform. *IEEE/AIAA 30th Digital Avionics Systems Conference*, Seattle, WA, USA.

Lauer, M. and Ermont, J. (2011). Latency and freshness analysis on IMA systems. *Main Technical Program at IEEE ETFA*.

Lee, Y.-H., Kim, D., Younis, M., and Zhou, J. (2000). Scheduling tool and algorithm for integrated modular avionics systems. *19th DASC. 19th Digital Avionics Systems Conference*, Philadelphia, PA, USA.

Lo, V. (1988). Heuristic algorithms for task assignment in distributed Systems. *IEEE Trans. Comput.* 37 (11): 1384–1397.

Manolios, P., Vroon, D., and Subramanian, G. (2007). Automating component-based system assembly. *Proceedings of the 2007 International Symposium on Software Testing and Analysis*.

Minot, F. (2010). Method for optimisation of an avionics platform. US Patent No. 2010/0292969 A1, issued 18 November 2010.

NASA (2007). *NASA System Engineering Handbook NASA/SP-2007-6105, Rev 1*. US: NASA.

NASA (2016). Final Report of the NASA Technology Readiness Assessment (TRA) Study Team.

Nickum, J., Robert Cormier, Albert Herndon et al. (2002). Avionics integration: a process to optimise avionics components to meet desired operational capabilities. *AIAA's Aircraft Technology, Integration, and Operations (ATIO)*, Los Angeles, California.

PREEvision (2020). https://www.vector.com/int/en/products/products-a-z/software/preevision/#c1789 (accessed 11 August 2021).

Radaei, M. (2021). Mathematical programming for optimization of integrated modular avionics. SAE Technical Paper 2021-01-0009, 2021. https://doi.org/10.4271/2021-01-0009.

Saaty, T. (1980). *The Analytic Hierarchy Process*. New York: McGraw-Hill.

Sabatini, R. (2000). MB-339CD aircraft development: COTS integration in a modern avionics architecture. *RTO Symposium on "Strategies to Mitigate Obsolescence in Defence Systems"*, Budapest, Hungary.

Sagaspe, L. and Bieber, P. (2007). Constraint-based design and allocation of shared avionics resources. *26th Digital Avionics Systems Conference*.

Salomon, U. and Reichel, R. (2011). Automatic safety computation for IMA systems. *IEEE/AIAA 30th Digital Avionics Systems Conference*.

Salomon, U. and Reichel, R. (2013). Automatic design of IMA systems. *IEEE Aerospace Conference Proceedings*.

Salzwedel, H. and Fischer, N. (2008). Aircraft level optimisation of avionics architecture. *27th Digital Avionics Systems Conference*.

Seabridge, A. (2017). Lecture notes on avionics data networking, hardware intergration and testing. Cranfield University.

Seabridge, A. and Moir, I. (2020). *Design and Development of Aircraft Systems*, 3rde. London: Wiley.

Shi, X. and Zhang, C. (2016). Integration design tool for avionics system based on mathematical programming. *IEEE/AIAA 35th Digital Avionics Systems Conference (DASC)*.

Sousa, F., Louiz, R., Marconi, E., and J. Calos (2013). A tool for multidisciplinary design conception of spacecraft equipment layout. *22nd International Congress of Mechanical Engineering, Brazil*.

Triantaphyllou, E. (2000). *Multi-Criteria Decision Making Methods: A Comparative Study*. Boston, MA: Springer.

Wang, G. and Zhao, W. (2020). *The Principles of Integrated Technology in Avionics Systems*. London: Elsevier.

Weilkiens, T., Lamm, J.G., Roth, S., and Walker, M. (2015). *Model-Based System Architecture*. Hoboken, NJ: Wiley.

Wolsey, L.A. and Nemhauser, G.L. (1999). *Integer and Combinatorial Optimisation*. Canada: Wiley.

Yang, J., Chen, X., and Yao, W. (2016). A rectangular cuboid satellite module layouts method based on integer optimisation. *6th International Conference on Advanced Design and Manufacturing Engineering*.

Zhang, S. (2008.) Communication infrastructure supporting real-time applications. PhD Thesis, Technical University of Hamburg.

Zhang, C. and Xiao, J. (2013). Modelling and optimisation in distributed integrated modular avionics. *34th Digital Avionics Systems Conference*.

8

Summary and Future Developments

8.1 Introduction

The systems listed in Chapters 3, 4, and 5 have been described as individual or stand-alone systems. This will be suitable for readers who have been given responsibility for a single system – for its design, its maintenance, or its procurement, for example. The systems diagrams will help to explain that there are almost inevitably interactions with other systems, and therefore, no system is truly 'standalone'.

This situation leads to a number of aspects which must be considered in the design of an aircraft, quite apart from customer requirements, standards, and regulations. They make a significant contribution to the overall design and ultimately to fitness for purpose. Some of these aspects are discussed in this chapter and include the following:

- Systems of systems
- Integration
- Complexity
- Emergent properties
- Chaos

8.2 Systems of Systems

There are many circumstances in which one or more of these systems are combined to form a higher-level system. This system will synergistically combine the various functions of the individual systems to form a function which is more comprehensive than the sum of its parts. Such a combination is often referred

---

Aircraft Systems Classifications: A Handbook of Characteristics and Design Guidelines, First Edition.
Allan Seabridge and Mohammad Radaei.
© 2022 John Wiley & Sons, Inc. Published 2022 by John Wiley & Sons, Inc.

to as 'a system of systems'. Example of such 'systems of systems' include the following:

- Propulsion (refer to Chapter 3)
- Navigation (refer to Chapter 4)
- Antennas (refer to Chapters 4 and 5)
- Weapon system (refer to Chapter 5)
- Military fast jet cockpit
- Flight deck of commercial airliner

8.2.1 Example of the Flight Deck as a System of Systems

From the list of systems of systems above, the flight deck is presented as an example of the complexity of influences, and eventually in the complexity of the resulting design. The major influences are illustrated in Figure 8.1 and described in the following paragraphs.

These general influences can be considered in more detail and referred to the previous chapters as illustrated in Figure 8.2. Each of these influences will be briefly described, the descriptions are not intended to be complete but are to give an impression of the sort of analysis required that will result in a complete human machine interface – the interface between the crew, the systems, and the aircraft.

8.2.2 Customer Requirement

The customer may have specific preferences for the design of the flight deck, these may be 'cosmetic' and seen as deviations from a standard cockpit design or may indeed be revolutionary, based on their own preferences or to fit in with the design of one or more of the systems. An example of this is the preference by some aircraft designers for a side control stick rather than the centre control column.

8.2.3 Standards and Regulations

The design of the flight deck is governed by standards that may be national or international. There are some mandatory requirements and some are advisory, but all will provide a framework for health and safety, safety and comfort that provide aircrew with an environment that is fit for purpose.

8.2.4 Human Factors Aspects

- Comfort is essential for the well-being of the crew who will have to remain in a constrained position for many hours with no opportunity for exercise. Comfort includes seating, temperature, humidity, and good quality air. Seating should minimise the risk of circulation issues such as deep vein thrombosis.

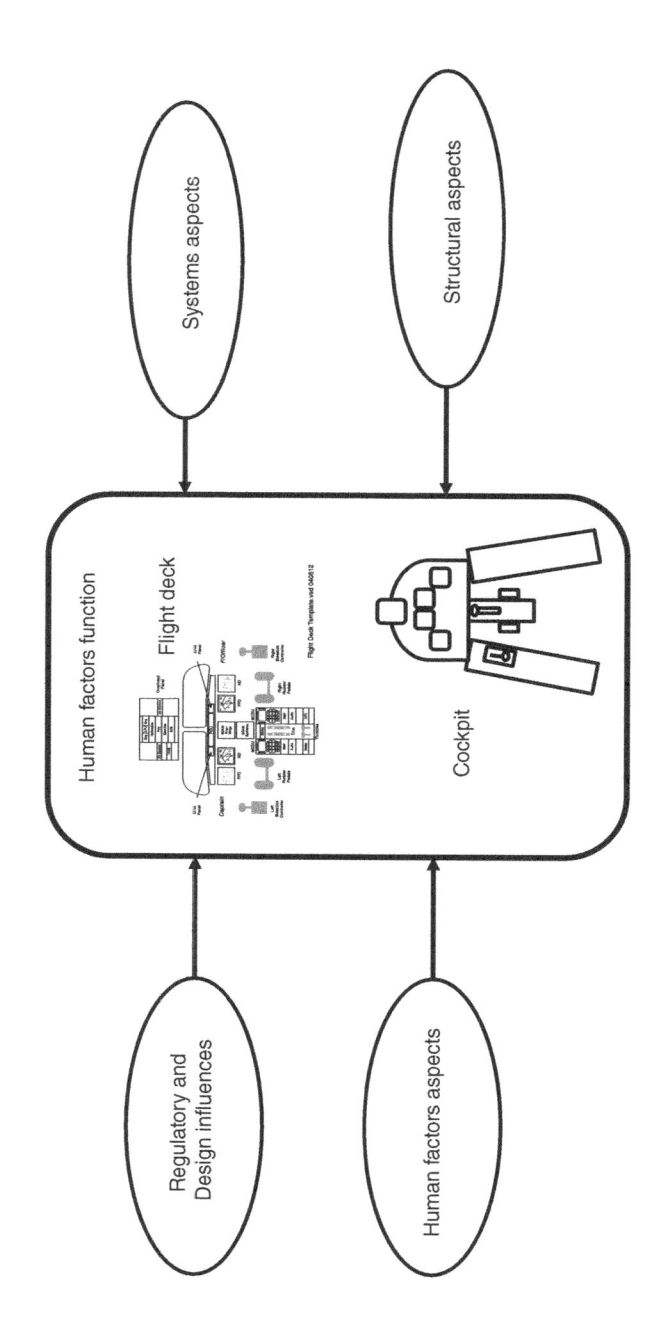

Figure 8.1 General influences on the design of the flight deck.

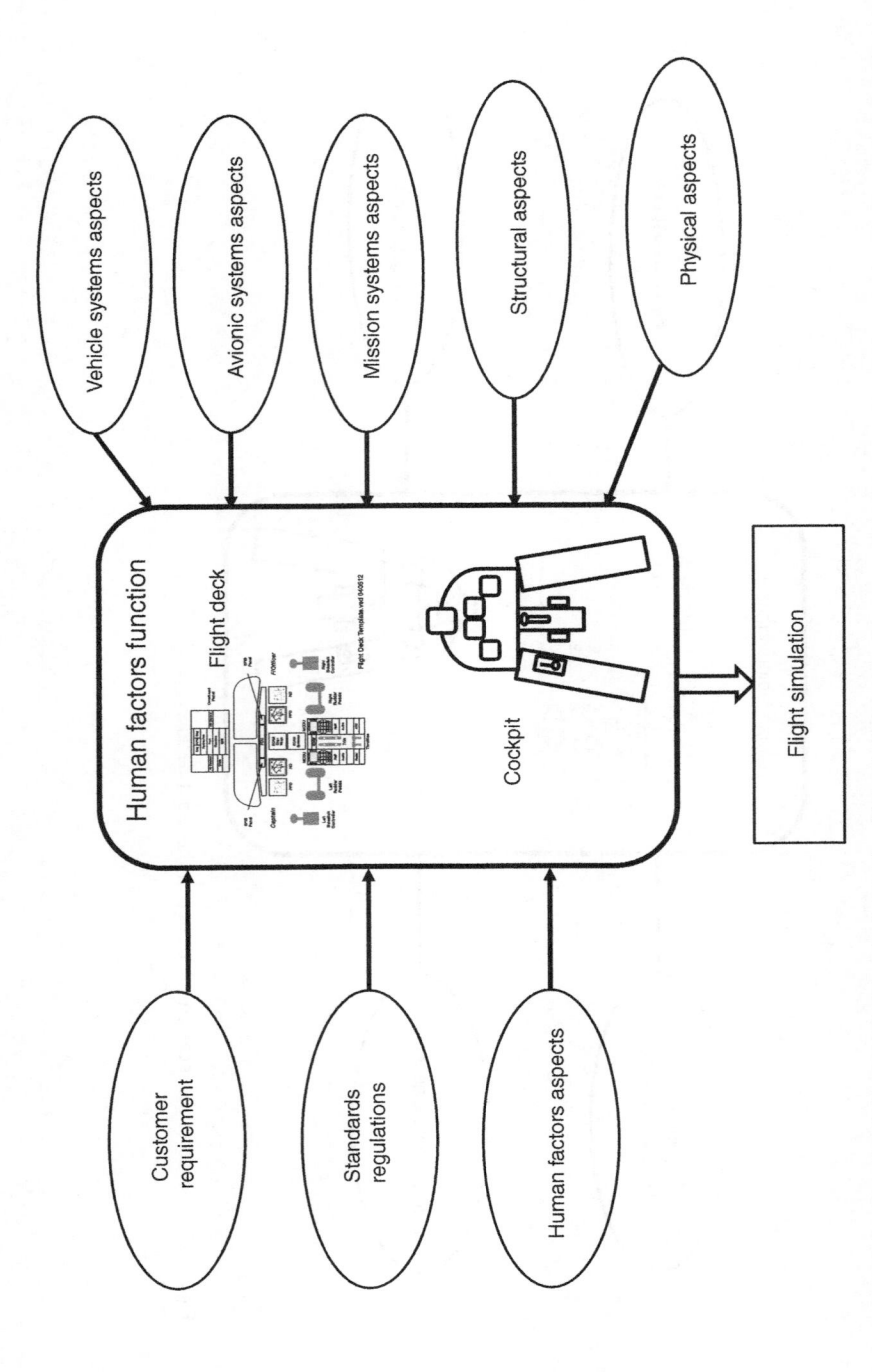

Figure 8.2 Detailed influences on the design of the flight deck.

- Viewing angles must be designed so that all displays and controls are visible by both pilots. In many cases, a pilot eye position is defined and viewing angles calculated with respect to that.
- External view is essential to allow the outside world to be viewed to the side and over the nose at all attitudes, during taxy, take-off, and landing approach as well as for scrutiny of airspace to avoid other aircraft. Fast jet pilots requires all round viewing which may be supplemented by external cameras.
- Reach affects the controls on the side consoles which must be reachable by the appropriate pilot. Controls on the centre console, main panels, and overhead are generally to be reachable by both pilots within the percentile range agreed for the project.
- Lighting must be balanced throughout the flight deck and must allow for reading of documents and for viewing of all displays with no precedence given to any single display. The lighting must take into account day, night, and high-altitude sunlight
- Consistent colour to be achieved throughout the furnishings to avoid distractions from the displays.
- Text and font size is to be to an agreed standard to ensure legibility without eyestrain. Critical parameters may be emphasised with colour or size of character or by flashing the message.
- Use of colour is selected to aid understanding and to provide emphasis and attention getting stimulus. The range of colours must be compatible with the lighting conditions and to obey standards and accepted preferences such as red, amber, green for warnings.
- Use of audio allows attention to be drawn to certain changes in conditions, especially when the crew may be engrossed in a particular task. The audio may take the form of one or more particular tones or the use of speech with clearly annunciated short phrases.
- Workload must be estimated by modelling the position of all items in the flight deck and all display and warning parameters and their implications on the ability of the pilot to perform the flying task. This estimate will be measured on flight deck rigs and on the final layout and confirmed in the flight simulator. Normal conditions and various failure modes must be modelled.

8.2.5 Vehicle Systems Aspects

Most of the vehicle systems will have an impact on the flight deck for display of status and for control actions. There are some controls that have traditionally occupied a prime position on the flight deck such as the throttles and control column and are almost fixed points around which the design is based. However, modern systems requirements and new technologies are challenging this supremacy. The following examples have a more profound on the design of the flight deck:

- Throttles (Section 3.1) are conventionally placed on the centre console of a commercial aircraft and on the left-hand console of a fast jet. The throttles must be reachable throughout their range of movement and there must be no features that impede the movement. The throttle handles are a convenient place to install switches to complement the switches in the control column handle.
- Control column (Section 3.7) is the primary input to the flight control system. It is conventionally placed in front of each pilot on commercial aircraft, but modern designs have seen it move to the side console. In the fast jet, it is conventionally placed in front of the pilot, although some designs have adopted a side stick on the right-hand console. The movement of the column has changed with time from a full fore and aft displacement to a limited displacement 'force' control. The control column handle is a convenient place to install switches to complement the switches in the throttle lever handle.
- Pressurisation (Section 3.5) demands that all opening windows and the canopy are sealed to prevent a loss of pressure. A slow loss of pressure leads to an insidious loss of capability to absorb oxygen and has led to loss of aircraft and life. Any items such as harnesses and ducts entering the flight deck or cockpit must therefore do so whilst respecting the pressure shell integrity.
- Pedals (Section 3.7) are used to control the attitude and trajectory of the aircraft through the flight control system and are also used to command the brakes and aerodynamic steering of the aircraft prior to the engagement of nose wheel steering on the ground.
- Equipment cooling (Section 3.12) is essential for some high dissipating equipment and ducts must be provided.
- Crew cooling (Section 3.12) is also essential and ducts and vents must be provided. The fast jet pilot must be provided with a high volume of cold air.
- Crew escape (Section 3.19) for fast jet pilots may require the installation of miniature detonating cord in the canopy material or devices to jettison the whole canopy. Notices must be displayed to warn ground crew of the hazardous nature of these devices.
- Ejection clearances (Section 3.19) must be observed with all fixed equipment mounted in the cockpit to ensure that the seat and the pilot can eject safely.

8.2.6 Avionic Systems Aspects

Most of the avionic systems will have an impact on the flight deck for display of status and for control actions:

- Displays and controls (Section 4.1) on legacy aircraft tend to be grouped or clustered, for example engine-related displays will be grouped together, as will navigation, fuel, etc. On multifunction displays, the same grouping may occur but on separate pages of display and will not necessarily be presented at all times, usually on demand or following a failure. Visibility is important as is brightness, legibility, and clarity.
- Communication control panels (Section 4.2) must be accessible to all crew-members.
- Navigation control panels (Section 4.3) will often be clustered on the centre console, close to the throttle levers.
- Electronic flight bag stowage and power connection (Section 4.16) is similar to carry on personal equipment. It must be stowed safely and may require a table or flat surface for a laptop device. A power supply cable must be provided but no other IT compatible cables to prevent connection to the aircraft systems.
- Lighting (Section 4.18) must be balanced to avoid any emphasis on particular areas of the flight deck, and it must be dimmable and must be compatible with night vision goggle requirements.

8.2.7 Mission Systems Aspects

The mission system of a military aircraft contains some elements of displayed parameters and control inputs. In many cases, these items need to be identified as belonging to the mission system and any control or switch input that could lead to weapon release must contain an interlock that makes inadvertent release extremely remote. This may take the form of covers or guards, or switches that require deliberate action to expose the switch.

- Head up display (HUD) (Section 5.4) is a key item of display in the cockpit and must be permanently situated to allow the pilot a direct view ahead with no reflections or obstructions. It may be necessary for the windscreen to be designed with a flat plane surface. On the commercial aircraft flight deck, the HUD can be sited on the front coaming or may be designed to fold down for landing.
- Weapon system (Section 5.14) demands that all controls and switches are clearly identified and may need an interlock such as a cover or wire locking. There will also be a requirement for wiring to be segregated from all other harnesses and additional shielding may be necessary.

8.2.8 General Aspects

There will inevitably be some items that do not fall easily into the discipline of the display system format for control and display demanded by multi-function display units. These items include individual switches, lamps, and indicators that connect directly to the system for integrity reasons. Nevertheless, the location and appearance of these items must be subject to the human engineering design philosophy.

- Warnings may need to be provided independently of the multifunction displays, for example if they are needed in the event of total display failure. If this is so, then they must be visible and unambiguous and must meet the flight deck design philosophy.
- Switches and controls are not always integrated with a control panel or part of the multi-function display. In such cases, a location must be found that is suitable for use and fits in with the flight deck design philosophy. Common locations are the overhead panel, the side consoles or the control column and throttle lever handles.
- Emergency oxygen cylinders and masks (Section 3.21) are provided for circumstances in which the pilots will need oxygen, for example after a loss of pressurisation, to recover the aircraft or effect a safe ejection. In both cases, pressurised oxygen cylinders are provided attached to the flight deck seats or to the ejection seat. Safety precautions must be taken to ensure that the cylinders are regularly checked for content and to reduce the risk of explosion.
- Emergency escape tools must be provided to assist in emergency evacuation of the aircraft and must be clearly labelled and stowed safely and easily accessible to flight crew and cabin crew.
- Personal items stowage must be provided to ensure safe stowage of carry-on items on the flight deck. This included personal items, manuals, and laptops (now mostly superseded by the electronic flight bag). These items must not obstruct escape routes and must be stowed to withstand crash conditions.

8.2.9 Structural Aspects

The structure plays a large part in the installation of the systems, not least in the flight deck or cockpit. The cockpit in particular is contained within its own pressure shell and built into the structure. The equipment in the flight deck must be securely mounted on rails or in enclosures that are connected to the structure and in most cases, must withstand the crash case deceleration. Connections between the cockpit pressure shell and the airframe such as wiring harnesses, air ducts, hydraulic pipes, and control rods or cables must be installed so that they do not weaken the structure or compromise the pressure shell.

- Equipment mounting is provided with standard rails and fastenings to enable panels and equipment to be securely attached and rapidly removed and replaced. Security of attachment must meet the crash case conditions to ensure safety of the occupants.
- Cooling must be provided for some equipment and most certainly for the pilots. For commercial aircraft pilots, a comfortable shirt sleeve environment is preferred. Fast jet pilots will require cooling to provide comfort in high-altitude solar radiation, wearing of flying suit, immersion suit, and g protection clothing. Much of the equipment mounted in the cockpit will dissipate heat and make a contribution to the heat load.
- Seat rails enable the seats to be adjusted independently to allow for different sizes and reach preferences for pilots.
- Sound insulation provides attenuation of external engine and aerodynamic noise to allow communications and systems status messages to be heard.
- Aerodynamics may be affected by the shape of the fuselage to accommodate the flight deck or cockpit – for example in fast jet trainers the fuselage shape around the cockpit may be longer or wider than a single-crew version.
- Escape or evacuation facilities must be provided to allow the crew to depart the aircraft on the ground and in the case of fast jet military aircraft in flight at a wide range of speeds, height, and attitudes.
- Canopy installation for a fast jet aircraft must provide a means of ingress and egress under normal conditions, clear all round vision, and to allow the pilot to escape in emergency.
- Pressure seals are required to seal all opening windows and an opening canopy to maintain the integrity of the pressure vessel.

8.2.10 Physical Aspects

There are aspects of a general nature that need to require special attention because they cause incursions into the pressure cabin on a military fast jet aircraft and therefore need specialist connections to be formed.

- Power supplies routed to the flight deck or cockpit will enter via pressure bungs or connectors and may be separated from other wiring. Some items of equipment may require a circuit breaker panel to enable breakers to be reset.
- Cooling ducts must be provided through the partitions or floors with appropriate adjustable vents and fans. In the fast jet cockpit, the ducts will enter the pressure shell.

- Wiring harnesses through pressure structure in a fast jet must be provided by means of rubber bungs or suitable connectors.
- Security of access to protect against unauthorised access and forced entry is necessary as a result of the terrorist threat. This will take the form of a strengthened door and frame, an eye hole/lens, a key pad, and secure locks.

8.2.11 Flight Simulation

Flight simulation is included in this chapter because it is so intimately related to the design of the flight deck or cockpit. The flight simulator is no longer merely a training aid to allow pilots to convert to type and familiarise themselves with the flight deck. It is used during the development of a new aircraft as a means of confirming that its fidelity to the design is robust. It is used regularly to extend the pilot's expertise under normal and abnormal conditions of flight and in the military world, it is used to develop and perfect new operational techniques. The simulator is strongly influenced by the development of the human machine interface and this influence is two way – it is used to reflect and confirm any changes identified which are fed back into flight deck improvement.

8.3 Architectures

The systems architecture is a representation of the conceptual shape and form of a system which can be visualised quite independently of any physical implementation. It is an invaluable device for making a simple and easy to understand representation of a system using a block diagram format as a convenient shorthand notation. This simple visualisation allows a concept to be represented clearly and acts as a mechanism for promoting discussion between various engineering disciplines to reach agreement on interfaces, functional allocations, and standards. From such simple basic representations, it is possible to develop the architectures further without the need to move to excessive detail of wiring interconnections or detailed components. This is true for physical and functional representations in terms of software and hardware building blocks. It is especially useful for setting and agreeing boundaries and interfaces. Figure 8.3 shows a generic system architecture and the way in which it can be layered to illustrate top-level architectures, major and sub-system and even the architecture of individual items of equipment.

Apart from allowing design decisions to be made, systems architectures are an ideal tool to assist in identifying candidates for early trade-offs and simple models using spreadsheets to perform cost, benefit, and performance comparisons between different architecture designs, see Chapter 7.

System architectures are discussed further in Chapter 5 of Seabridge and Moir. 2020.

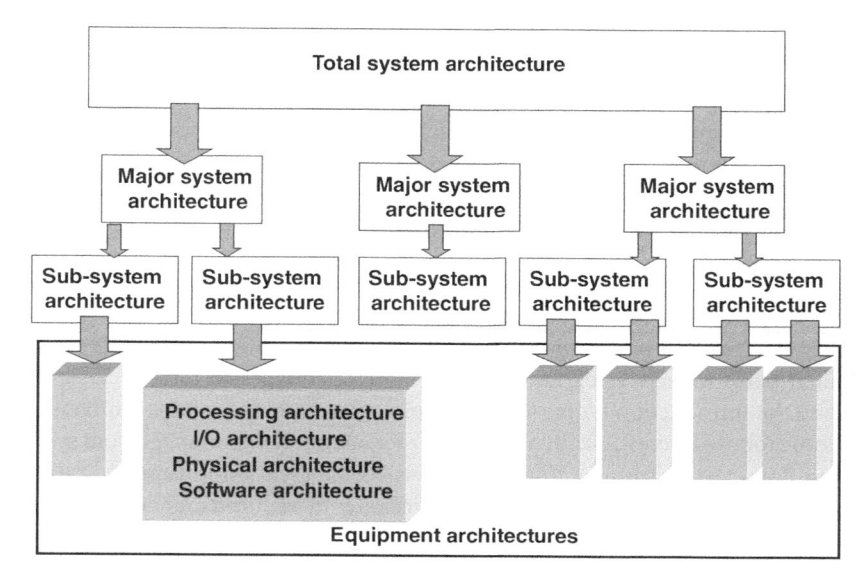

Figure 8.3 Example generic system architecture.

8.4 Other Considerations

There are other considerations that are less easy to determine and model. Some of these appear as issues when systems are being tested or during flight test. In extreme cases, they appear when the aircraft is in service, when issues are costly to rectify. Some example considerations are discussed below, but the cautious system engineer will try to identify others.

8.4.1 Integration

Integration can be characterised by the performance of many functions by a small number of units. It is more commonly applied to avionics type systems, but is also found in major structural and mechanical systems such as aircraft powerplant and propulsion systems and even the airframe itself. In an integrated system, the individual systems interact with each other and display forms of 'interconnectedness' that makes the larger system more than the sum of its constituent sub-systems. Interconnectedness of systems is achieved by

- Data communication networks conductive or optical.
- Direct physical interconnections by design, e.g. welded, fastened, or glued connections.

- Direct mechanical interconnections by drive shafts, air motors, or actuators.
- Indirect physical interconnections, e.g. thermal, vibration, and electrical conduction.
- Physical amalgamation at component level.
- Functional interconnection in processors by shared computer architectures, memory locations, or data use.
- Serendipity – emergent properties or unexpected behaviours in one system that affect one or more others.

There is little doubt that integration has brought advantages to the design of aircraft systems. There is better inter-system communication and greatly improved functional performance. Integration has brought advantages to aircraft design, including the introduction of innovative technologies. Data is available on-board in real time for use by crew, accident data recorder, maintenance management system, and flight test instrumentation. There has been a great improvement in the design of cockpits and flights decks with a consequently reduced crew workload.

System integration has been achieved by design and has been implemented by the application of readily available technologies by means of which functionality has been incorporated using a combination of

- Computing systems based on distributed computers
- High-order languages
- High-speed data buses
- Multi-function displays and controls
- Touch, voice, and switch controls

All this has been enhanced over the years by the use of colour screens, high-resolution images, easy access to on-board databases for intelligence and maintenance data, access to ground-based systems by data link and incorporation into the air transport management system. The cockpit or flight deck is now roomy, uncluttered, comfortable and a safe and clean place in which to work. The pilot is now integrated with the whole aircraft and its systems. This looks likely to continue with predicted technologies such as gesture control, synthetic vision and neural networks.

Some mechanical transfer functions such as those previously provided by mechanical devices such as cams and springs or hydro-mechanical devices have been incorporated in software and electrical effectors, thereby blurring the distinction between mechanical and avionics functions. There are issues associated with each of these characteristics that need to be understood:

- **Computing systems based on distributed computers:** the functions in individual systems may be conducted in real time in a cluster of computers and are

part of a complex network of more clustered computers. All of these will generally operate asynchronously.
- **High-order languages:** the early choice of languages such as Pascal and Ada was based on their inherent rigid structure and the use of limited instruction sets to form a deterministic structure essential for the acceptance of safety critical software. As time progressed, this structure was seen as limiting in terms of speed, even cumbersome in its inherent inelegance of software design, and the computer games industry began to spawn languages more suited to graphics, visual effects, and high-speed operations. The favoured language became C++ which is still prevalent today and has been adopted for aerospace applications, after early resistance. There are potential issues with its use (unless a limited instruction set is mandated) which can lead to non-deterministic structures.
- **High-speed data buses:** a similar sequence of events led from the initial choice of MIL-STD-1553B because of its simple and deterministic nature, towards the widespread use of high-speed data bus structures which can also exhibit non-determinism if not correctly designed.
- **Multi-function displays and controls** have released panel space in cockpits and improved pilot awareness by the use of colour and easily understood formats for data and information. Much of this advance has come from the world of desktop computing and games. The perceived need to give aircrew information only when it is needed for a particular phase of flight means that data may only be accessed after several pages of information (screens) have been viewed.
- **Touch, voice, and switch controls** have increased the opportunity for overlaying visual screen information with other cues such as voice and aural tones to signify events. They have also increased the opportunity for instinctive reactions by the pilot in response to these cues. In a controlled environment, this is seen as an advantage, but in emergencies, it may result in an overload of information which is not permanently displayed – it needs to be remembered. Instinctive reaction may also lead to a response being less considered than a switch press action where the switch is on a panel and can be checked by a second pilot, rather than being a 'twitch of the finger'.

The result of this is a perceived obscuration of the end-to-end behaviour of the system. This may be seen as an inability to be able to trace the route of an input signal to an output effector in a simple manner – once the way in which system behaviour was checked. There are severe implications to this obscuration:

- It is difficult to envisage the correct behaviour of any one individual system and the whole system for an observer and, more seriously, for a reviewer checking and signing-off for correctness of operation.

- It is difficult to design a series of tests that go deeper than merely checking that an appropriate combination of inputs gives rise to an expected output and does not give rise to an unexpected output.
- It is almost impossible to verify the soundness and segregation of systems and to detect any unexpected interactions.
- It is difficult to determine the correct operation of the whole system – not simply the fact that functions have been performed correctly, but in understanding the mechanics of this.
- It is very possible that serious issues will remain dormant until a particular combination of events, demands, data structure, and software instructions leads to their initiation.

This all leads to uncertainty in the complete design unless something is done to improve the visualisation and understanding of the design, its implementation and the results of testing.

8.4.2 Complexity

It certainly appears that modern aircraft systems have become more complex, both as individual systems and as total systems. Some of the reasons for this include the following:

- Advances in technology have made it easier to implement more functions in software in a high-speed processing architecture.
- The desire for engineering advancement: the simple need for engineers and designers to incorporate new functions and to apply new technology.
- The need for more automation in a bid to reduce crew workload.
- A desire for autonomous operations for unmanned vehicles and in the event of further reductions in flight deck crew complement.
- The desire to get more functional performance from fewer items of equipment.

As computing technology makes it easier to include new functions, so customer requirements are enhanced by the results of this and their needs become more complex. This may be spurred on by pilots flying new types and learning new techniques, thus making more demands based on their experience. Engineers are usually more than happy to entertain this scenario, as their natural inclination is to apply new technologies.

This desire is not solely restricted to the flying side of operations. Airline and military fleet maintenance and logistics support are also under pressure to improve turn-around times. One method of doing this is to improve the accurate location of failures and to provide advance notification to the ideal repair and replacement bases. This has led to the introduction of more sophisticated algorithms to enable monitoring, on-board analysis together with data link download of information. It has also led, incidentally, to scenarios in which faults can be 'carried' in redundant systems until the most cost-effective repair base is reached.

This has the result of making the systems associated with control of the aircraft and mission to become closely connected to the maintenance function, further increasing complexity. Complexity is rarely a good thing in engineering. Despite the process of requirements capture, analysis, careful design, and robust reviews, many projects fail because of cost and time over-runs. Much of this has to do with the impact of complexity. Complexity adds time to the design and test process as re-work is required and because the task has been under-estimated. It is important to try to assess or measure any trends towards increasing complexity and to focus attention on their solutions or to amend the estimate.

Complexity results in an increase in relationships between elements of the system. This is bad enough if the complexity is confined to a bounded system, but when the impact of integration blurs the boundaries of individual systems then the situation becomes difficult to control. Hence, the widening scope of integration to encompass external systems such as maintenance and air traffic management requires more diligence in design.

It is well then to make an assessment of a new system design to see if it has become more complex than its predecessors. This assessment can be used for a number of purposes:

- To determine what additional design and test work will be required.
- To determine what further investment will be needed in new design tools.
- To determine what new or additional skills will be required.
- To determine what training needs will be needed.
- To ensure that any work or cost estimates carried forward from a previous project are appropriately factored.
- As an aid to determining the feasibility of the project.

8.4.3 Chaos

The current situation of aircraft systems is that they have reached an advanced state of integration in which the 'core' or essential systems of the aircraft are themselves tightly integrated and are becoming an integral part of advances in navigation and air traffic management systems aimed at providing efficient navigation. In part, this is being driven by environmental issues aimed at reducing fuel burn and potential pollution of the atmosphere.

Figure 8.4 takes this further to illustrate how advances in external systems, many of them ground-based and in continuous operation are also embracing the aircraft. In this way, the controlled operation of systems, with regular and relatively

8 Summary and Future Developments

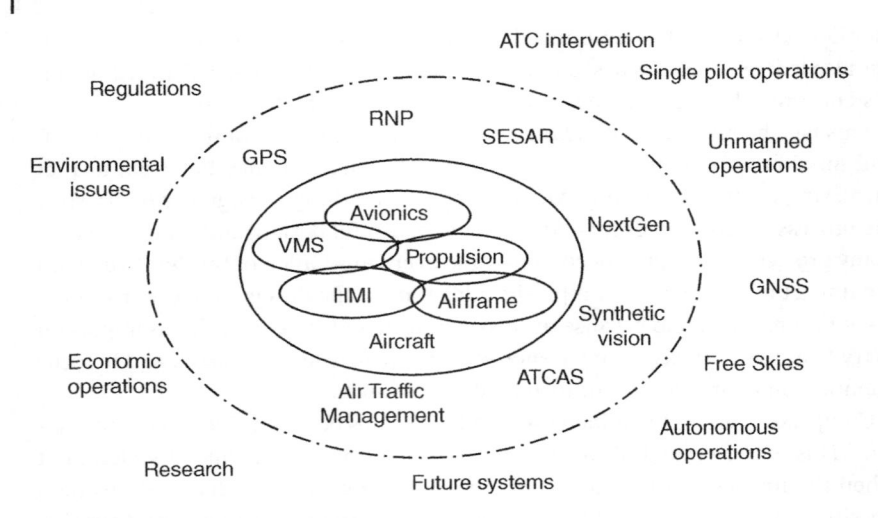

Figure 8.4 Increasing integration.

frequent power up and power down phases are now a part of a continuously operating system. Thus, there are fewer opportunities for system clocks to be reset and for memories to be refreshed.

The 'core' systems can be represented by their own individual sub-system architectures and can be visualised as a combined architecture. This illustration can be extended to encompass a complete aircraft architecture to give an example of the complexity of modern systems. This extension can be illustrated as a federated architecture or an integrated modular architecture. Each functional block in the extension should be considered as performing one or more functions which can be implemented in individual avionics boxes or integrated into some form of computing architecture with interconnecting data bus links. The functions are implemented in software and interconnected with some form of data bus. The message is the same no matter how it is implemented – it is complex. The system should be perceived as many functions performed in many computers, with instructions and data in software, and inter-functional data embedded in data bus messages. There the systems will probably run asynchronously, and there will be many data items and many non-linearities. It is likely that some processing structures and some data bus mechanisms will be non-deterministic which may lead to variability in system timing. The system is also the recipient of random inputs from the human operators and maybe from external sources.

This situation leads one to pose a question about complex real-time systems: 'are there any conditions in which the system can become chaotic?' Is there a risk that the whole system can enter a state of chaos?

8.4 Other Considerations

At one time, the aircraft systems operated for short periods of time only in any one flight and were then powered down, and thus all conditions were re-set. For fast jet military jets, the power up time might only be an hour or so, for long-haul commercial airliners, maybe eight hours. Now with air-to-air re-fuelling even fast jets are powered up for many hours, and airliners routinely remain powered up for days at a time. This situation arises on long-haul flight of up to twenty hours where the systems are not shut down during the aircraft turn-around interval. They are operating within an air transport management system that itself operates continuously.

As a result of some unexplained behaviours, it has been necessary to impose a mandatory re-boot for the B787 and A350. This was reported in MRO-Network.com under the header 'EASA orders periodic reset of A350 internal clock'. The report described the issue of a European union Aviation Safety Agency (EASA) airworthiness directive AD 2017-0129 mandating that operators power-down the aircraft systems after a continuous period of operation of 149 hours. This was issued as a response to reports of loss of communications between avionic systems and networks. According to EASA 'different consequences have been observed by operators from redundancy loss to a specific function hosted on the remote data concentrators and core processing I/O modules'. Shutting down and powering up (power re-boot) has had some success in previous incidents. In 2015, a Federal Aviation Authority (FAA) directive directed a re-boot of all B787s that had been powered up continuously for more than 248 days to prevent a computer internal counter overflow. A year later, software issues were reported to have re-surfaced when flight control modules were found to reset automatically after 22 days of continuous power up.

There are a number of reasons that give rise to the question about chaos: aviation systems have moved from relatively short time scale durations to a situation in which some parts of the systems are in use continuously, and now airborne systems have moved to long duration power up. In this was they have closely approached the operating scenarios for ground-based commercial systems – which have been known to produce unusual results.

The complexity of systems means that it is difficult to guarantee that some or all of the precursor conditions for chaos will not arise. In its 'normal' state, the total system operates with many transactions, iterations, rates, and asynchronous conditions. Inputs to the system will be provided by the operators on board such as first pilot, second pilot, cabin crew. There are situations in which some use of non-deterministic software and data bus applications may arise. When installed in the aircraft there are many interactions in software and data bus, and many systems contain non-linearities. These are all conditions in which chaos can arise.

- Is it possible for a transition to occur that will disturb the 'normal' state?
- Could this lead to unexpected system behaviour?

8.4.4 Emergent Properties

One issue associated with complexity and chaos is that of unexpected behaviours or emergent properties. This phenomenon has been seen in large ground-based systems where a re-boot, control/alt/delete or power down is feasible to reset the system.

There are many definitions of the term 'emergent property' to be found. Many of them include phrases similar to 'behaviour exhibited by a collection of systems that is not exhibited by the constituent systems on their own'. This means that even if systems are individually tested rigorously, they may not behave as a collection of systems. It is to be expected then that a system of systems, integrated systems or even interconnected systems may produce unexpected behaviours. Some of these will be benign, but the risk is that some may lead to an accident.

Rigorous testing on the ground and in the air may reveal some of these behaviours. However, the test environment is carefully controlled and many of the tests are based on expected behaviours. Therefore, a system under test will not necessarily experience all the conditions that an in-service system will.

It will be interesting to see if a model can be devised using artificial intelligence and synthetic reality to discover emergent properties. A major issue here is that a model, like the testing process, is a controlled environment.

8.4.5 Fitness For Purpose

The whole design and development process is intended to produce an aircraft that meets the customer's requirement throughout its life and in all defined environmental and performance conditions – in other words it is fit for purpose. The process of the development lifecycle, including testing, trials, and initial in-service experience, is intended to demonstrate this fitness for purpose. However, if any unexpected behaviours or fitness shortfalls are found, they will often be discovered late in the lifecycle where their rectification is costly.

It is important to fully understand what fitness for purpose means for the whole aircraft, and what contribution each of the systems makes to this. This is where the issues of emergent properties, unexpected behaviours, and chaos can influence the demonstration of fitness for purpose. Some form of predictive model will be helpful in identifying any weaknesses in the design and development process that may indicate a deviation from the path to demonstration of fitness for purpose.

8.5 Conclusion

This book has provided an illustration of aircraft systems as recognised in many types of aircraft in the aerospace industry. The industry is currently dominated by the need to deliver hard products, many of them complex interactions of airframe, components, human operators, and systems – both hardware and software. These products are provided to customers as a part of their armed forces or airline infrastructure, which in turn may be part of a wider national or international entity. This increasingly complex nature of products has led to an approach to dealing with them as complex systems.

An understanding of what constitutes a system is important. There is an increasing tendency for domain-specific engineers to take a broader view of their system, a state of mind that is stimulated by the increasing integration of systems in the modern aircraft. Thus, individual systems are perceived as being sub-systems of larger integrated systems existing in complex environments.

Chapters 3, 4, and 5 introduced the systems of an aircraft individually classified as vehicle, avionics, and weapon systems respectively. In the system diagrams used to describe individual systems, some aspects of interactions with other systems are shown. This must not be taken as definitive – specific solutions will contain different interactions. Chapter 6 briefly introduced some external systems to which aircraft are routinely connected. Systems architectures are a convenient method of visualising emerging concepts in both functional and physical form. The block diagram is a convenient notation for identifying the form of systems and is used as a medium for brainstorming, debate, and discussion. Chapter 7 briefly introduced this topic and presented and a method of modelling architectures to aid selection of the most appropriate solution.

8.6 What's Next?

The aerospace industry has a long history of innovation and new opportunities will arise from new scientific discoveries. These will be turned into new technologies by an industry that has great enthusiasm for new technology. Technology advance will continue to provide a focus for implementations to gradually improve the performance of aircraft or to provide solutions for specific problems. A brief scan through current literature reveals the following topics that will be developed over the next decade:

The flight deck or cockpit will undoubtedly transform from a physical entity into a virtual environment. This can be accomplished in the fast jet cockpit using sensors and display in the helmet, whereas the flight deck can be realised

8 Summary and Future Developments

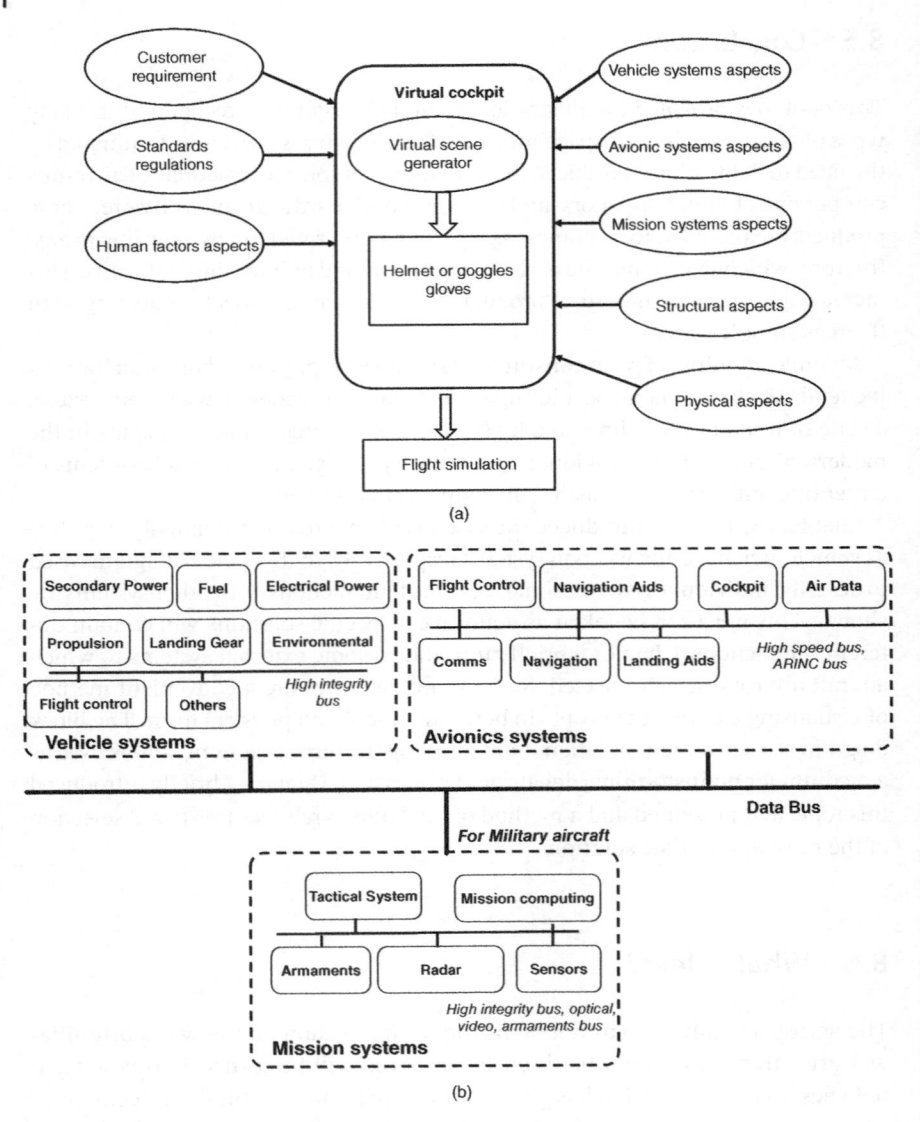

Figure 8.5 The virtual flight deck or cockpit.

with goggles, now familiar in a number of training aids to provide a virtual environment. An example of this is shown in Figure 8.5.

There are safety issues with this approach which must to be considered in the design. The most obvious example is the redundancy offered by multiple physical display surfaces. This allows the continuation of the flight with the loss of one display surface, and allows cross-checking by pilots of each other's displays.

There will be more changes as systems become more integrated with the development of extended reality, virtual controls, and displays and the increase in AI driving more automation and autonomy.

Computing has been at the heart of most advances in aircraft systems since the emergence of microprocessors suitable for use in aircraft applications in the 1970s. Changes in technology have resulted in faster computers and denser memories, and the application of these technologies in the commercial field of home computing, games, and Internet applications have been put to good use in aircraft systems. Artificial intelligence algorithms in design and in real time on-board applications will benefit from neuromorphic computing for fast and energy-efficient processing. This will have an impact on real-time pattern recognition, speech processing, and image classification. This will affect the design of intelligence gathering, sensor processing, human factors design and navigation systems in future aircraft and may pave the way towards unmanned passenger aircraft with real autonomy. The main benefit of neuromorphic devices is their lower-energy demand and the potential to incorporate more processing power into smaller volumes. This is predicted to improve over the next three decades, exceeding the predictions of Moore's law.

One use of this advance in computing will be the ability to move towards synthetic vision in cockpits and flight decks. To some extent, this has been applied in head-up displays and in cameras mounted to provide all round vision from within the aircraft – a perceived spherical view by the pilot. Artificial intelligence (AI) and learning software techniques will lead to the provision of such facilities as object recognition and avoidance, techniques being developed for autonomous road vehicles. There is a danger that this can lead to an overload of information in the cockpit unless robust human factors approaches are applied. Techniques such as only displaying the required contextual information to an operator at a point in time and managing certain tasks on the operator's behalf will be applied to control the workload. Emerging analysis models such as trusted reasoning and trusted AI will be used to increase the certainty in which systems can be understood and tested.

There will always be a need for high-energy systems to control the attitude of air vehicles, and there will continue to be a move towards electrical rather than hydraulic or pneumatic effectors. This will stimulate a demand for electrical actuation and electro-hydrostatic systems with a need for electric motors with materials capable of operating at higher temperatures, higher magnetic field strengths, and rotational speeds using in-built failure prediction software which will greatly improve flight and engine control systems. This is expected to improve power and rate performance of control surfaces at reduced mass.

It has been shown that power demand on successive generations of aircraft is growing. Despite the promises of lower power requirements of succeeding generations of semi-conductors, this always off-set by an increasing demand for more processing power. To reduce generated power and large heavy generators connected to the main engines, it is expected that next generation thermo-electric generators will be used to scavenge waste and rejected heat sources to convert thermal energy directly into electrical energy. This can be used to capitalise on waste engine heat from engines, auxiliary power units, brakes, and avionics sources. There have been studies on how to use heat generated by the brakes to contribute to the energy needed to taxy an aircraft. A slight dilemma arises here, in that one of the major heat generating systems on an aircraft is the hydraulic system, and of course, its replacement by electrical systems is already being predicted.

On the subject of electrical power, it is expected that electrical generation components are likely to be embedded in the shafts of future engines. Engine manufacturers have been pursuing this for some time, and it is now very likely to happen.

Composite materials have been changing the approach to design and manufacturing over the past 30 years to great advantage in commercial and military aircraft types. New materials such as graphene will pave the way for novel approaches to structure design and for the design of electronic equipment housings which may require alternative approaches to screening and bonding.

A serious interest in environmental issues is at last moving away from simply banning the use of materials and substances and towards a creative use of technology and techniques to reduce carbon emissions. This has resulted in studies for different ways of operating aircraft and for different shapes and propulsions solutions. This, in turn, has led to consideration of novel sources of fuel being sought for cleaner combusting engines.

In the avionic systems of the aircraft the electrical wiring, generally copper with shielding and insulation is a contributor to weight, and it needs a large volume of the fuselage interior for harnesses or bundles of wire (many 10s of kilometres), connectors are always a contributor to faults and the whole assembly is costly. Testing and repair are further costs. Despite the widespread use of data buses including some fibre-optics, there is still a lot of wiring in an aircraft. In the military data-centric world, the introduction of new mission computers, real-time video capture, high definition cockpit displays, and advanced electronic warfare systems means extensive re-wiring for updates. Future systems will need to make greater emphasis on optical networks and signal multiplexing, novel mechanisms for incorporating wiring or optical fibres into the airframe structure and using transceivers. The commercial aircraft industry is leading the way with fibre optical backbones in the A380 and B787.

The scope of unmanned aircraft has expanded since their inception. Full-size military vehicles are common with remote piloting and now the concept of a single aircraft accompanied by a swarm of 'colleagues'. The role of vehicles has also expanded from military applications to commercial and civilian uses for peace-keeping, policing, and purely civil uses. As yet, the carriage of passengers in unmanned vehicles has not happened, but with autonomous road vehicles and trains, it can't be far away. This will have an impact of the design of the vehicle. Without the mass of a pilot and the volume required, the 'flight deck' or 'cockpit' functions will remain on the ground.

There will be many changes and the world of 'systems' will develop to provide the functions of vehicle systems and avionic systems in a completely different way.

Exercise

8.1 What sort of properties will be required of a system model to enable it to predict emergent properties?

Index

a

AADL-based modelling of IMA 265
ACARS 246
accident data recording 172
acoustics 205
AFCS 156
AFCS architecture 283
AFDX topology model 272
airborne data acquisition system 242
airborne early warning 20
aircraft as set of systems 6
aircraft roles 5, 14
aircraft types, classification 1
air data 168
airframe 6, 8
airframe loads 8
air superiority 16
air-to-air refuelling 2
allocation modelling 292
alternative fuels 48
altitude lighting facility 131
antenna configuration, mission 186, 235
antenna modelling 234
antenna placement 186, 234, 235
antennas 185, 234
anti-spin parachute 110

approximation algorithms 276
architecture, aircraft systems 29
architecture modelling 260
architectures 314
architecture, top level 28
ARIANE federated architecture 274
ARINC 653, 265
ARINC 664, 32
ARINC 429 data bus 30, 31, 92
ARINC 629 data bus 31, 92
arrestor hook 105
artificial intelligence 25
assignment and mapping 266
ATEX 46
atmosphere, impact 11
atmosphere impact on airframe 8, 9
audio, attention getting 309
automated architecture generation 268
automated model driven methodology 257
automatic architecture optimization 260
automotive industry experience 274
avionics architecture modelling 263
avionics equipment list 285
avionics integration constraints 294

Aircraft Systems Classifications: A Handbook of Characteristics and Design Guidelines, First Edition.
Allan Seabridge and Mohammad Radaei.
© 2022 John Wiley & Sons, Inc. Published 2022 by John Wiley & Sons, Inc.

Index

avionics integration optimisation software system   289
avionics LRUs assignment   294
avionics operational capability assessment   285
avionics requirements   279, 281
avionics software architecture   289
avionics system architecture   285
avionics system integration and architecting   277
avionics technical specifications   285
avionic system architecting   258
avionic systems aspects   310
axiomatic design   260

b

batteries   52
battlefield surveillance   19
biological & chemical protection   102
bleed air   56
brake parachute   108
braking, generic description   71
branch and bound algorithm   277

c

cabin and emergency lighting   122
camera system   218
CANbus   32
canopy   313
canopy jettison   97
chaos   319
civilian agencies   25
classification of systems   25
CoBaSa   269
cockpit image recorder   248
cockpit voice recorder   248
colour, use of   309
comfort   306
COMINT, as part of ESM   214
commercial roles   13
communications   131
communications control panels   311
comparison of modelling approaches   275
comparison of optimisation methods   276
complexity   318
component-based system assembly   269
connectivity constraints   293
control column   310
control function   12
conversion to military role   16
cooling, crew   310, 313
cooling ducts   313
cooling, equipment   310, 313
Covid-19 impact   79
CPLEX   271, 274
Cranfield University projects   277
crash case   10
crew escape   93, 310

d

data bus   29, 30
data link   224
decision-analysis tool   273
defensive aids   209
Def Stan 00-18   30
design structure matrix   260
displays and controls   127, 311
distributed computers   316
DME/TACAN   152
Dormant   26

e

ECCM   215
Eclipse EMF   271
ECM   215
ejection clearances   310
electro-hydraulic pump   63

electronic flight bag   174
electronic flight bag stowage   311
electronic support measures   200
electronic warfare   21, 215
electro-optical sensors   197, 198
electro-optics   197
ELINT, as part of ESM   214
emergency oxygen   312
emergency power   61
emergency power unit   62
emergency tools stowage   312
emergency use   26
emergent properties   322
engine, types of   38
environmental control   76
environmental issues   326
equipment mounting   313
escape and evacuation   313
ETOPS   40, 61
evolutionary algorithms   276
exact algorithms   276
experimental aircraft   243
external lighting, generic description   86
external view   309

f

fire protection   79
fitness for purpose   322
flight controls   65
flight deck   12, 307, 308
flight deck, system of systems   306
flight management   138
flight simulation   314
flight test data analysis   243
flight test instrumentation   244
fuel   44
fuel cell   62
functional decomposition   280
function/means tree   260, 262

g

galley   112
general aspects, flight deck design   312
general aviation   14
generic system architecture example   315
genetic algorithm   269
get-u-home display   128
glideslope   165
ground arresting   105
ground attack   17
ground systems   12
GUROBI   271, 274

h

hardware layer   263
head up display   220
helmet mounted systems   222
HEPA filter   79
heuristic algorithms   276
high order languages   317
high speed data buses   317
house of quality   261
HUD   312
human factors aspects   306
hydraulic accumulator   63
hydraulics   53

i

ice and rain   9
ice detection   83
ice protection   85
IEEE 1398 data bus   92
IMA   183
IMA architecture optimization   268
IMA, mathematical-based modelling   265
IMINT   217
inertia switch   40

Index

in-flight entertainment 117
installation layer 264
integer programming 293
integrated development environment 262
integration 315, 320
integration of airborne and ground systems 242
internal lighting 181
ionising radiation effects 9

j
JTIDS 225

l
landing aids 163
landing categories 163
landing gear 68
law enforcement role 25
lighting 309, 311
link 11, 225
link 16, 225
localiser 164
logistics ground based system 247
long haul 15

m
magnetic anomaly detector 202
maintenance management system 246
maritime patrol 18
marker beacon 165
microwave landing system 166
military roles 16
MILP 272
MIL-STD-1553 data bus 30, 92
mission computing 207
mission critical 26
mission displays and controls 230, 231
mission management ground system 251
mission support system 250
mission system, illustration 193
mission systems aspects 312
model based development 259
model based systems engineering 259
modelling techniques/tools 262
modelling and analysis techniques 260
modelling of optimisation 292
model management 262
multi criteria decision making method 277
multi-function displays and controls 317

n
navigation 134, 135
navigation control panels 311
night vision goggle compatibility 130
NILE 225
non-aerospace industry experience 273
non-critical 26
non-ionising radiation effects 9

o
on-board oxygen generation 101
one-shot battery 62
operational capability assessment 288
oxygen 99

p
Pareto front 271
particle swarm optimisation 269, 277
passenger evacuation 115
π-calculus 265
pedals 310
performance constraints 293
permanent magnet generator 62
personal items stowage 312

Index | 333

photographic reconnaissance 21
physical aspects 313
physical constraints 293
physical integration 258
power supplies 313
precision DME 152
pressurisation 310, 313
probe heating 89
production acceptance testing 242
prognostics and health monitoring 178
propulsion 38, 40
prototype aircraft 243

r

rad alt 160
radar 194
radar modes 192
radome 195
ram air turbine 61, 63
reach, pilot 309
regional 15
resource constraints 293
RF spectrum 186

s

safety critical 26
safety involved 26
scatterD tool 269
seat rails 313
secondary power 59
security of access, flight deck 314
segregation/collocation constraints 293
SIGINT, as part of ESM 214
simple additive weighting method 287
sonobuoys 206
sound insulation 313
spacecraft models 273
special roles 24
stakeholder diagram 27

stakeholders, aviation industry 26
stakeholders, project 28
station keeping 213
steering 73
strategic bombing 18
structural aspects 312
structure interface 11
switch control 317
switches and controls 312
synthetic vision 325
system architecting 258
system behaviour, end to end 317
system boundary 11
system characteristics 12
system diagram, explanation 10
system integration 258
system layer 263
system of systems 305
systems health data 246

t

TAWS 149
TCAS 146
technology alternative comparison 288
technology readiness level 287
text and font size 309
throttles 310
toilet and water waste 119
touch control 317
training aircraft 23
transponder 143
troop and materiel transport 22
turbulence 9

u

UAV ground control system 252
unmanned air vehicles 23, 326
unmanned air vehicles, classification 24
UV radiation effects 9

V

VAPS 131
vehicle management 92
vehicle systems aspects 309
viewing angle 309
virtual flight deck 323
voice control 317
VOR, generic description 154

W

warnings 312
weapon system 227, 312
weather radar 140
wiring harnesses 314
workload 309

X

XML database 271